# GENERAL CHEMISTRY *for* ENGINEERS AND BIOLOGICAL SCIENTISTS

FIRST EDITION

# GENERAL CHEMISTRY *FOR* ENGINEERS AND BIOLOGICAL SCIENTISTS

BY BRIAN DIXON AND LORI NOBLE

*Massachusetts Maritime Academy*

cognella

SAN DIEGO

Bassim Hamadeh, CEO and Publisher
John Remington, Executive Editor
Michelle Piehl, Senior Project Editor
Alia Bales, Production Editor
Emely Villavicencio, Senior Graphic Designer
Stephanie Kohl, Licensing Coordinator
Natalie Piccotti, Director of Marketing
Kassie Graves, Vice President of Editorial
Jamie Giganti, Director of Academic Publishing



Printed in the United States of America.

cognella® | ACADEMIC PUBLISHING
3970 Sorrento Valley Blvd., Ste. 500, San Diego, CA 92121

# CONTENTS

CHAPTER 1

# Atoms and Compounds

## What Is Chemistry?

Chemistry is the study of the composition, properties and behavior of ***matter*** in all of its forms. Matter is anything that has mass and takes up space. The book you are reading, the air you are breathing and the chair you are sitting in are all composed of matter. In order to study the properties and behavior of matter, an understanding of is composition is important.

Chemistry is an important part of all aspects of life. The various branches of chemistry each deal with a specific component of chemistry. Chemistry can explain how human beings acquire energy from sugars and fats while plants produce their own energy using sunlight. It can even explain why some fireworks produce a red burst while others produce blue or green colors. For many years chemistry's application to medicine has found cures for life threatening illnesses. Hence an understanding of chemistry can provide great insight into various aspects of life. In order to understand all of these applications of chemistry, it is important to understand matter, what it is made of, and why it behaves the way it does and that is where this text begins.

## 1.1 What Is Matter?

### Atoms, Elements and Molecules

Atoms are the building blocks of matter. There are over 100 unique types of atoms and each unique atom is called an ***element***. Some elements occur naturally on Earth while others have to be synthesized in a laboratory and may exist only for a matter of seconds or much less. Iron is the most abundant, naturally occurring element on Earth, and is very stable. A new element first synthesized in 2010 called ununseptium exists for only microseconds. The properties and behaviors of each form of matter is determined by its unique composition, combination and arrangement of atoms (Figure 1.1).

The simplest forms of matter are comprised of atoms from one element. Pure gold is comprised only of gold atoms. A sample of pure oxygen is comprised solely of oxygen atoms. However, in its pure form, two atoms of oxygen are attached by a chemical bond. This is different than gold where atoms of gold are independent of each other and not

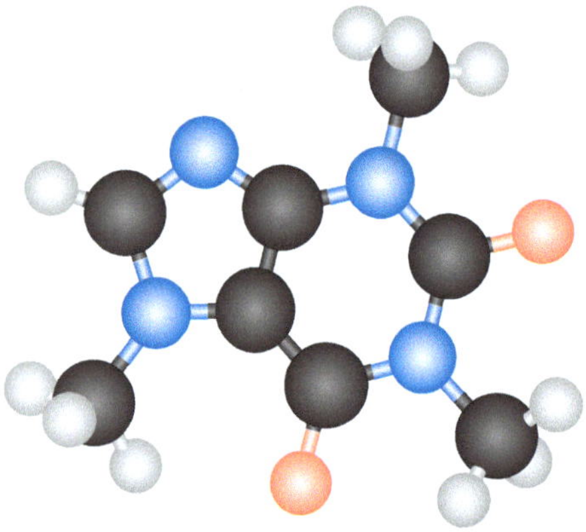

FIGURE 1.1. Molecules/Atoms.

held together by chemical bonds. When two or more atoms are held together by a chemical bond, they form a ***molecule***. Many elements exist as molecules in their pure form (Figure 1.1). These elements are hydrogen, nitrogen, oxygen, fluorine, chlorine, bromine and iodine. Any sample that contains only one type of atom, whether as a molecule or individual atoms, represents matter in its simplest form.

The periodic table lists all of the known elements in an organized manner and will be fully discussed in the next chapter. For now, a brief discussion of this mysterious table is in order. Each element is represented by a unique letter symbol. The table is arranged in groups shown as (columns, Figure 1.2a) vertical, and periods (rows, Figure 1.2b) that are horizontal. The elements are arranged in a manner that groups elements with similar properties together. The table also offers important information about each element.

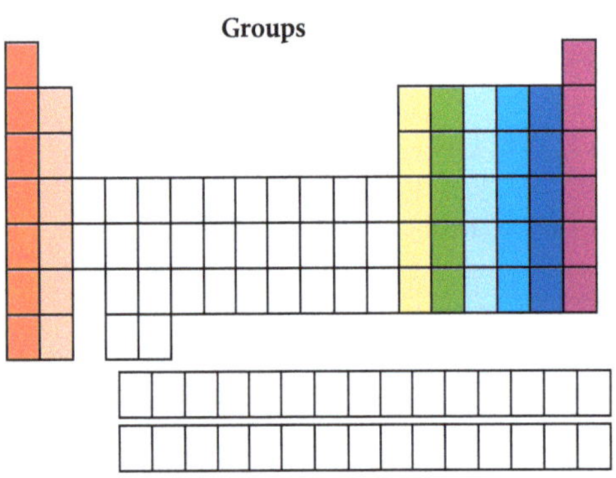


FIGURE 1.2A. Periodic Table: Groups.

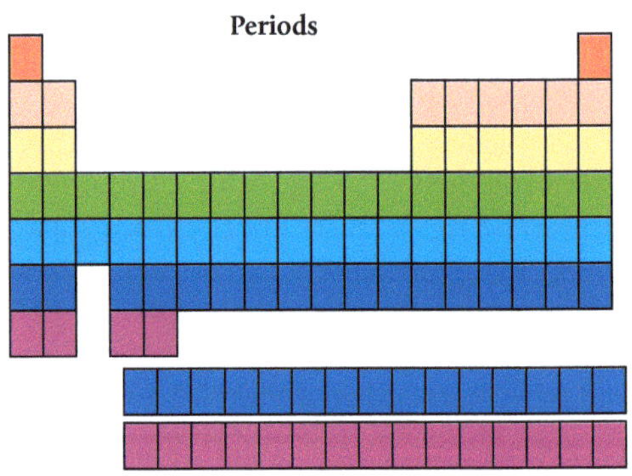


FIGURE 1.2B. Periodic Table: Periods.

## Compounds

Two or more atoms from different elements can come together to form molecules as well. When molecules contain atoms from different elements, they are called compounds. Water is a compound comprised of an oxygen atom and two hydrogen atoms held together through *chemical bonds*. The same number and ratio of atoms can create two completely different types of matter with different properties depending on their arrangement. There are two forms of a compound thalidomide that have different three-dimensional arrangements of the atoms. One version, r-thalidomide, is an effective sedative while s-thalidomide is a teratogen and causes severe birth defects if taken during pregnancy. Modest changes in the arrangements of the atoms in a compound can lead to drastic changes in the properties of different substances.

The composition of a compound is the same regardless of its source. The ***law of constant composition*** (also referred to as the law of multiple proportions or Proust's Law) states that the elemental composition of a given compound is always the same.

It was first stated by Joseph-Louis Proust near the turn of the 19th century and was based mostly on his studies of iron oxide. This means that a compound, when broken down into the elements it is composed of, will always produce the same ratio of elements. This law can be applied to water which contains hydrogen and oxygen. The composition of water when broken down to these two elements is 89% (by weight) oxygen and 11% hydrogen. This law is true for all compounds. No matter what the source of the sample, so long as it is pure, the composition of the elements in the compound will always be the same.

**PRACTICE** Classify each of the following as a compound or element: water, copper, sugar, chlorine, oxygen, and octane.

## Pure Substances and Mixtures

Nature most often consists of matter that is not pure in a chemical sense. A ***pure substance*** is matter with properties and composition that do not differ from sample to sample. Matter in nature often contains many different molecules, compounds and elements mixed together. The air we breathe is not pure as it contains oxygen, nitrogen, carbon dioxide, and other gases as well. In order to obtain a substance that is pure, different forms of matter have to be separated from one another. Sugar and table salt are examples of pure substances. To obtain these materials; however, they first had to be separated from the other materials they were mixed with originally. Once a substance is pure, its composition should be consistent throughout and should not differ sample to sample. The law of constant composition will apply to all substances that are pure.

Most materials in nature exist as ***mixtures.*** Mixtures can vary in their composition from sample to sample. Take salt water for example. Some mixtures of salt water contain lots of salt while others contain a little salt. These are not pure substances but mixtures. In this case, the law of constant composition will not apply because the composition of the mixture can vary from sample to sample.

Some mixtures contain the same composition and properties throughout a single sample. In the case of salt water, two different samples will have a different composition, but each individual sample will have a uniform composition throughout. Salt water is considered a ***homogenous mixture*** as individual samples have constant composition throughout. Other mixtures have an inconsistent composition and their properties may vary within an individual sample. Sand and granite are examples of mixtures with an inconsistent composition in a given sample or a ***heterogeneous mixture.*** Both samples vary in texture and composition in different portions of a sample. A very easy way to discern between homogeneous and heterogeneous mixtures is visually. One can see the variation in heterogeneous mixtures. The colors and shapes in a single slab of granite can have huge variations. In a sample of salt water, one cannot discern between the salt and the water it looks the same throughout the entire sample (Figure 1.3).

**PRACTICE** Classify each of the following as pure substance, homogenous mixture or heterogeneous mixture: rice pudding, feldspar, air, brass, sand, iron, and ethanol.

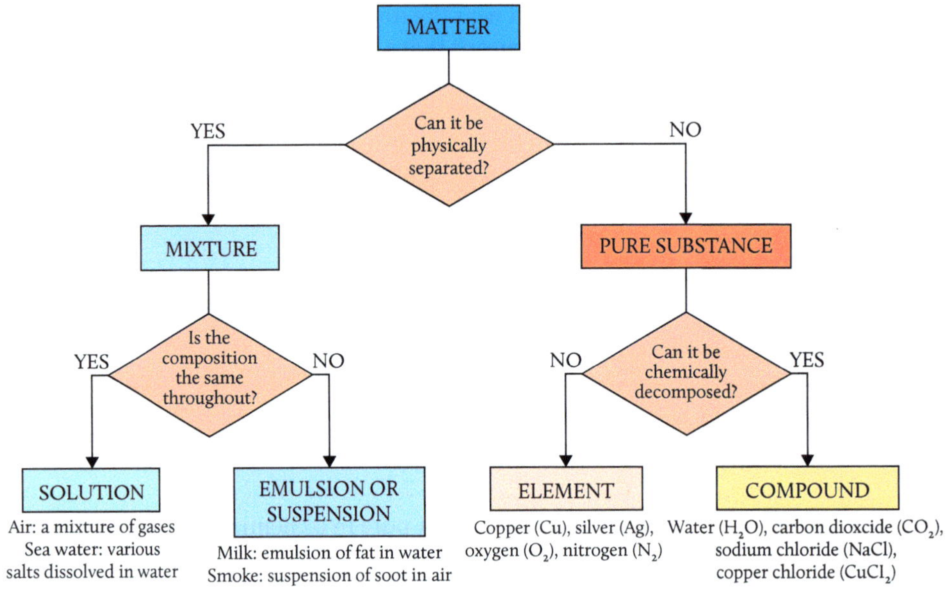


FIGURE 1.3. Classification of Matter Flow Chart.

# 1.2 The Structure of the Atom

## Atomic Theory

The word atom comes from *atomos* a Greek word meaning *uncuttable.* Ancient philosophers theorized that the world was made up of tiny particles that could not be divided (or cut) any further. It is remarkable how perceptive these philosophers were given the lack of experimental evidence for their ideas. A theory on the atom was formalized by John Dalton in his atomic theory. His theory contained the following postulates:

1. All matter is made up of indivisible and indestructible atoms.
2. All atoms of a given element are identical.
3. Atoms of different elements are different.
4. Atoms combine to form compounds.
5. Atoms cannot be created or destroyed during ordinary chemical reactions.

At the time, Dalton's atomic theory aligned with many of the chemical observations made to that point. However, technology during Dalton's time did not allow for direct evidence of the existence of atoms. Scientists later were able to study his postulates experimentally as new experimental methods were advanced. While some of Dalton's postulates hold true today, others have been modified based on experimental observations (Figure 1.4).

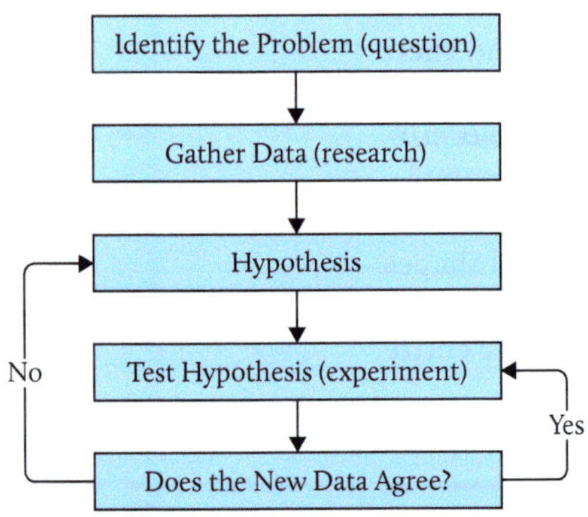


FIGURE 1.4. Scientific Method.

This procedure is called the ***Scientific Method.*** Today we are well-aware that atoms contain ***subatomic particles,*** thus disproving

Dalton's theory of atoms being "indivisible". Around the turn of the 20th century, J.J. Thompson performed a series of studies on cathode ray tubes and made some important observations. First, he observed that no matter the source of the atoms, the composition of the cathode rays was the same. This is significant when we consider Dalton's theory that atoms of different elements are different. Thompson's observation suggests that while the atoms are different, some part or component of them must be the same.

Thompson later discovered that he could deflect the cathode rays using magnetic fields. The beam would deflect away from negative magnetic fields and towards positive magnetic fields, suggesting that the cathode rays had a negative charge (Figure 1.5). These experiments are generally credited with the discovery of the ***electron*** and the fact that the atom can be divided into smaller parts.

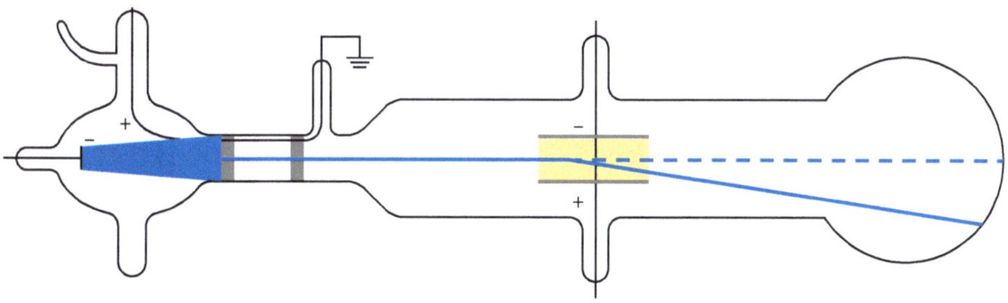


FIGURE 1.5. Cathode Ray Tube.

Based on his observations, Thompson developed a model of the atom that became known as the plum pudding model (Figure 1.6). He theorized that the atom was a sphere of positive charged electric field with electrons embedded in the field (similar to raisins in plum pudding). While he did not have direct evidence of the positively charged field, he knew something had to be present to balance out the negatively charged electrons he discovered. Just imagine the chaos if nature was only composed of negative charges! Based on experiments performed later he also concluded that electrons did not contribute much mass to the atom and were small relative to the atom itself.

The next scientist to further study the structure of the atom was a student of Thompson's, Ernest B. Rutherford. He wanted to provide evidence supporting the plum pudding model of the atom so he designed an elegant experiment to prove the atom was comprised mostly of empty space. This experiment is known as the gold foil experiment.

Rutherford theorized that if the atom was made mostly of empty space, and electrons were small he should be able to shoot something at atoms and it should travel straight through undeterred similar to a bullet passing through a bag of sand. The experiment used a very thin sheet of gold foil (the sand), positively charged particles (the bullets) to shoot at the gold foil and a photographic screen surrounding the foil to detect the particles that passed through the foil (Figure 1.7).

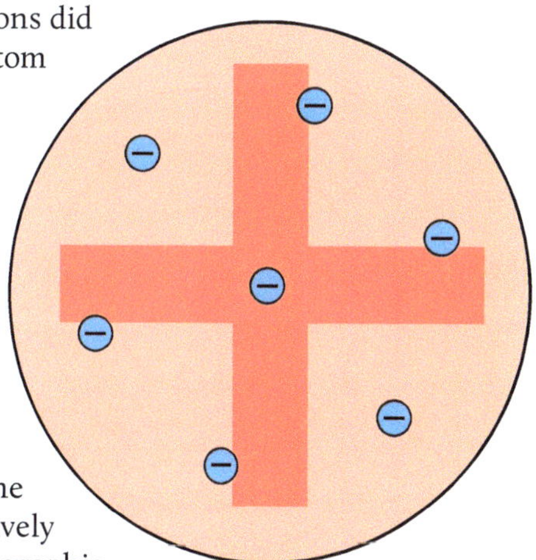

FIGURE 1.6. Plum Pudding Model.

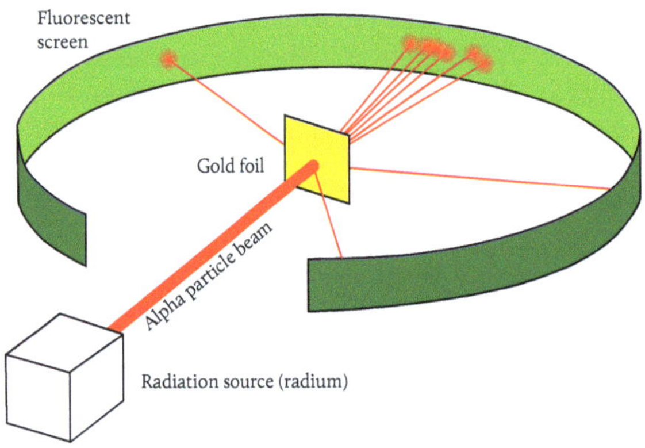


FIGURE 1.7A. Rutherford Experiment.

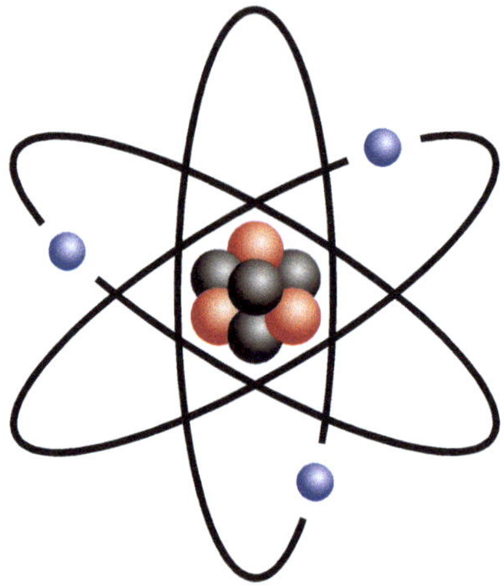

FIGURE 1.7B. Rutherford Experiment.

While Rutherford hypothesized that all of the particles should shoot straight through the foil, this was not what he observed. True to his hypothesis, a majority of the particles passed straight through the foil not encountering anything large enough to alter the trajectory. However, a small fraction of the particles were deflected at a slight angle and an even smaller fraction collided with something that sent them back to the source of the particles. This disproved Thompson's "plum pudding model".

Rutherford's observations led to a new model for the atom which more closely aligns with the modern view of the atom. Seeing that most of the particles shot straight through the gold foil, Rutherford concluded that atoms are comprised mostly of empty space. In addition the observation that a small fraction shot straight back at the source of the particles led to the conclusion that the atom contained a small but very dense center of mass known as the ***nucleus.*** Rutherford also concluded the cause of the deflection of some of the particles was due to the center of mass containing positive charge. Finally, Rutherford concluded that the electrons filled the empty space surrounding the nucleus. While his experiment disproved Thompson's plum pudding model it did provide evidence of a positive charge present in the atom.

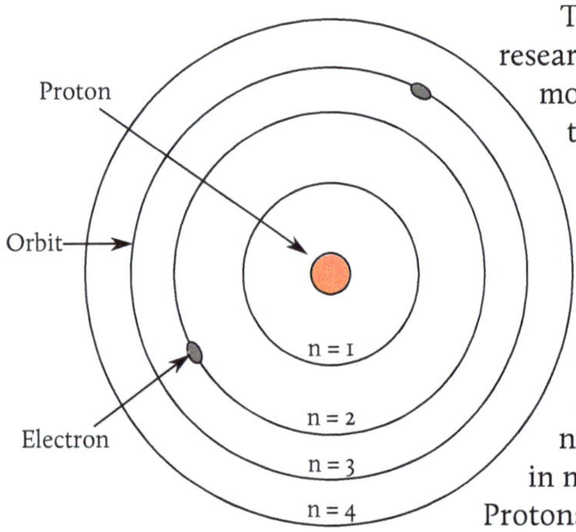


FIGURE 1.8. Bohr Model.

The modern view of the atom has been enhanced through the years of research by many different scientists. New particles were discovered, and more precise, complicated descriptions of the structure of the atom exist today. To simplify the structure of the atom for the purpose of this book, we will focus on three subatomic particles: protons, neutrons and electrons. Electrons are negatively charged and contain very little mass, so little mass that it does not have a significant bearing on the mass of an atom. The charge they are assigned is a charge of –1. Protons are positively charged with a charge of +1. The mass of a proton does contribute ~1 amu (atomic mass units) to the mass of an atom. Finally, atoms contain neutrons. As their name implies, neutrons are neutral with respect to charge. Neutrons are about equal in mass to protons and also contribute ~1 amu to the mass of an atom. Protons and neutrons are located in the nucleus and electrons are found surrounding the nucleus in discreet energy levels as part of an "electron cloud" (Figure 1.8). Understanding this simplified model of the atom will aid in your understanding of the topics that are discussed in this text.

## 1.3 Atomic Numbers, Isotopes, and Atomic Weights

### Atomic Number

The fundamental construction of the atoms of an element determines its physical and chemical properties. Of particular importance is the number of protons contained in the atom's nucleus and the corresponding number of electrons that circle the nucleus as an electron cloud. The number of protons present is called the ***atomic number*** and is unique to any given element. For example, the nucleus of carbon (symbol C) contains 6 protons, hence it is atomic number 6. A carbon atom cannot have say 8 protons. An atom with 8 protons would be the element oxygen, not carbon.

As mentioned earlier, each proton has a positive charge, of +1 and each electron a charge of –1. For an element that is electronically neutral (that is, no net charge) the number of electrons must equal the number of protons. So for our example of carbon there are six protons whose +6 charge is perfectly balanced, or neutralized, by six electrons which contribute a total charge -6. When combined, 6 positively charged protons and 6 negatively charged electrons created a neutral atom.

### Isotopes

Besides protons the nuclei of atoms may contain neutrons, well named since they have no charge. Neutrons have the same mass as protons do, and are far more massive than electrons are. As opposed to protons the number of neutrons in a given element can vary over a very wide range from 0 up to almost 300! Hydrogen (H) usually has no neutrons, but occasionally is found with 1 (deuterium) or 2 (tritium). These variations of atoms with the same number of protons but a different of neutrons are known as ***isotopes.*** The only difference in these atoms is their masses. Their other properties will remain the same including their behaviors in chemical reactions. Figure 1.9 depicts the isotopes of hydrogen.

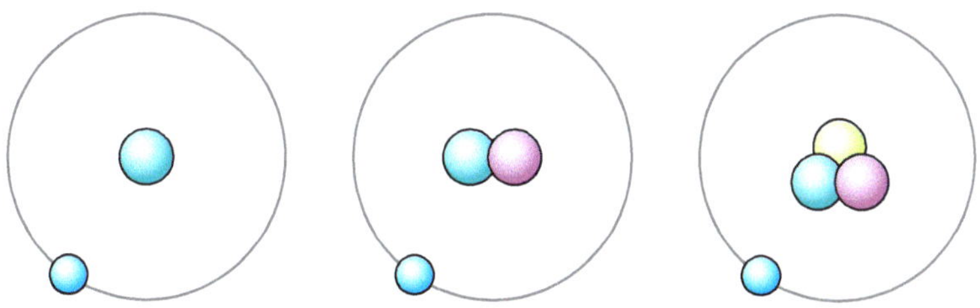

Protium ($^1H$) Deuterium ($^2H$) Tritium ($^3H$)

FIGURE 1.9. Carbon Isotopes.

The sum of the number of protons and neutrons in an atom is known as the ***mass number.*** Conventional notation shows the atomic number of an element as a subscript on the elemental symbol, and the mass number as a superscript as shown below for carbon, nitrogen and oxygen, three of the most important elements:

| | | |
|---|---|---|
| Carbon 12, 13, 14: | $^{12}_{6}C$ | 6 protons + 6 neutrons = 12 |
| | $^{13}_{6}C$ | 6 protons + 7 neutrons = 13 |
| | $^{14}_{6}C$ | 6 protons + 8 neutrons = 14 |
| Nitrogen 14, 15: | $^{14}_{7}N$ | 7 protons + 7 neutrons = 14 |
| | $^{15}_{7}N$ | 7 protons + 8 neutrons = 16 |
| Oxygen 16, 17, 18: | $^{12}_{6}C$ | 6 protons + 6 neutrons = 12 |
| | $^{13}_{6}C$ | 6 protons + 7 neutrons = 13 |
| | $^{14}_{6}C$ | 6 protons + 8 neutrons = 14 |

Once again, if the number of protons in an atom of an element changes the identity of element changes. Also, as mentioned previously the mass of an electron is vanishingly small compared to a proton or neutron. In fact, the difference is about five orders of magnitude.

### Atomic Weights

Naturally occurring elements are usually found as mixtures of isotopes. Chlorine exists in nature predominantly as chlorine-35 and chlorine-37. The 35 isotope has 17 protons and 18 neutrons while the 37 isotope has 17 electrons and 20 neutrons. Seeing as the isotopes occur in nature as a mixture, the effective atomic weight is a mathematical average of the relative abundances of the two isotopes. For our example of chlorine the calculation can be completed based on the abundances of the two isotopes mentioned. While other isotopes of chlorine do exist in nature, their abundances are negligible when calculating atomic mass. The isotope $^{35}Cl$ naturally occurs 75.8% of the time, and $^{37}Cl$ 24.2% of the time. This means the effective atomic mass, more commonly referred to as the atomic weight, is:

$$(0.758)(35) + \mathbf{(0.242)(37) = 35.5}\ \boldsymbol{amu}\ \mathbf{(}\boldsymbol{atomic\ mass\ units}\mathbf{)}$$

**PRACTICE** Copper occurs as two isotopes 63Cu at 69.2% and 65Cu at 30.8%. What is the effective atomic weight of copper?

## 1.4 Ionic Compounds

It is rare for elements to occur in a pure form. That is, atoms occur naturally in forms in which they have reacted with other atoms to form ions and molecules. This fact alone accounts for the impressive richness of chemistry as a science. There are two broad classes of molecules, ionic and molecular compounds. These classes will now be discussed.

Compounds are formed when atoms of elements react with each other. We can think of our reaction process as a system of reactants and products. Energy optimization is the driving force that makes these reactions either go or not go. So long as the energetics of the reaction is favorable, it will proceed. For the following discussion it will be useful to refer to the periodic table presented earlier. Here we will focus on a few of the groups at the horizontal extremes, more specifically Groups, 1(1A), II (IIA) and 14 (4A) through 17 (17A).

## Ion Formation

As mentioned previously, a change in the number of protons in an element's nucleus changes the identity of the element. For example, sodium ($_{11}$Na) has 11 protons in the nucleus. Adding another proton creates the element magnesium ($_{12}$Mg), an element with similar, but also substantially different physical and chemical properties. This proton addition is one type of nuclear reaction, very different from, but related to, chemical reactions.

The number of protons in a given element must remain constant, but the number of electrons may or may not do so. Any element in its pure, unreacted state possesses an equal number of electrons and protons. Upon undergoing a chemical reaction the number of electrons can change, either increasing or decreasing in number. The loss of electrons from a neutral (uncharged) atom creates a ***cation,*** one with a positive charge because the atom lost a negative charge. Similarly a gain of electrons creates an ***anion,*** one with a negative charge because the atom gained a negative charge. If an atom begins with a positive charge and loses an electron (shown as **e**$^-$) the positive charge increases by one. A negatively charged atom that gains an electron will become more negative by one.

Examples are:

| | | | |
|---|---|---|---|
| Li | ⟶ | $Li^{+1}$ + 1e$^-$ | $Li^{+1}$ is a cation |
| $Mg^{+1}$ | ⟶ | $Mg^{+2}$ + 1e$^-$ | $Mg^{+1}$ and $Mg^{+2}$ are both cations |
| F + 1e$^-$ | ⟶ | $F^{-1}$ | $F^{-1}$ is an anion |
| $O^{-1}$ + 1e$^-$ | ⟶ | $O^{-2}$ | $O^{-1}$ and $O^{-2}$ are both anions |

## Ionic Compounds

By definition ***ionic compounds*** are formed by a ***complete electron transfer*** from one element (or polyatomic entity) to another (Figure 1.10). An electron is a negative charge so its loss leaves behind a positive charge of exactly the same magnitude. The descriptor word complete above should be emphasized. As a metaphor consider the nucleus and an electron as a child walking with a candy bar. Consider this is the neutral atom before electron transfer. Now another child comes along and takes the candy bar away. The candy bar transferred completely from one child to the other.

By way of chemical example consider a sodium (Na) atom reacting with a chlorine (Cl) atom. Na is from group IA, Cl from group 7A in the periodic table.

$$Na + Cl \longrightarrow Na^+ + Cl^-$$

The process can be broken down into two steps as follows:

$$Na \longrightarrow Na^+ + 1e^-$$
$$Cl + 1e^- \longrightarrow Cl^-$$

The sodium is losing or giving away an electron to chlorine. Chlorine is accepting the electron and gaining a negative charge. Sodium gets a positive charge as it lost a negatively charged electron. Group 1A members (for example Li, Na, K) are all extremely reactive metals that upon reaction lose or give away one, and only one, electron (e$^-$). Group 2A metals (Mg, Ca) only ever give up two electrons. Group 7A elements (F, Cl, Br, I)

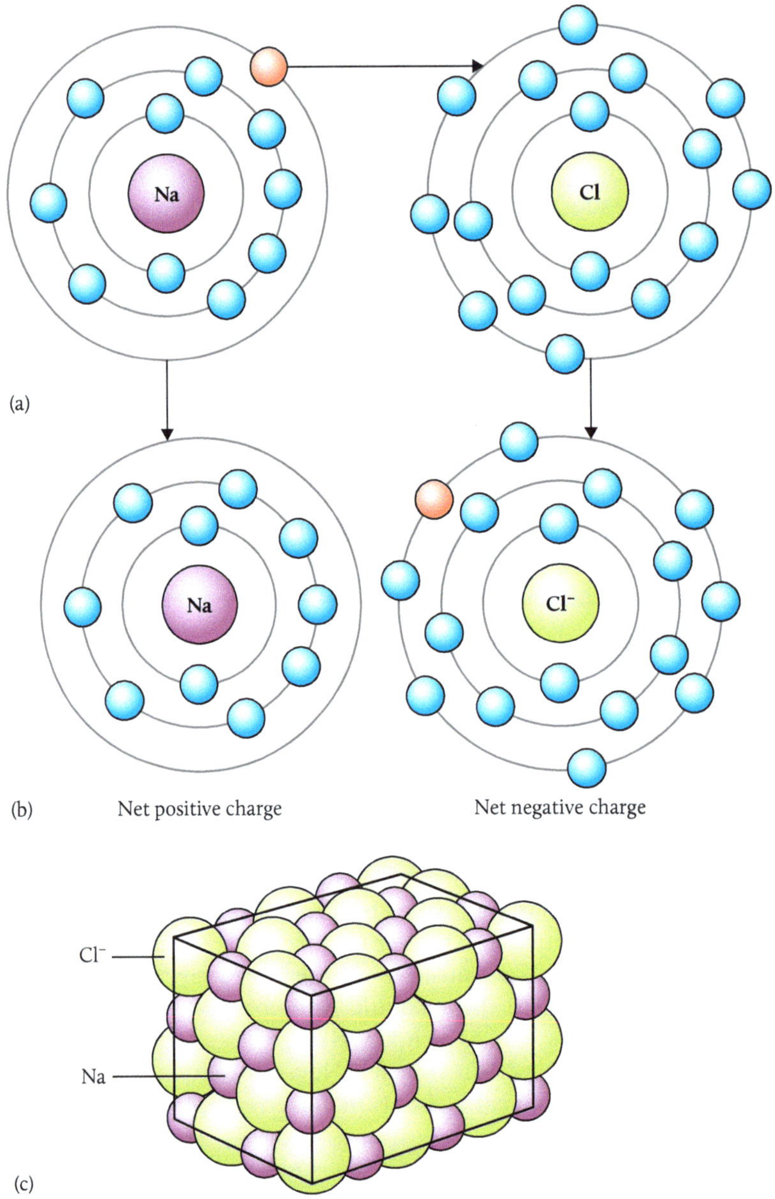


**FIGURE 1.10.** Ionic Compound.

generally do the reverse, that is, they accept an electron. Another complete reaction example is:

| | | | | | | | |
|---|---|---|---|---|---|---|---|
| K | $\longrightarrow$ | $K^+$ | + | $1e^-$ | | | |
| Br | + | $1e^-$ | $\longrightarrow$ | $Br^-$ | | | |
| *combined:* | **Br** | + | **K** | $\longrightarrow$ | $\mathbf{K^+}$ | + | $\mathbf{Br^-}$ |

In the last step the two equations are combined with the electrons, common to both sides of the equation, being removed.

**PRACTICE** Magnesium (Mg) is a group 2A metal that readily gives up two electrons. Oxygen (O) is a non-metal from group 6A that readily accepts two electrons. Write out the two half reactions and full reaction for the process involving electron transfer from Mg to O. Use the example giving above from K and Br above as a model keeping in mind that the latter example involves a transfer of just one $e^-$ rather than two.

## Predicting Charges, Ionic Formulas, and Ionic Compound Formation

A key skill in chemistry is the ability to predict chemical formula and structures based upon the elements involved, and where they occur in the periodic table. Again the discussion is now focused upon just a few groups and elements of the periodic table, but these turn out to also be some of the most important ones industrially, commercially, and especially, biologically. Here are the groups and elements that will be focused on:

Group 1A: Li, Na, K
Group 2A: Mg, Ca
Group 5A: N, P
Group 6A: O, S
Group 7A: F, Cl, Br, I

Remember that the groups are arranged vertically in the periodic table and atoms in the same group share similar properties. Lithium (Li), sodium (Na) and potassium (K) are very reactive metals that always give away just one electron to take on a +1 charge as cations. Magnesium (Mg) and calcium (Ca) give away only two electrons and take on a +2 charge, also as cations. On the other side nitrogen (N) and phosphorous (P) can readily accept three electrons, oxygen (O) and sulfur (S) two electrons and fluorine (F), chlorine (Cl), bromine (Br) and iodine (I) one electron. The reasons for these abilities will be discussed at a later time but knowing these trends allows us to predict ionic compound formulas.

Now let us assume we wish to form a compound between Ca and F that is neutral, that is, has no net electrical charge.

Here is the equation for Ca:

$$Ca \longrightarrow Ca^{+2} + 2e^-$$

For fluorine (F):

$$F + 1e^- \longrightarrow F^-$$

Combining them yields:

$$F + Ca + 1e^- \longrightarrow Ca^{+2} + F^- + 2e^-$$

But the number of electrons on the opposite sides of the last equation are unequal and will not cancel out as required. However, if we multiply the second equation through by two and then combine it with the first the electrons will cancel out:

$$2F + 2e^- \longrightarrow 2F^-$$

*Combined with Ca:* $2F + Ca + 2e^- \longrightarrow Ca^{+2} + 2F^- + 2e^-$

A simple formula for this ionic compound is $\mathbf{CaF_2}$. The calcium atom gave up two electrons with one electron transferring to a fluorine atom and the second electron transferring to a second fluorine atom. By convention, the two fluorine atoms are denoted in the $\mathbf{CaF_2}$ formula is by the subscript 2 beside the F, *not* Ca2F or CaF2. By reading a chemical formula, one can determine not only the types of atoms present, but the number of each atom as well. ***It is essential that you learn these formulaic conventions in order to understand the language of chemistry.***

Similarly we can predict what products would be formed by the reactions of the other elements listed above.

---

**PRACTICE** Predict the formulas of the ionic compounds that can be formed between the cations of Li, K, and Mg and the anions of P, S, Br, and I.

---

One more important point should be made concerning ionic compounds. At the instant the electron transfer(s) occur, the generation of the positive cations and negative anions also produces an electronic attraction between the opposite charges. This attraction can be quite strong and referred to as an ***ionic bond.*** A simile for this is the attraction of the north and south poles of a magnet. Also like magnets, charges of the same sign will strongly repel each other as such, cations do not combine with other cations and anions do not combine with other anions.

There are 118 known elements (gulp) presented on a modern periodic table, of greater or lesser importance. We will get to know the relatively few that are the critical and essential ones. As discussed further in the next section not all compounds are formed using complete electron transfer chemistry. Can we predict when they will? Yes, and two general rules will serve us well:

***If a metal cation is present there is an ionic bond present.***

***If the elements come from the extreme left and right of the periodic table, an ionic bond is likely present.***

Like much of chemistry there are exceptions to the rules. In this case group 8A (18) on the far right hand side contains the so-called noble gases that are not of much interest because, for practical purposes, they do not undergo chemical reactions. So we can ignore them, at least for today.

# 1.5 Molecular Compounds

## Covalent Bonding

As described above ionic compounds are characterized by a complete transfer of one or more electrons from one atom to another. Molecular compounds are characterized by an ***incomplete electron transfer.*** More specifically the electrons are ***shared*** between atoms rather than being completely transferred. The sharing of electrons between atoms forms a ***covalent bond*** (Figure 1.11). Based on this bond formation, molecular compounds are often referred to as covalent compounds. Because electrons are shared in molecular compounds and not transferred, they do not contain a charge. Going back to our candy bar example from before, instead of one child taking the candy bar away, in this case the two children would share the candy bar.

As with ionic compounds, molecular compounds are represented by a chemical formula using the elemental symbols and numerical subscripts to represent the type and number of each atom present. As mentioned earlier, there are seven common diatomic (two atoms) *elemental molecules* which have no ionic character whatsoever. They are $O_2$ (dioxygen), $N_2$ (dinitrogen), $H_2$ (dihydrogen), $F_2$ (difluorine), $Cl_2$ (dichlorine), $Br_2$ (dibromine), and $I_2$ (diiodine). The first two in this list comprise close to 100 percent of the earth's atmosphere. The reader will notice that the last four are members of Group 7 (17) in the periodic table.

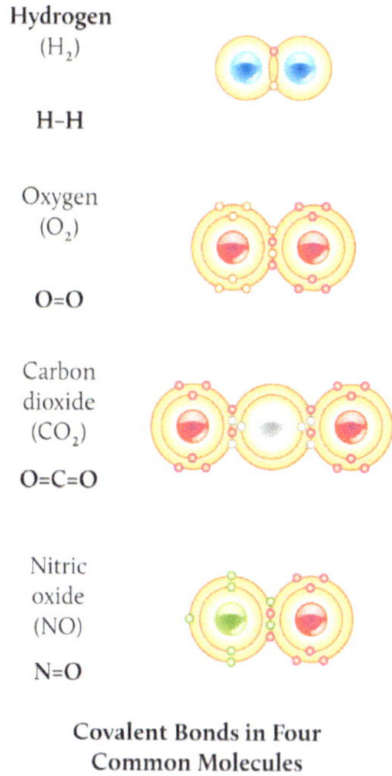


**FIGURE 1.11.** Molecules.

## Molecular Compounds

***Molecular compounds*** form between atoms of *different* elements. They are simply chemicals that contain more than one type of atom and the atoms are held together by electrons that are shared in a covalent bond. There are endless possible combinations of atoms to form unique molecular compounds and this explains the complexity and diversity of compounds found on Earth. Here a few important examples:

**$CO_2$: Carbon Dioxide** contains one carbon bonded to two oxygen atoms. The formula demonstrates an important point, if there is no subscript after an element's symbol, one (1) assumed.

**$CH_4$: Methane** is better known as natural gas. It has one carbon bonded with four hydrogens.

**$C_{12}H_{22}O_{11}$:** This compound is best known as a **sugar!** It contains 12 carbon atoms, 22 hydrogen atoms and 11 oxygen atoms.

Molecular compounds are generally formed from combinations of elements contained in the upper right hand corner of the periodic table plus hydrogen. Once again the noble gases in group 8 are unreactive and exist in their elemental forms. This is due to the fact that elements on the right side of the periodic table (with the exception of the noble gases) are more likely to share electrons with each other rather than transfer electrons from one atom to another. Applying the rules we discussed in the previous section, when trying to discern whether a compound is ionic or molecular, look to see if a metal atom is present. If there is a metal atom presence chances are extremely high that the compound is ionic.

We will learn later that molecular compounds and ionic compounds not only differ in the way in which they are chemically combined, but also in the behavior in terms of physical properties. These behaviors are rooted in the manner in which the atoms in the compounds are held together. Moving forward it will be important to have the ability to differentiate between ionic and molecular compounds simply by looking at the formula given. It will also be important to understand the difference between an ionic bond and a covalent bond.

# Chapter 1 Summary

**Chemistry** is the study of the composition, properties and behavior of ***matter*** in all of its forms. Matter is anything that has mass and takes up space.

**Atoms** are the building blocks of matter. There are over 100 unique types of atoms and each unique atom is called an **element.**

**Molecules** are formed when two or more atoms are held together by a chemical bond.

**Compounds** form when two or more atoms from different elements are joined by chemical bonds.

**Law of constant composition** states that the elemental composition of a given compound is always the same.

**Pure substances** is matter with properties and composition that do not vary from sample to sample. The can be made up of an element or a compound.

**Mixtures** vary in composition from sample to sample and are formed when more than one element or compound are added to a sample.

***Homogeneous mixture*** as individual samples have constant composition throughout a sample while **heterogeneous mixtures** have variable composition throughout a sample.

**Dalton's Atomic Theory** states that:

1. All matter is made up of indestructible atoms.
2. All atoms of a given element are identical.
3. Atoms of different elements are not identical.
4. Atoms combine to form compounds.
5. Atoms cannot be created or destroyed.

**Covalent bonds** form when two atoms share electrons and are found in molecular compounds.

**Ionic bonds** form when an electron is transferred from one atom to another atom and form ionic compounds.

The **nucleus** is the dense center of an atom and contains positively charged **protons** and uncharged **neutrons.** Negatively charged **electrons** surround the nucleus arranged in discrete energy levels.

**Isotopes** are different atoms of the same element that vary in their number of neutrons.

The **atomic number** (Z) of an atom represents the number of protons. Each element has a unique atomic number. Changing the number of protons changes an atom to a different element.

The **mass number** (A) is the number of protons and neutrons contained in an atom's nucleus. Atoms of the same element can have different mass numbers and represent different isotopes of an element.

# CHAPTER 1 EXERCISES

1. Classify each of the following as a pure element, pure compound, homogenous mixture or heterogeneous mixture:

   a. Granite countertop

   b. Aluminum foil

   c. Ice water

   d. Baking soda

2. What is an isotope?

3. Is a glass of orange juice with pulp a pure substance? If not, what type of substance is it?

4. What happens when the number of protons of an atom changes?

5. Determine the formula for the ionic compound when the following elements combine:

   a. sodium and fluorine

   b. chlorine and chlorine

   c. potassium and oxygen

   d. sulfur and magnesium

6. Bromine has two naturally occurring isotopes. Bromine-79, has a mass of 78.92 amu and an abundance of 50.69%. The other isotope is bromine-81 has a mass of 80.92 amu.

   a. What is the percent abundance for bromine-81?

   b. What is the average atomic mass for bromine?

7. Explain how an ionic bond is formed.

8. What is difference between an ionic bond and a covalent bond?

9. Chromium has four naturally occurring isotopes. Chromium-50 (49.95 amu) has an abundance of 4.35%, chromium-52 (51.94 amu) with an abundance of 83.79%, chromium-53 (52.94) with an abundance of 9.50% and chromium-54 (53.94 amu).

   a. What is the abundance of chromium-54?

   b. What is the average atomic mass of chromium?

10. Name the different subatomic particles.

11. What happens when an atom gains and electron?

12. What happens when an atom loses an electron?

13. What is the identity of the element with an atomic number of 48? How many protons in this element?

14. Caffeine has 8 carbon atoms, 10 hydrogen atoms, 4 nitrogen atoms, and 2 oxygen atoms. What is the molecular formula of caffeine?

15. Calculate the atomic mass of an atom has 24 neutrons and a mass number of 44. How many protons in this atom? What is the identity of this atom?

16. An oxygen atom contains 8 neutrons and 8 electrons.

    a. How many protons does this atom have? Explain how you determine this number.

    b. What is the mass number? Explain how you determine this number.

    c. What is the charge on the atom? What determines the charge of an atom?

17. Classify each of the following as an ionic or molecular compound.

    a. NaCl

    b. $CH_4$

    c. $H_2O$

    d. $NH_3$

18. Classify the substances depicted below as an element, compound, or mixture:

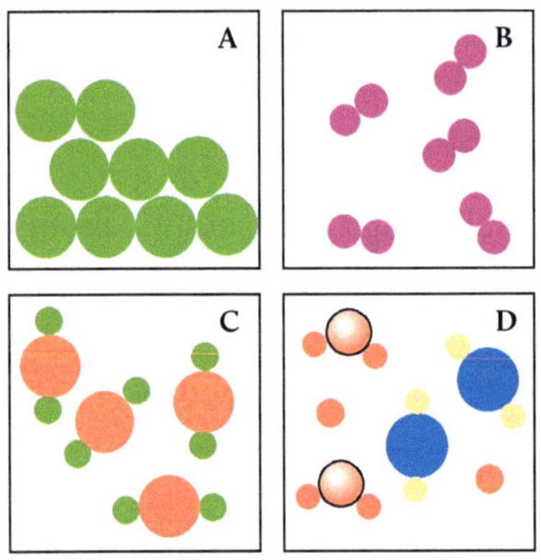


IMAGE 1.1.

19. Using your knowledge of isotopes, complete the following table.

| Element | Symbol | Atomic Number | Mass Number | # of Protons | # of Electrons | # of Neutrons | Charge |
|---|---|---|---|---|---|---|---|
| | | | 199 | | | 120 | +5 |
| | | | | 35 | 36 | 46 | |
| | | 54 | 136 | | | | 0 |
| Chromium | | | | | 18 | 29 | |

# Credits

Fig. 1.1: Copyright © 2006 by Mstroeck, (CC BY-SA 3.0) at https://commons.wikimedia.org/wiki/File:Caffeine_Molecule.png.

Fig. 1.2a: Source: http://www.chem4kids.com/files/art/elem_pertable3.png.

Fig. 1.2b: Source: http://www.chem4kids.com/files/art/elem_pertable2.png.

Fig. 1.3: Copyright © 2014 by Siyavula Education, (CC BY 2.0) at https://www.flickr.com/photos/121935927@N06/13579731025.

Fig. 1.4: Copyright © 2014 by Michael Fullerton, (CC BY-SA 4.0) at https://commons.wikimedia.org/wiki/File:The_Scientific_Method_Blue.png.

Fig. 1.5: Source: https://commons.wikimedia.org/wiki/File%3AJJ_Thomson_Cathode_Ray_Tube_2.png.

Fig. 1.6: Copyright © 2014 by Kurzon, (CC BY-SA 4.0) at https://commons.wikimedia.org/wiki/File%3APlum_pudding_model.svg.

Fig. 1.7a: Copyright © 2014 by Kurzon, (CC BY-SA 3.0) at https://commons.wikimedia.org/wiki/File:Geiger-Marsden_experiment.svg.

Fig. 1.7b: Copyright © 2007 by Indolences, (CC BY-SA 3.0) at https://commons.wikimedia.org/wiki/File:Stylised_atom_with_three_Bohr_model_orbits_and_stylised_nucleus.svg.

Fig. 1.8: Copyright © 2010 by Sharon Bewick and Adrignola, (CC BY-SA 3.0) at https://commons.wikimedia.org/wiki/File%3ABohr's_model.svg.

Fig. 1.9: Copyright © 2013 by OpenStax College, (CC BY 3.0) at https://commons.wikimedia.org/wiki/File:204_Isotopes_of_Hydrogen-01.jpg.

Fig. 1.10: Copyright © 2013 by OpenStax College, (CC BY 3.0) at https://commons.wikimedia.org/wiki/File:207_Ionic_Bonding-01.jpg.

Fig. 1.11: Copyright © 2016 by BruceBlaus, (CC BY 4.0) at https://commons.wikimedia.org/wiki/File:Covalent_Bonds.png.

CHAPTER 2

# The Periodic Table

## (is not nearly as confusing as it looks)

The information contained in the periodic table underlies all of chemistry. Although there are 118 known elements this number is a bit deceiving. For example, only the first 94 are found in nature, many in vanishingly small quantities. The remainder of the elements, numbers 95–118, are manmade, and not of much interest from a practical standpoint. Just as importantly the first 94 elements vary widely in their relative importance. We will focus on the elements of greatest importance and usefulness.

## 2.1 Periodic Table Organization

### The Basics

Fortunately for all of us the chemical and physical properties of the elements are related to each other in systematic and predictable ways. The periodic table is organized based on repeating patterns in the properties of elements. It can thus be segmented or mentally organized based upon some fundamental properties of the elements, and groups of elements. Let us look at some of these relationships as summarized in Figure 2.1.

The vertical columns of elemental cells are known as ***groups*** while the horizontal rows are called ***periods***. At the top of the table is a set of numbers identifying the various groups. There is a second set of numbers as follows: 1A, 2A, 3A, (3B, 4B, 5B, 6B, 7B, 8B, 1B, 2B) 4A, 5A, 6A, 7A, 8A is an older American system that has been in usage for a long, long time. In this system the A groups represent the main elements, main meaning those of particular interest biologically and commercially. The B groups contain the transition elements; so-called because there is an orderly transition in characteristics from left to right across the table. The second system is simpler, and it numbers the groups from 1 to 18. The periods are simply numbered 1–7 in either system. We will use the newer systems of numbering in this book.

The reader will note that element numbers 57–70 (*), and 89–102 (**) are physically separated from the whole of the periodic table. These periods are known as the lanthanide and actinide series, respectively. They are relatively rare, and will generally not be involved in our further discussions of chemistry here. The reason they are separated out is merely one of practicality. Doing so allows us to look at (and print out) the periodic table on a single 81/2 × 11 page.

| Group→ / ↓Period | 1 | 2 | | 3 | 4 | 5 | 6 | 7 | 8 | 9 | 10 | 11 | 12 | 13 | 14 | 15 | 16 | 17 | 18 |
|---|---|---|---|---|---|---|---|---|---|---|---|---|---|---|---|---|---|---|---|
| 1 | 1 H | | | | | | | | | | | | | | | | | | 2 He |
| 2 | 3 Li | 4 Be | | | | | | | | | | | | 5 B | 6 C | 7 N | 8 O | 9 F | 10 Ne |
| 3 | 11 Na | 12 Mg | | | | | | | | | | | | 13 Al | 14 Si | 15 P | 16 S | 17 Cl | 18 Ar |
| 4 | 19 K | 20 Ca | | 21 Sc | 22 Ti | 23 V | 24 Cr | 25 Mn | 26 Fe | 27 Co | 28 Ni | 29 Cu | 30 Zn | 31 Ga | 32 Ge | 33 As | 34 Se | 35 Br | 36 Kr |
| 5 | 37 Rb | 38 Sr | | 39 Y | 40 Zr | 41 Nb | 42 Mo | 43 Tc | 44 Ru | 45 Rh | 46 Rh | 47 Ag | 48 Cd | 49 In | 50 Sn | 51 Sb | 52 Te | 53 I | 54 Xe |
| 6 | 55 Cs | 56 Ba | * | 71 Lu | 72 Hf | 73 Ta | 74 W | 75 Re | 76 Os | 77 Ir | 78 Pt | 79 Au | 80 Hg | 81 Tl | 82 Pb | 83 Bi | 84 Po | 85 At | 86 Rn |
| 7 | 87 Fr | 88 Ra | ** | 103 Lr | 104 Rf | 105 Db | 106 Sg | 107 Bh | 108 Hs | 109 Mt | 110 Ds | 111 Rg | 112 Cn | 113 Uut | 114 Fl | 115 Uup | 116 Lv | 117 Uus | 118 Uuo |
| | | | * | 57 La | 58 Ce | 59 Pr | 60 Nd | 61 Pm | 62 Sm | 63 Eu | 64 Gd | 65 Tb | 66 Dy | 67 Ho | 68 Er | 69 Tm | 70 Yb | | |
| | | | ** | 89 Ac | 90 Th | 91 Pa | 92 U | 93 Np | 94 Pu | 95 Am | 96 Cm | 97 Bk | 98 Cf | 99 Es | 100 Fm | 101 Md | 102 No | | |

**FIGURE 2.1.** Periodic Table.

## Cell Information

Each cell in the periodic table contains fundamental information about the element. First is the elemental symbol. For example rubidium (Group 1, period 5) has been given the symbol **Rb**. The symbol contains a maximum of two letters, and the first letter is capitalized. As far as naming goes the elements have been named in many different ways. The name rubidium is derived from the Latin meaning red. Of particular note here is that some of the most important elements have unexpected names as follows:

| Element | Symbol | Name |
|---|---|---|
| Sodium | Na | Natrium |
| Potassium | K | Kalium |
| Copper | Cu | Cuprous or Cupric |
| Iron | Fe | Ferrous or Ferric |

In this regard, maybe the most confusing is the symbol **P** which stands for phosphorus, not potassium, which is **K**.

Also contained at the top of each cell is the atomic number which identifies the number of protons and electrons present in an element's pure (unreacted) state. Underneath the elemental symbol is the atomic masses. Recall this number is the statistical average of the various isotopes of an element as discussed in Chapter 1. Periodic tables often contain much more information about a given element, but the simplified one shown here contains enough for now.

Let's briefly look at the most important groups from left to right across the table. The focus will be upon the most important elements within each of these groups.

***Group 1 (red): Alkali Metals:*** The most reactive elements. Upon reaction these elements *always* take a charge (valence) of +1. In all biological systems a couple of these elements are of critical importance.

Most important members: Sodium (Na) and Potassium (K)

***Group 2 (pink): Alkaline Earth Metals:*** Reactive elements. Upon reaction *always* take a charge (valence) of +2. Biologically these are of critical importance also.

Most important members: Magnesium (Mg) and Calcium (Ca)

***Groups 3-12 (more pink left to right): Transition Metals:*** All are metals which constitute the majority of the periodic table in terms of numbers, but not importance.

Some very common and important members: Iron (Fe), Copper Cu), Zinc (Zn), Nickel, (Ni), Mercury (Hg), Silver (Ag), and Gold (Au).

***Groups 13-15 (purple): Post Transition Metals:*** Somewhat of an orphan group in terms of properties, but one that includes three important, and well-known metals: Aluminum (Al), Lead (Pb), and Tin (Sn).

***Groups 13-16 (green-brown): Metalloids:*** This interesting set of elements possesses properties that are in between those of the metallic elements to their left and the non-metallic elements to their right. Virtually all electronic devices in use today depend upon one or more of these elements to function correctly, or to function at all.

Especially important member: Silicon (Si)

***Groups 14-16 (greens): Non-metals:*** Along with hydrogen (alone, top left) five of these elements (C, N, O, S, P) are easily the most important elements in the periodic table. All life as we know it is dependent upon these six elements.

Critically important members: Carbon (C), Nitrogen (N), Oxygen (O), Sulfur (S), Phosphorous (P)

***Group 17 (green): Non-metals:*** Also called the halogens, these are especially important in inorganic chemistry, but are also biologically active as well.

Important members: Chlorine (Cl), Bromine (Br), and Iodine (I)

***Group 18 (blue): Non-metals:*** These gases are not of much general interest as they are very unreactive, hence their common moniker the noble gases. They do not form useful compounds.

Important members: Helium (He), Argon (Ar)

Now that we have partitioned the periodic table into related sectors let us look at each sector in more detail, but before we do, let us consider the subject of **chemical reactivity** as it relates to **valence electrons**.

## 2.2 Valence Electrons

***It is critical that the chemistry student understand the concept of valence electrons.*** As described in Chapter 1 each atom in an element consists of protons contained in the nucleus that is surrounded by a cloud of electrons that swirl around it at blindingly fast velocities. Prior to any chemical reaction the number of electrons (each with a –1 charge) and protons (each with a +1 charge) is equal, so the atom has no net electrical charge (–1 + 1 = 0). Upon reaction the situation gets much more interesting as electrons can be transferred or shared between different atoms. This is how new chemicals and materials are formed regardless of whether these events occur in nature or are made to occur synthetically by man. Some would argue the difference between the two processes negligible.

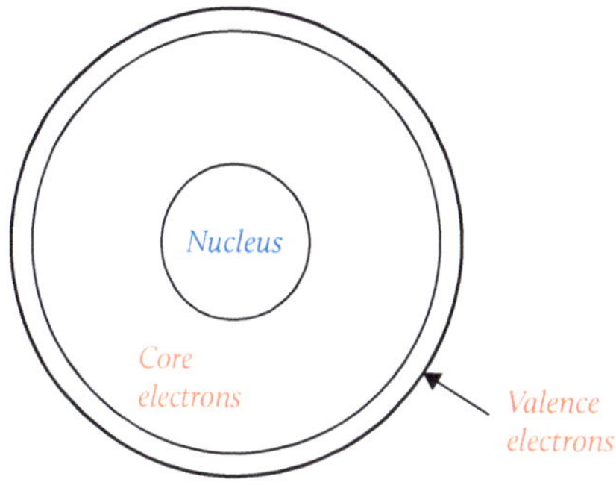


FIGURE 2.2. Generic Atom.

Hereto it is fortunate for all of us that the electrons in an atom are not created equally. First let's assume the nucleus to be stationary (a good approximation), but having an attractive positive force for the electrons that is directly proportional to the number of protons present. A good analogy is our solar system with the sun in the middle, and the planets circling around. Similar to gravitational attractions, the electrons closest to the nucleus experience the strongest nuclear attractive forces, and thus are the most difficult to separate (pull) away. Looking at Figure 2.2 we have the nucleus, followed by the ***core electrons***, and finally the outer or ***valence electron(s)***. It turns out that the electrons exist in energy shells sort of like the layers of an onion. It also turns out that for stability reasons most elements much prefer to have 2 or 8 electrons in their outermost electron shell. Why this is so is a topic we will revisit in Chapter 3. For now please accept this fact at face value.

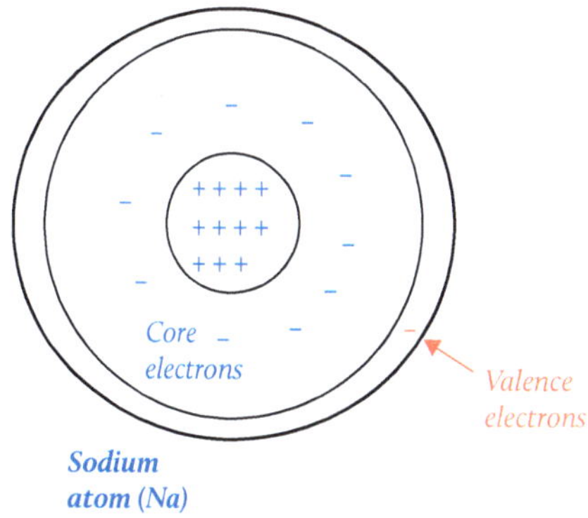


FIGURE 2.3. Na Core-Valence E–.

Now let us look at a real and very important example of our atomic "onion" for the case of sodium (Na). Na occupies the periodic table cell at Group 1, Period 3 (Figure 2.3). Sodium is atomic number 11 so, prior to reaction, it contains 11 protons and 11 electrons. The eleven protons are unchanging, but the electrons are split into two basic categories, a set of 10 electrons, known as **core electrons**, that are between the nucleus and the lone outer **valence electron** (the 11$^{th}$ electron). The inner 10 electrons are found in shells 1 and 2 while the valence electron is found in shell 3, the outermost shell.

*It turns out that in most cases the reaction of an element can be predicted simply by looking at the number of valence electrons present.*

Upon reaction sodium will give up its single valence electron to form a sodium cation with 11 protons, 10 electrons and a +1 charge as follows:

$$Na \longrightarrow Na^{+1} + 1e^{-}$$

The core electrons are not shown in the chemical equation as they have no bearing on the reaction. An important aside is that Na is from group 1 and always forms a +1 charge cation upon reaction. The same is true for the rest of the elements in group 1. This makes the number of valence electrons easy to remember.

How about group 2? Virtually the same, but two valence electrons are now present. Take calcium:

$$Ca \longrightarrow Ca^{+2} + 2e^{-}$$

Group 2 elements, two valence electrons. The same equation holds true then for the rest of the group 2 elements Mg, Sr and Ba as well.

Let's consider group 17, also known as the halogens. In this case the the atoms in this group have 7 valence electrons. Fluorine's atomic number 9 so in its unreacted state it possesses 9 protons in the nucleus and 9 electrons, just 2 of which are core electrons, the remaining 7 are valence electrons. To achieve the desired 8 valence electrons, F has two realistic options. One is to lose 7 electrons leaving the core of 2 electrons:

$$F \longrightarrow F^{+7} + 7e^-$$

The other option is to steal an electron from some other element:

$$F + e^- \longrightarrow F^{-1}$$

The latter option is simpler, and the one that does occur. Where does this electron come from? Fluorine will gain this electron from number of possibilities here, but say sodium is available:

$$Na + F \longrightarrow Na^+ + F^-$$

This is exactly what would occur if we brought these two elements together to form the compound sodium fluoride - sodium will give up one valence electron to fluorine. This would be an explosive reaction, also something it turns out we can predict! These are valuable chemical tools. All of a sudden we will be able to design and synthesize compounds with beneficial properties. Just as importantly we can assess potentially hazardous chemicals and synthetic procedures.

**Alchemy**

It is important to understand the difference between reaction chemistry and nuclear chemistry. In a typical chemical reaction electrons exchange or sharing of electrons. Although the atomic nuclei are important in terms of a reaction's course, they do not change composition. That is, the number of protons and neutrons present in the nuclei do not change, during a chemical reaction.

In a nuclear reaction the numbers of protons and neutrons can change. In essence this is alchemy, that is, the changing of one element into another. Here is a simple example:

Li (3 protons) He (2 protons) + 1 proton + tremendous energy release

Lithium is atomic number 3. If it is forced to lose a proton helium (atomic number 2) is formed and a phenomenal amount of energy is released. It also takes a phenomenal amount of energy to initiate this reaction. Neutron nuclear reactions act similarly. Such processes involve orders of magnitude larger amounts of energy than do chemical reactions and form the bases for nuclear power systems and weapons.

Let us now look at some key properties of the periodic table as a whole, and then some general properties of the especially important groups and elements contained therein.

## 2.3 Periodic Property Trends

### Effective Nuclear Charge, Zeff

It should have been noted already that as we go across a period changing elements a proton and an electron are added with each step (cell) to the right as we travel along. As the protonic (+) charge increases the attractive force the nucleus imparts to the electrons also increases. Recall too that the electrons are not randomly distributed within an electronic cloud, but rather only occur at energy levels set at discrete distances away from the nucleus. For the time being let us focus on period 2 which contains elements Li, Be, B, C, N, O, F, and Ne; atomic numbers 3,4,5,6,7,8,9, and 10. Li has 3 protons and three electrons and so on up to Ne with 10 protons and 10 electrons.

Consider our onion metaphor again where the proton-containing nucleus is at the center of a sphere of electrons distributed into layers of energy shells containing electrons. It turns out that the first layer can only contain two electrons, the second layer just 8 at a maximum. For both cases the inner shell of electrons is nearer the nucleus and thus feels a stronger attractive pull from the nucleus. As the atomic number increases this pull also goes up. So the attraction between the neon nuclei and its inner shell electrons is stronger than that of lithium's nucleus to its inner electrons because neon has more protons and all its electrons are sheilded but the same number of electrons as lithium. For neon this has the effect of pulling in the electrons and shortening the distance between the nucleus and the electrons.

The remainder of the electrons are 1 valence (3 total with 2 inner or core electrons that can shield) for lithium and 8 valence (10 total and 2 core electrons that can shield) for neon. These valence electrons go into the second layer of our onion, one that is further away from the nucleus and thus held in by weaker attractive forces. Figure 2.4 shows the overall effect for the elements contained in periods 2 and 3.

For the inner or first electron layer as the atomic charge increases the effective nuclear charge Zeff, a measure of the strength of the attraction between + and – charges, also goes up in a linear manner. After filling the first layer with those two electrons, the next ones go into the next higher energy layer or shell. This is at some further distance away from the nucleus; thus the attractive force between the electrons and the nucleus is weaker, recall the solar system analogy discussed earlier in this chapter. The further away an electron is from the nucleus, the weaker the attraction it will experience towards the nucleus.

There is another factor affecting Zeff as well, the inner electrons. The first shell of electrons shields the outer electrons from the nucleus. As a result, the outer electrons are not only further away from the nucleus but they also feel less of a pull as the inner electrons block the pull of the nucleus. The upshot of these two effects is shown in Figure 2.4. As electrons 3-10 are added in the effective charge drops dramatically at first, but then increases once again in a linear fashion. As expected these effects occur repeatedly as we add more protons and electrons to make the higher atomic weight elements.

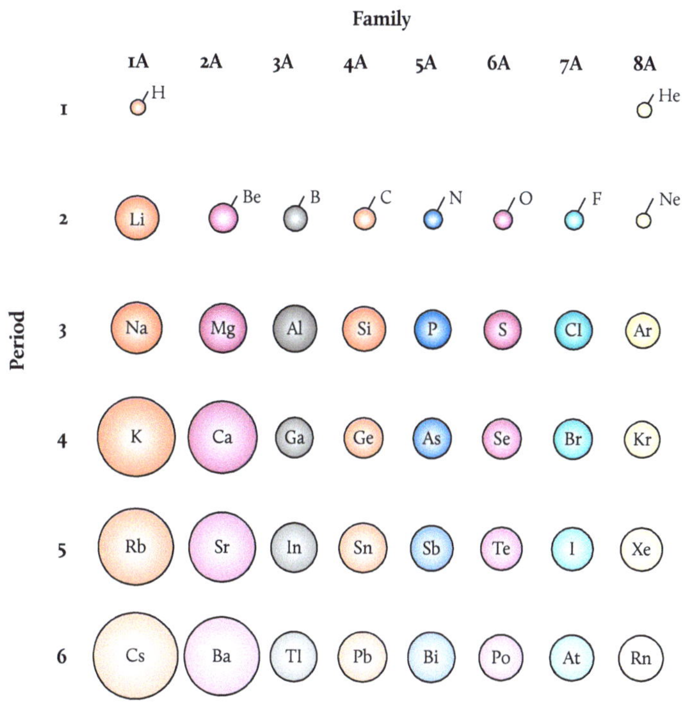


**FIGURE 2.4.** Atomic Sizes.

The reader might have made an important observation in the discussion above. When the first energy shell is filled the element helium (He) will have been formed. As the second is filled with eight more electrons neon (Ne) results. These are both well-known and important inert gases. They are very unreactive. This is no coincidence. Why is this so? Can we then predict reactivity based upon electron distributions?

**PRACTICE** How many protons and electrons does the element sulfur (S) have?

## Atomic and Ionic Diameters

Two rules to keep in mind here:

1. Proceeding from left to right across any period results in the one at a time addition of protons to the nucleus and electrons into the same energy shell. The Zeff increases.
2. Proceeding down a group an electronic shell or layer is added on with each period. Each additional layer is further removed from the nucleus. The Zeff decreases.

### Atomic Radius

As just described the effective nuclear charge increases from left to right across any period, but the layers of electrons remains unchanged. As a result, negatively charged electrons experience a stronger attraction to the positively charged nucleus. This results in a decrease in the atomic radius moving from left to right across a period. This effect is also reflected in an increase in the amount of energy needed to remove an electron, known as the ionization energy: This is correct as shown in Figure 2.5.

Here the vertical and horizontal arrows show the directions of decreasing atomic diameter. An upshot of the decreasing diameter is to make the related charges more

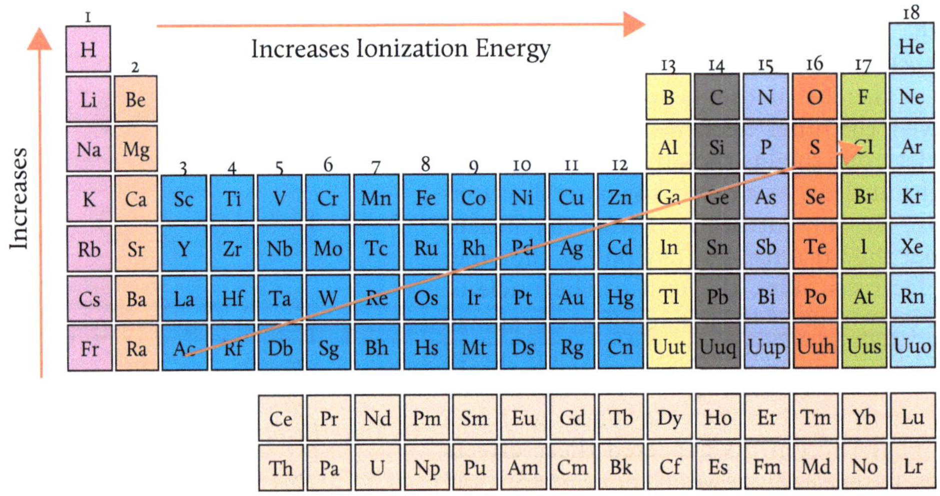


**FIGURE 2.5.** Ionization Energies.

intense. Again thinking of the atom as a sphere more and more positive and negative charge is being packed into a smaller and smaller volume. In a strange way this basic concept is key to another topic—nuclear reactions. In summary, moving across a period from left to right the size of atoms will decrease. Moving from the top of a group to the bottom, the size of the atoms increase. The largest atom is cesium and the smallest is fluorine. Size is dictated by the Zeff discussed earlier and will have a great impact and allow for the prediction of properties yet to be covered in this text.

### Ionic Radius

As mentioned already the number of protons of a given element cannot be changed since this act would change the identity of an element. However, we are allowed to add and subtract electrons as described in Chapter 1. What happens to the atomic size if we do so? Say the 3rd electron of Li, located by itself in the second energy level, is removed leaving $Li^+$ as a cation. Here again intuition suggest that the removal of the second layer, just like an onion, will decrease the diameter. A secondary impact is that the effective nuclear charge becomes greater with the electron's loss. This intuition is also correct (see Figure 2.6).

V · T · E **Periodic table of electronegativity using the Pauling scale**

**→ Atomic radius decreases → Ionization energy increases → Electronegativity increases →**

| Group→ / ↓Period | 1 | 2 | 3 | 4 | 5 | 6 | 7 | 8 | 9 | 10 | 11 | 12 | 13 | 14 | 15 | 16 | 17 | 18 |
|---|---|---|---|---|---|---|---|---|---|---|---|---|---|---|---|---|---|---|
| 1 | H 2.20 | | | | | | | | | | | | | | | | | He |
| 2 | Li 0.98 | Be 1.57 | | | | | | | | | | | B 2.04 | C 2.55 | N 3.04 | O 3.44 | F 3.98 | Ne |
| 3 | Na 0.93 | Mg 1.31 | | | | | | | | | | | Al 1.61 | Si 1.90 | P 2.19 | S 2.58 | Cl 3.16 | Ar |
| 4 | K 0.82 | Ca 1.00 | Sc 1.36 | Ti 1.54 | V 1.63 | Cr 0.66 | Mn 1.55 | Fe 1.83 | Co 1.88 | Ni 1.91 | Cu 1.90 | Zn 1.65 | Ga 181 | Ge 2.01 | As 2.18 | Se 2.55 | Br 2.96 | Kr 3.00 |
| 5 | Rb 0.82 | Sr 0.95 | Y 1.22 | Zr 1.33 | Nb 1.6 | Mo 2.16 | Tc 1.9 | Ru 2.2 | Rh 2.28 | Pd 2.20 | Ag 1.93 | Cd 1.69 | In 1.78 | Sn 1.96 | Sb 2.05 | Te 2.1 | I 2.66 | Xe 2.60 |
| 6 | Cs 0.79 | Ba 0.89 | * | Hf 1.3 | Ta 1.5 | W 2.36 | Re 1.9 | Os 2.2 | Ir 2.20 | Pt 2.28 | Au 2.54 | Hg 2.00 | Tl 1.62 | Pb 1.87 | Bi 2.02 | Po 2.0 | At 2.2 | Rn 2.2 |
| 7 | Fr 0.7 | Ra 0.9 | ** | Rf | Db | Sg | Bh | Hs | Mt | Ds | Rg | Cn | Uut | Fl | Uup | Lv | Uus | Uuo |

| | | | | | | | | | | | | | | | |
|---|---|---|---|---|---|---|---|---|---|---|---|---|---|---|---|
| *Lanthanoids | La 1.1 | Ce 1.12 | Pr 1.13 | Nd 1.14 | Pm 1.13 | Sm 1.17 | Eu 1.2 | Gd 1.2 | Tb 1.1 | Dy 1.22 | Ho 1.23 | Er 1.24 | Tm 1.25 | Yb 1.1 | Lu 1.27 |
| **Actinoids | Ac 1.1 | Th 1.3 | Pa 1.5 | U 1.38 | Np 1.36 | Pu 1.28 | Am 1.13 | Cm 1.28 | Bk 1.3 | Cf 1.3 | Es 1.3 | Fm 1.3 | Md 1.3 | No 1.3 | Lr 1.3 |

FIGURE 2.6. Ionic Diameters.

How about adding an electron instead? Fluorine (F, group 17, atomic number 9) has 2 core and 7 valence electrons in its elemental state. The addition of one more electron gives it 8 and a charge of –1. Its electronic structure is now the same as neon's (atomic number 10), but with one fewer protons! As can be seen in the Figure 2.6 fluorine's ionic size is larger than the pure element. Why? This is due to an effect we have not yet discussed, electron-electron repulsion. The negative (–1) charge on each electron repels other nearby negative charges, much like two north or south poles on magnets repel each other. The result of this is that as electrons are added to an atom they will occupy more space, hence the ionic diameters are larger for anions As electrons are removed, repulsion decreases and as such so does the ionic size when cations are formed.

# 2.4 Ionization Energies

## Ionization Energies (also referred to as Ionization Potentials)

For either an element or an ion electrons can be removed; however this act requires an input of energy since the attractive forces existing between the positive nucleus and the negative electron must be overcome. The attractive forces vary substantially from element to element and ion to ion. The energy needed for the removal of an electron is known as the **ionization energy**.

The units involved here are kilojoules (energy) per mole (amount of substance); both units to be defined further along. The relative magnitudes are all that is important to understand this concept.

As you should recall, sodium (Na, atomic number 10) has one valence electron and ten core electrons as defined previously. The removal of the one valence electron is very easy and requires just 496 kJ/mole. This called the first ionization energy. Removing the valence electron leaves the Na nucleus surrounded by the core electrons and possessing a +1 charge. Can another electron be removed?

The answer is yes, but it takes almost an order of magnitude more energy to do so at 4552 KJ/mole. There are a number of good reasons for this difference in required energy. First is that the inherent stability of the core electron structure is being broken which requires a substantial extra energy input. Second, is that the removal of the first electron's negative charge intensifies the attraction between the positively charged nucleus and the remaining 10 core electrons (due to less repulsion between electrons). Also, core electrons are found in an electron shell that is closer to the nucleus so these electrons feel a greater attraction to the nucleus. Removing more electrons compounds these effects.

Ionization energies tend to increase across a given period. They tend to decrease as one goes down a given group. Notice that this trend follows the trend in effective nuclear charge. As effective nuclear charge increases, the ionization energy also increases. This makes sense—the stronger the nucleus "pulls" on electrons in an atom, the harder they will be to remove from an atom.

Once again this is logical. When proceeding across a period the number of shells (onion layers) of electrons remains constant, but the number of protons steadily increases. Thus the nucleus pulls the electrons in tighter and tighter. In going down a group additional layers of electrons are added, each one substantially further removed from the nucleus.

## Electron Affinities

Electron affinity is basically the reverse of the ionization energy. It is a measure of the energy associated with adding an electron to an atom or ion. The greater the electron affinity, the more the element or ion likes having electrons added to it. The elements on the upper right hand corner of the periodic table (excepting group 18) have the greatest electron affinities. Once again, this trend can be explained by considering the trend in effective nuclear charge. The greater the effective nuclear charge of an atom, the better is will be at drawing electrons to the atom.

Although numerical values have been assigned to electron affinities, and ionization energies, in general all that is important here is to have a feeling in a qualitative sense for what these terms mean.

What this leads to is one of the most important elemental properties associated with the periodic table: ***electronegativity***.

## Electronegativity

Electronegativity can be defined in a number of different ways:

The desire of an element to draw in a pair of electrons within a bond.

The pull of electrons towards an element or group of elements at the expense of surrounding or nearby elements.

Within the chemical bonds of a compound a measure of how the electrons are distributed among various atom types.

Electronegativity also has a numerical value associated with it as shown in Figure 2.6 above. For lack of confusion, we will consider electronegativity as it pertains to electrons shared between two atoms in a covalent bond. Atoms with high electronegativity values attract electrons more than those with lower values. This will have many consequences on chemical reactions and properties yet to be covered in this text.

In Figure 2.7, fluorine (F at the upper right hand corner of Figure 2.6) has the highest electronegativity of all of the elements at 3.98. Cesium (Cs at 0.79, lower left) has the lowest. The trends in electronegativity across the periodic table is very important to understand but the absolute numerical values are not. When electronegativity values are close between two atoms, this indicates electrons in a bond between them are more likely to be shared equally. *Why is the concept of electronegativity so important?* In a nutshell if you know where the electrons spend most or all of their time you can predict chemical reactivity with a great deal of accuracy.

We will examine bonds and electron sharing in a later chapter, but let's do a couple of arbitrarily chosen examples of bonds between atoms, and

**FIGURE 2.7.** Fluorine in the Solid State.

look at the qualitative importance of the element electronegativities (numbers from Figure 2.6):

| | | |
|---|---|---|
| **Oxygen - Carbon** | O - C | 3.0 – 2.5 = 0.5 |
| **Bromine - Bromine** | Br - Br | 2.8 – 2.8 = 0.0 |
| **Fluorine - Cesium** | F - Cs | 4.0 – 0.7 = 3.3 |
| **Nitrogen - Chlorine** | N - Cl | 3.0 – 3.0 – 0.0 |

Here the absolute values of the numbers are not especially important. What is important is the *difference* between the numbers. For O – C the difference is 0.5 with the oxygen being more electronegative. What does this mean? It means that oxygen has a greater desire for and pull upon the electrons than carbon does. This negative charge pull results in a charge distribution of a negative end of the bond on the O, and a positive end on the C.

Contrast this result with the Br to Br bond. Here both atoms have the same electronegativity value thus there is no net charge distribution. Put another way there is no positive and negative end to this bond, and the electrons are shared equally.

How about the Cs to F bond? The electronegativity here is the largest possible for any two elements in the periodic table. One might expect a very negative fluorine end and a very positive cesium end. That is the case.

The final example is shown to demonstrate that the elements do not have to be the same to for a bond to have no negative and positive ends. Here nitrogen and chlorine coincidentally have the same electronegativity values just like the Br – Br case.

It is not shown here, but a large enough difference in electronegativity of atoms will result in a full transfer of electrons between two atoms. You should recall that a transfer of electrons will result in the formation of ions and the attraction between the ions will form an ionic bond. As such, electronegativity will also allow for the prediction of ionic bond formation.

A thorough understanding of electronegativity, and even more basically how and why electrons are distributed as they are in molecules, is fundamental to grasping the science of chemistry.

**PRACTICE** Based upon the electronegativity values in Figure 2.6 where is the negative end of each of the following bonds: Si – Al; S – I; Na – K. What about carbon dioxide ($CO_2$) which has a linear structure that can be represented by O – C – O.

## 2.5 Properties of the Most Important Periodic Table Groups

Now that we have a broad set of rules, let's look at the physical and chemical properties of the most important groups, as well as some of the other essential elements, mainly transition elements, that are scattered around the periodic table. Shown below in Figure 2.8 is crystalline silicon, an element that is essential to all of modern electronics.

Before we do let us examine three broad classes as follows:

***Metals:*** Most of the elements are metals. Metals are:

1. Ductile and malleable. Easily formed into shapes.
2. Shiny and with surfaces that are reflective of heat and light.
3. Electrically and thermally conductive.
4. Solids at ambient temperatures (except mercury).

***Nonmetals:*** All biological systems are critically dependent upon some of the nonmetals. Nonmetals are:

1. Solids, liquids or gases.
2. As solids they are often brittle, and fracture rather than bend.
3. Are electrical and thermal insulators.

***Metalloids:*** Just a very few elements are metalloids; however, they form the basis of all modern electronics: Metalloids are hybrids in the sense that they combine some of the properties of metals with other properties that are characteristic of nonmetals. Metalloids:

FIGURE 2.8. Crystalline Silicon.

1. Are solids at ambient temperatures.
2. Are semiconductors that conduct electricity and heat, but orders of magnitudes more poorly than do metals.
3. Are brittle, and fracture like nonmetals.
4. Have a metallic-like luster, but are not as reflective or shiny.

Now a discussion of some of the most important groups and elements.

## Hydrogen (H)

This element is put up front for a couple of reasons. It is one of the (arguably) two most important elements, along with oxygen, with which it is often combined. H is a nonmetal that is essential to all life as we know it. Hydrogen is here presented by itself because it shows a wide range of properties depending upon the chemical and physical site it inhabits. As element atomic number 1 it is as simple as elements get with just one proton and one electron in the pure state. It is a gas that is extremely flammable, and that does not occur in nature in any great quantity. Upon losing its lone electron a bare proton remains with a +1 charge. This $H^+$ ion forms the basis for much of the acid-base chemistry we will discuss shortly.

## Group 1: Alkali Metals

In their elemental state these metals are extremely reactive, and can be explosively dangerous. They do not occur in pure forms naturally. As metals they are electrically and thermally conductive, but are soft metals that can be easily cut with a knife. As one proceeds down this group the elements become more and more reactive.

As ions Na and K are especially important in all biological systems. Li is essential in the batteries that power a lot of electronics including cell phones, laptops, tablets, and the most powerful electric cars.

## Group 2: Alkaline Metals

The alkaline earth metals are similar to the alkali metals, but not as reactive. They conduct electricity and heat well. Magnesium and calcium are crucial in biological systems, much like Na and K of Group 1.

Magnesium is also an important structural metal, and as such has the properties of ductility and high strength. It is also flammable and chemically reactive. Also like Group 1 as one proceeds down the group the chemical reactivity increases. This property limits usage of these elements in many applications where water and oxygen are present.

## Transition Metals

Most of the elements in the periodic table are transition metals, well named since from left to right the elements transition from metals to nonmetals in a more or less seamless manner. The transition metals vary greatly in their relative importance. We can arbitrarily describe two basic areas where transition elements are important; biological systems, and inorganic or materials based systems. Another important property of these metals is that many can have variable ionic charges which has a impact on how these metals behave when forming compounds.

Biologically many transition metals are crucial components of a vast number of processes, often in vanishingly small quantities. So, even though many are not abundant on earth naturally this fact does not matter. A partial list of biologically important transition metals include: Fe, Cu, Zn, Mo, Co, Cr, Ni, Mn, and V. There are almost surely more yet to be identified. In addition, there are many more transition elements that are biologically active in an environmental, and in a toxicological sense. These topics are covered in more detail in the latter chapters of this book.

**PRACTICE** The previous paragraph contains many elemental symbols. What are the names of the biologically important elements?

Transition metals are also important and essential in the non-biological world, that is, the world that includes all kinds of materials with an almost infinite number of uses and applications. By way of example, without the metals iron, aluminum, and copper (for starters) our world would be drastically different. We are all familiar with many of the uses of the pure, unreacted metals. The list of uses becomes geometrically larger when these elements are combined with other elements to form myriad compounds and alloys.

Except for those that are radioactive, and thus toxicologically and environmentally dangerous, almost all of the transition elements have important uses. Even those that are radioactive, in particular U and Pu, are important and used in large scale power generation plants all over the world. Once again we will revisit some of these topics in more detail in later chapters.

## Group 14 (4A): Carbon, Semimetals, and Metals

FIGURE 2.9. Crystalline Sulfur.

Group 4 is arguably the most interesting group on the periodic table. At the top carbon (C) is one of the six elements essential for organic chemistry-based life as we know it. Its properties also vary over a very wide spectrum. It is a nonmetal, but unusual in that it is electrically conductive, but also brittle rather than ductile. Diamond, graphite and coal are three forms of carbon that are in widespread existence and usage.

The next two group 4 members, silicon (Si) (Figure 2.9) and germanium (Ge), are semiconductors meaning they are electrically conductive, but only poorly so. It is this property that makes them useful and essential in virtually all modern solid state electronics. They are brittle like nonmetals, and have a dull luster or shine like metals.

The last two members of importance in this group are tin (Sn) and lead (Pb), both valuable metals known since antiquity. They are widely available in the earth, and have many uses. They are true metals in that they are ductile, shiny, as well as electrically and thermally conductive.

## Group 15 (5A): Non-Metals: N and P

Nitrogen (N) and phosphorous (P) are two more of the six elements crucial for all biological systems to function. They are also two of the big three nutrients required by plants, with potassium being the third. Nitrogen gas comprises almost 80% of the earth's atmosphere. Phosphorous is a pyrophoric white or red solid in its pure forms, and exists primarily in the form of phosphate rock in the earth. Besides biology both of these elements find widespread use in many different applications in compounds.

The next two elements, arsenic (As) and antimony (Sb) are semiconductors, and similar to that of Si and Ge, are used in electronics. Of course arsenic is a well-known poison that functions as such because of its position directly beneath P in Group 15. Because of arsenic's chemistry being similar to phosphorus it can mimic the latter's biology with devastating results. Bismuth (Bi) is a metal with fairly brittle properties, and limited usefulness.

## Group 16 (6A): Nonmetals: O and S

This group is comprised mainly of nonmetals although tellurium's properties are metallic in many ways. Oxygen and sulfur are the last two of the six elements that form the basis of organic chemistry. The six are:

### Hydrogen, Carbon, Nitrogen, Phosphorous, Oxygen, and Sulfur

Without any one of these six elements there would be no plant or animal life as we know it on earth. The situation is just as simple as that.

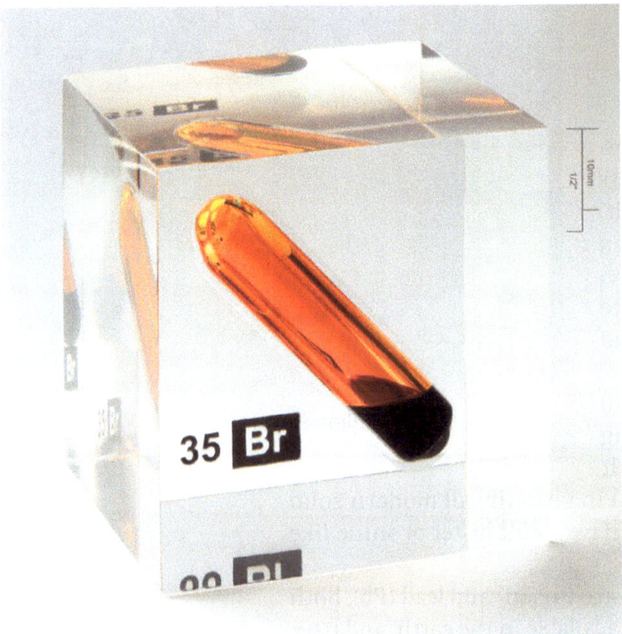


FIGURE 2.10. Bromine; Liquid and Gas.

Oxygen is ubiquitous in our world in compounds of an almost infinite number, but especially combined with hydrogen as water. In gaseous form as the pure element (~20% of the atmosphere) its value is obvious. The important chemical processes and reactions that do not involve oxygen are in a distinct minority.

Besides its biological uses sulfur (Figure 2.9) is used to make many different compounds and in many different processes. Its primary use is in the manufacture of sulfuric acid, $H_2SO_4$, a very strong acid that is the chemical manufactured in the largest quantities worldwide.

## Group 17 (7A): Non-Metals. The Halogens

Group 17, known as the halogens, occur commonly in many different important compounds, both organic and inorganic classes. They are reactive and when pure in the following forms. They occur in nature in compounds, not pure.

**Fluorine:** Colorless gas. The most electronegative element, and very reactive.

**Chlorine:** Pale yellow-greens gas. High electronegativity. Dangerously reactive. Original WWI war gas. Commonly used in synthetic herbicides and pesticides.

**Bromine (Figure 2.10):** Volatile red liquid gas. High electronegativity, and reactive.

**Iodine:** Purple solid that sublimes readily. Electronegative, and reactive.

Fluorine steals electrons from anything it comes in contact with, even water where the reaction can run out of control and generate large quantities of heat. Chlorine also reacts with water, but relatively slowly. The products of this reaction are excellent disinfectants. This latter property is also true for $Br_2$ and $I_2$.

All four halogens form salts with metals, many of which are very important; for example NaCl comprises over 3% by weight of ocean waters worldwide, a tremendous quantity of material.

## Group 18 (8A): Noble Gases

The noble gases are special in that they are almost totally unreactive. As such they are of relatively little interest here; however, they do have important applications. Besides lighter than air applications, such as dirigibles, helium is essential for cryogenic applications including MRI.

Argon comprises about 1% by weight of the earth's atmosphere. Krypton and xenon have similar properties to argon, and argon especially is useful as inert atmospheres for preventing explosive situations, for example in oil tankers. Radon is radioactive and carcinogenic, and as such its presence is commonly monitored in homes as a precaution.

# Chapter 2 Summary

*Periodic Table Organization:* The periodic table is arranged into **groups** (vertical) and **periods** (horizontal) such that elements with similar properties are close to each other. From left to right, following the periods, the groups vary from the most active metals through the transition metals, metalloids (semiconductors), non-metals, and finally the noble gases on the far right.

*Valence and Core Electrons:* In essence **valence electrons** are those that are primarily involved in chemical reactions. They reside in the outermost electron shells that surround the nucleus. Between the latter and the valence electrons are the **core electrons**. In general core electrons are not directly involved in chemical reactions, but have the effect of shielding the valence electrons from the nucleus which significantly affects an atom's reactivity.

*Periodic Property Trends:* There are predictable property trends as one proceeds across a period or down a group. These include:

*Effective Nuclear Charge:* A key to the chemical and physical properties of any element is the strength of the attraction between the positively charged nucleus and the valence electrons. The more protons in a nucleus for atoms with electrons added to the same energy shell increases the Zeff. The larger the distance between the two entities the weaker the attractive forces. In essence this distance is determined by the intermediating core electrons. The more layers of core electrons, the larger the distance is which results in a diminished, effective nuclear charge.

*Atomic Diameters (Sizes):* The atomic diameters of uncharged atoms decreases as one proceeds across a period since the number of electronic shells remains the same, but the number of protons increases. This results in a greater effective nuclear charge and increased ionic attraction. A reverse trend occurs as one proceeds down a group as with each step down a new layer of electrons is added.

*Ionic Diameters (Sizes):* Cations possess diameters smaller than the corresponding neutral atm. Anion diameters are larger.

*Electron Affinity:* Is a ranking of an elements inherent attraction for electrons. Progressing across the periodic table from lower left to upper right shows a general increase in electron affinity.

*Ionization Energy:* Is a measure of the ease with which electrons can be removed from an atom or ion. The trend here is to become more difficult as one moves from lower left to upper right of the periodic table.

*Electronegativity:* Is a ranking of an elements inherent attraction for electrons shared in a bond. Progressing across the periodic table from lower left to upper right shows a general increase in electronegativity.

## Periodic Table—Major Regions

*Metals:* Most of the elements are metals. Metals are:

1. Ductile and malleable. Easily formed into shapes.
2. Shiny and with surfaces that are reflective of heat and light.

3. Electrically and thermally conductive.
4. Solids at ambient temperatures (except mercury).

*Nonmetals:* All biological systems are critically dependent upon some of the nonmetals. Nonmetals are:

1. Solids, liquids or gases.
2. As solids they are often brittle, and fracture rather than bend.
3. Are electrical and thermal insulators.

*Metalloids:* Just a very few elements are metalloids; however, they form the basis of all modern electronics: Metalloids are hybrids in the sense that they combine some of the properties of metals with other properties that are characteristic of nonmetals. Metalloids:

1. Are solids at ambient temperatures.
2. Are semiconductors that conduct electricity and heat, but orders of magnitudes more poorly than metals.
3. Are brittle, and fracture like nonmetals.
4. Have a metallic-like luster, but are not as reflective or shiny.

## CHAPTER 2 EXERCISES

1. How do the following variables change as one proceeds across the periodic table?

   a. electronegativity

   b. cationic size

   c. anionic size

   d. atomic size

2. What is the difference between a core electron and a valence electron? Which is more important in regards to an atom's (element's) reactivity?

3. Locate each of the following elements in the periodic table; then identify it by name and atomic number, and indicate whether it is a metal, metalloid, or nonmetal:

   a. Si

   b. S

   c. Mg

   d. Na

   e. Mn

   f. P

   g. Ar

3. How many protons, neutrons, and electrons are present in the following atoms?

   a. $^{43}Ca$

   b. $^{29}Si$

   c. $^{109}Ag$

   d. $^{37}Cl$

   e. $^{56}Fe$

   f. $^{238}U$

4. In compounds aluminum virtually always has a charge of 3+. What chemical formulas you would expect for compounds formed between aluminum and (a) oxygen, (b) iodine, (c) nitrogen, (d) phosphorus

5. Which of these spheres represents K, which represents $K^+$, which represents Br, and which represents $Br^-$?

6. Predict the charges of the ions of the following elements based upon their location in the periodic table:

   a. K

   b. Sr

   c. N

   d. I

   e. S

7. Using only the periodic table, arrange each set of atoms in order from largest to smallest:

   a. K, Li, Cs

   b. Pb, Sn, Si

   c. F, O, N

8. a. Which neutral atom is isoelectronic with each of the following ions? $Al^{3+}$ $Ti^{4+}$, $Mn^{7+}$, $Cl^-$, $Sn^{2+}$? b. After forming the ions what is the common feature of the resultant electronic structures?

9. Write equations that show the processes that describe the first, second, and third ionization energies of an aluminum atom. Which process would require the least amount of energy? Which the most? Why?

10. Predict which atom of the following pairs of elements will have the smaller first ionization energy:

    a. I, Xe

    b. Mg, Ca

    c. K, Co

    d. S, Ge

    e. Sn, Te

11. Arrange each of the following sets of atoms and ions, in order of increasing size:

    a. $O^{2-}$, $S^{2-}$, O

    b. $Co^{3+}$, $Fe^{3+}$, $Fe^{2+}$

    c. Ca, $Ti^{4+}$, $Y^{3+}$

    d. $Be^{2+}$, $Na^{+}$, Ne

12. a. Why is calcium generally more reactive than magnesium? b. Why is calcium generally less reactive than potassium?

13. Iodine is essential for the thyroid to function correctly. One can predict its properties based upon where it occurs in the periodic table. (a) Do you expect the element to be a gas, liquid, or solid at room temperature? Explain. (b) Would you expect I to be a metal, nonmetal, or metalloid? Explain. (c) What is the chemical formula of the compound it forms with Na?

14. What is the difference in reactivity between valence electrons and core electrons?

15. Estimate the density of gallium based upon the densities of Ge, Zn, In and Al.

16. Valence electrons are shielded from the attractive forces of the nucleus by the core electrons. What effect does this shielding have upon an elements reactivity?

17. Why do atomic radii decrease from left to right within a period in the periodic table?

18. Arrange each of the following sets of atoms in order of decreasing atomic radii:

    a. the alkali elements

    b. the group VII (17) elements

    c. the first five elements in the fifth period

    d. C, Si, Sn, and Pb.

19. Explain why the second ionization energy is always greater than the first ionization energy?

20. Arrange the following sets of cations in order of increasing ionic radii:

    a. $K^{+}$, $Ca^{+2}$, $Ga^{+3}$

    b. $Ca^{+2}$, $Be^{+2}$, $Ba^{+2}$, $Mg^{+2}$

    c. $Al^{+3}$, $Sr^{+2}$, $Rb^{+}$, $K^{+}$

    d. $K^{+}$, $Ca^{+2}$, $Rb^{+}$

21. What is electronegativity? Why is this property important?

22. Arrange the following elements in order of increasing electronegativities:

    a. Pb, C, Sn, Ge

    b. S, Na, Mg, Cl

    c. P, N, Sb, Bi

    d. Se, Ba, F, Si, Sc

# Credits

Fig. 2.1: Copyright © 2015 by Sandbh, (CC BY-SA 4.0) at https://commons.wikimedia.org/wiki/File:14LaAc_periodic_table_IIa.jpg.

Fig. 2.4: Copyright © 2010 by CK-12 Foundation, (CC BY-SA 3.0) at https://commons.wikimedia.org/wiki/File:Comparative_atomic_sizes.png.

Fig. 2.5: Copyright © 2011 by Cdang and Adrignola, (CC BY-SA 3.0) at https://commons.wikimedia.org/wiki/File:Ionization_energy_periodic_table.svg.

Fig. 2.6: Copyright © 2013 by DMacks, (CC BY-SA 3.0) at https://commons.wikimedia.org/wiki/File:Periodic_table_Pauling_electronegatvity.jpg.

Fig. 2.7: Copyright © 2007 by Giovanni Dall'Orto, (CC BY-SA 2.5 IT) at https://commons.wikimedia.org/wiki/File:9104_-_Milano_-_Museo_storia_naturale_-_Fluorite_-_Foto_Giovanni_Dall%27Orto_22-Apr-2007.jpg.

Fig. 2.8: Source: https://commons.wikimedia.org/wiki/File:Silicon_wafer_with_mirror_finish.jpg.

Fig. 2.9: Source: https://commons.wikimedia.org/wiki/File:Sulfur.jpg.

Fig. 2.10: Copyright © 2009 by Alchemist-hp, (CC BY-SA 3.0 DE) at https://commons.wikimedia.org/wiki/File:Bromine_vial_in_acrylic_cube.jpg.

CHAPTER 3

# Compounds and Chemical Bonding

As discussed in Chapter 2, elements in the same group on the periodic table have similar properties. Different atoms, in their elemental form, have a unique number of protons and a certain number of electrons to maintain neutrality. During chemical reactions however, electrons can be added to or removed from atoms. When atoms react, they often do so to achieve an electron configuration that is more stable. This leads to the formation of different compounds. When different elements react and form compounds, the atoms are held together by chemical bonds. In this chapter, you will learn about different types of chemical bonds and when they will form. You will also learn how to use chemical bonds to predict the three-dimensional structure of a compound.

## 3.1 Octet Rule and Lewis Symbols

As discussed earlier in this textbook, atoms contain a positively charged nucleus surrounded by negatively charged electrons. Changing the number of protons in an atom changes the identity of that atom. Looking at the periodic table, Sodium has 11 protons. If a proton were to be taken from this atom, its identity would then change to Argon (10 protons). Unlike protons, electrons can be added to or removed from an atom to form an ***ionic bond***. It is also possible for two atoms to share protons to form a ***covalent bond***.

So what determines whether an atom will gain, lose or share electrons? Furthermore, why do some atoms lose or gain multiple electrons? Atoms react in such a way to achieve eight electrons in their valence shell. The phenomenon is known as the ***octet rule***. Electrons can be transferred between atoms or shared in order to achieve the desired octet of electrons. It is possible to predict the manner in which atoms react by considering how many electrons each atom has and whether gaining, losing or sharing electrons will achieve a total of 8 in the atoms valence shell.

Why eight electrons? Looking at the periodic table, the last column represents the Noble Gases. Remember, these element are very inert or unreactive. They prefer to remain independent of other atoms and are unlikely to form compounds of any type. Why? Noble gases have a very stable electron configuration with 8 electrons in their valence shell. Seeing as Noble Gases contain 8 electrons in their valence shell already, they do not need to react with other atoms in order to achieve an octet of electrons. This being the case, these elements are inert.

So why is it important to appreciate the inert nature of the Noble Gases? The manner in which all other elements react during chemical reactions is governed by their desire to achieve a Noble Gas electron configuration. Let's consider some examples. Sodium (Na) has an atomic number of 11. As you already learned, this tells us that a neutral (uncharged) Na atom contains 11 electrons and 11 protons. To determine how Na will behave to achieve a stable electron configuration, examine its periodic table position. There are two options here, one option is that sodium can gain seven electrons to have a total of 18 electrons like Ar, a Noble Gas. It can also lose one electron giving it 10 electrons or the electron configuration of Ne. What option do you think is more likely or, better put, easier? Na will lose 1 electron in a chemical reaction to achieve an electron configuration of Neon simply because it is very easy for sodium to lose one electron and more difficult to gain seven. This is why Na, when not in its elemental form, it found to have a charge of +1. In fact, this +1 state is so stable it is the form of Na found in nature in different ionic compounds including NaCl and $NaNO_3$.

Now let us consider Sulfur (S). It contains 16 electrons (10 core electrons, 6 valence electrons). Sulfur can either lose 6 electrons to have an electron configuration like Neon or it can gain two electrons to have an electron configuration of Argon. Based on what was discussed with Na, will Sulfur gain 2 electrons or lose 6? In this case, it is much easier for Sulfur to gain two electrons rather than lose 6. So Sulfur will gain 2 electrons in a chemical reaction to form a −2 charged anion. Two common, naturally occurring, ores of sulfur are galena (PbS) and pyrite ($FeS_2$). In both cases, Sulfur is in its stable −2 anion state.

Like sodium and sulfur other atoms also seek to achieve a Noble Gas configuration. Whether they form anions or cations will depend on their whether it is easier to gain or lose electrons to achieve a stable configuration. This will also dictate when atoms gain or lose multiple electrons. Using this logic, you should be able to predict common charges for the main group elements on the periodic table.

**Lewis symbols** are used to represent an atom of an element and its valence electrons. The type of atom is represented by its atomic symbol and the *valence* electrons are represented by dots surrounding the symbol. This is a convenient representation of an element as it only shows the valence electrons and these are the electrons that are primarily important for chemical reactions.

A Lewis symbol is drawn by first recording the atomic symbol for the representative element. Simply by giving the atomic symbol, the element of interest is shown and also the number of core electrons are known by default. Next, to represent the valence electrons present, dots are placed around the symbol. Visualize a box surrounding the atomic symbol. Begin adding dots around the symbol one at a time to each side. Then begin to pair up the electrons until all of the valence electrons are represented. Note that only one or two electrons are placed on each side. Below are examples of some Lewis symbols.

Note that for Carbon, which has four electrons, the electrons are not paired. They remain as single electrons on four different sides of the chemical symbol. When placing the dots around the symbol, it is important to pair off electrons after adding a single electron to one of the four "sides" of the symbol first (Figure 3.1a, b, c, and d).

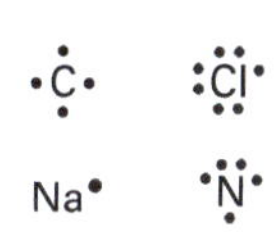


**FIGURE 3.1A-D.** Examples of some lewis structures.

Lewis symbols can be very useful in chemistry. First, they reveal the number of valence electrons in a given element. Secondly, by looking at the symbol one can determine the number of electrons needed to satisfy the octet rule. Finally, they reveal how many electrons are available for chemical bonding. Each unpaired electron in a Lewis symbols can be used in chemical bond formation.

# 3.2 Ionic Bonding

## Ion Formation

During chemical reactions two atoms can form chemical bonds with each other by sharing electrons or transferring electrons from one atom to another. A ***chemical bond*** is a force of attraction occurring between two atoms in a compound. Only valence electrons are involved in bond formation. When forming bonds, atoms seek to achieve an octet of electrons to achieve a stable electron configuration.

One type of bond formation involves a complete transfer of an electron from on atom to another and is known as an ***ionic bond***. This type of bond occurs in ionic compounds which were discussed at length in Chapter 1. So when will atoms form an ionic bond? In order to answer this question, one must consider the two atoms involved. In general, an ionic bond will form between an atom that has a low ionization energy and low electron affinity (willing to give up an electron) and another atom that has a high ionization energy and high electron affinity (willing to gain an electron). The greater the difference between these properties in the two atoms, the more likely is an ionic bond to form.

You will recall from the previous chapter, atoms on the left side of the periodic table (metals) have low ionization energies and low electron affinities. Conversely, atoms on the right side of the periodic table have high ionization energies and high electron affinities (nonmetals). This indicates that metals (left side of the periodic table) are likely to give up and electron while nonmetals (right side of the periodic table) are likely to gain electrons. Thus, it is common to see ionic bonds form between a metal and a nonmetal. In fact, in almost all cases,you can expect a bond between a metal and a nonmetal will be ionic (Figure 3.2).

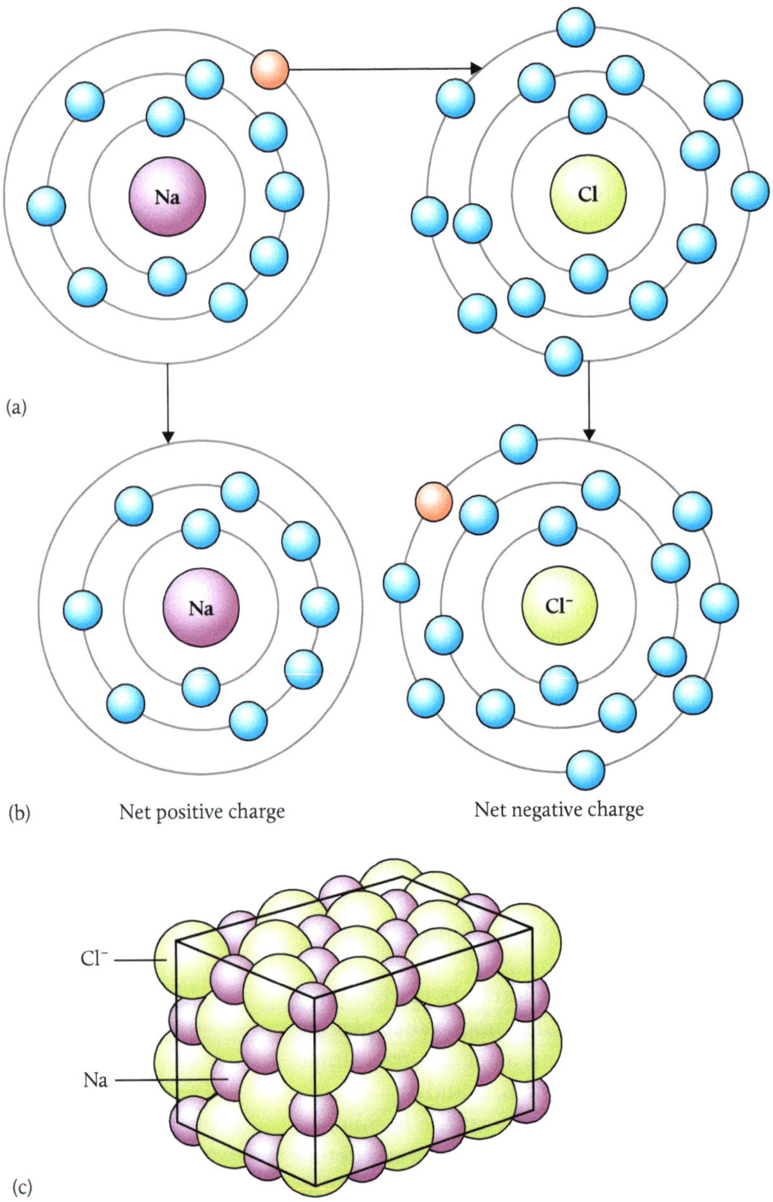


**FIGURE 3.2.** Ionic Bond Formation.

## Ionic Bond Formation

As discussed in Chapter 1, when a metal gives up an electron(s) a cation is formed while the nonmetal forms an anion when it gains electron(s). The positive charge of the cation and negative charge of the anion create an electrostatic interaction between the two ions. This attraction is very strong and is the ionic bond between the atoms.

In some cases, a metal atom will give up a single electron. In other cases, a metal atom may give up more than one electron. This will depend on how many valence electrons are present in the atom. Likewise, some nonmetals will only accept a single electron while other nonmetals will accept more than on electron. This will also depend on the number of valence electrons present in the atom. In both cases, the atom will give up or gain electrons to follow the octet rule.

Let's consider some examples of ionic compounds by looking at their chemical formulas.

1. NaCl: In this case, Na has one valence electron and chlorine has seven valence electrons. So, one Na atom will give up one electron to one Cl atom. This will achieve a 1:1 Na to Cl ratio as indicated by the formula.
2. $CaCl_2$: Here Ca has two valence electrons and Cl has its seven valence electrons. In this case, one Ca atom will give up two valence electrons. Two different Cl atoms will each accept one valence electron creating a 1:2 Ca to Cl ratio.
3. CaO: Here Ca has two valence electrons and O has six valence electrons. One Ca atom will give up two electrons to one O atom. This creates a 1:1 Ca to O ratio.

In the examples above, the number of valence electrons for a particular atom did not change. What changed was how one atom reacted with different atoms so each of the atoms involved in the reaction could follow the octet rule. In some cases an even exchange can occur between two atoms as was the case with NaCl and CaO. In other cases, one atom may have to interact with more than one atom to properly follow the octet rule. We see this in the case with $CaCl_2$ where two Cl atoms are required to accept the two electron Ca gives up. An important take home message here is that an electron must go from one atom to another it cannot simply disappear.

## Ionic Compound Structure

When ionic compounds are in the solid state, a crystal lattice structure is formed. This optimizes interactions between positively charged cations and negatively charged anions. Let us consider NaCl in three dimensions (Figure 3.2, bottom image). One sodium atom will be surrounded on all sides by chlorine atoms. One chlorine atom will be surrounded on all sides by sodium atoms. The ratio of Na to Cl atoms is 1 to 1 but this does not mean we will only find only *pairs* of Na atoms and Cl atoms interacting with each other in the solid state instead a maximum number of interactions is optimized between Na and Cl while maintaining a 1 to 1 ratio (Figure 3.3).

## Ionic Nomenclature

Chemical compounds are given unique names through the process known as ***nomenclature.*** Naming allows one to identify the type and number of each atom present

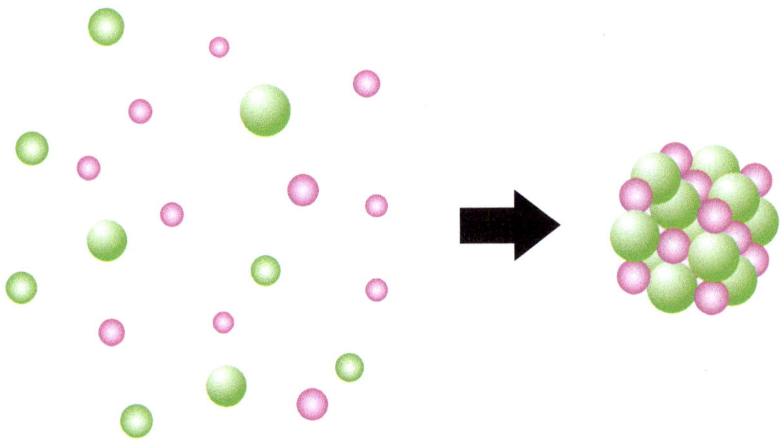

FIGURE 3.3. Formation of Sodium Chloride.

in certain compound. It is important that a system be used so that names given to compounds are consistent.

To name a compound it is important to examine the formula of the compound that indicates the type and number of each atom present.

**NaCl** **$MgCl_2$** **$Li_2O$**

For each of the compounds listed above we know the type of each atom and the number of each atom present simply by looking at the formula for the compound. Now it is necessary to assign a name to each compound. The name must be simple and unique to each compound and must also indicate the type of atoms and number of atoms present. Let's review the process for naming ionic compounds.

When naming ionic compounds it is important to remember they are made up of cation and an anion. First we will learn how to name binary ionic compounds or ionic compounds that are made up of just two atoms. These types of ionic compounds will fall into two categories based of the type of metal present: Type I or Type II. Type I ionic compounds are those ionic compounds containing metals from column 1, column 2 or Al, Zn and Ag. These are considered type one because the metals only have one valence state. Type II ionic compounds contain a transition metal and this category will encompass any metals that are NOT type I.

Steps for naming type I ionic compounds:

1. Name metal cation first followed by the name of the nonmetal anion.
2. The cation name is simply the metal name.
3. The nonmetal anion is named by changing the ending on the nonmetal name to *–ide*.

### Example: NaCl

The cation is sodium, the anion is chlorine. Chlorine gets changed to chlor*ide*. So the name for this compound is:

***sodium chloride***

The naming process for type II ionic compounds is more complicated because they involve transition metals. Type I metals only have one valence state and therefore the charge on these metal ions is always the same. We can name ionic compounds without indicating the charges on the metals. Transition metals can have different valence states or different charges so when naming these compounds, the charge on the metal must be noted in some manner. Consider copper, a type II metal. It has two common valence states of +1 and +2. Manganese has 5 common valence states of +2, +3, +4, +6 and +7! So it is very important to not only name the metal in this class of compounds but also to indicate the charge the metal has in the compound.

Steps for naming type II ionic compounds:

1. Determine the charge on the metal by looking at the formula and using the charge on the nonmetal.
2. The name of cation is simply the name of the metal.
3. In brackets or parenthesis the charge of the metal is indicated with the Roman numeral. (ie. A charge of +4 is indicated by [IV])
4. Name the anion by changing the nonmetal ending to ***-ide***.

#### Example: CuCl

To determine the charge we look to the charge on chlorine (−1). Looking at the formula, only one copper ion is present. To balance the −1 charge out, the charge on the copper ion must be +1. We will name the cation copper (I) and the anion will be chorine changed chloride again. That gives us the name:

***copper (I) chloride***

## Polyatomic Ions

At this point we have discussed binary ionic compounds at length. These again, are compounds with only two types of atoms present. There are also ionic compounds that can have 3 or more different atoms present. In these cases, a ***polyatomic ion*** is present (Figure 3.4). Polyatomic ions are molecular compounds that carry a charge because it either has too many or too few electrons. Due to their charge, polyatomic ions form ionic compounds by combining with other ions to balance out positive and negative charges—in a manner similar to binary ionic compounds. Figure 3.4 contains a list of polyatomic ions.

Here are two examples of compounds containing polyatomic ions:

1. $NaNO_3$ is a compound containing one sodium atom and one nitrate ion.
2. $Ba(NO_3)_2$ is a compound containing one barium atom and two nitrate ions. Notice that the nitrate formula is placed in brackets when indicating more than one of that ion is present.

To name compounds with polyatomic ions, it is first important to determine the type of cation present (type I metal, type II metal, polyatomic ion). Once the cation is identified the rules for that type of cation are applied. The polyatomic ion is named simply by naming the ion—no change to the ending of the ion is necessary. As with all ionic compounds, the name of the cation comes first, followed by the name of the anion.

#### Example: $NH_4NO_3$

Because this compound contains two polyatomic ions, it is simply named by naming the two ions present and as always the cation is named first followed by the anion.

***ammonium nitrate***

#### Example: $CuCO_3$

Copper is a type II metal so first, the charge on copper must be determined by looking at the charge on the anion. Carbonate carries a charge of −2. There is only one copper ion

| Name | Formula | Typical Compound |
|---|---|---|
| **Cation** | | |
| Ammonium ion | $NH_4^+$ | $NH_4Cl$ |
| **Anions** | | |
| Acetate ion | [a]$C_2H_3O_2^-$ | $NaC_2H_3O_2$ |
| Carbonate ion | $CO_3^{2-}$ | $Li_2CO_3$ |
| Hydrogen carbonate ion (or bicarbonate ion)[b] | $HCO_3^-$ | $NaHCO_3$ |
| Hypochlorite ion | $ClO^-$ | $Ca(ClO)_2$ |
| Chlorite ion | $ClO_2^-$ | $NaClO_2$ |
| Chlorate ion | $ClO_3^-$ | $NaClO_3$ |
| Perchlorate ion | $ClO_4^-$ | $KClO_4$ |
| Chromate ion | $CrO_4^{2-}$ | $K_2CrO_4$ |
| Dichromate ion | $Cr_2O_7^{2-}$ | $(NH_4)_2Cr_2O_7$ |
| Cyanate ion | $OCN^-$ | KOCN |
| Thiocyanate ion[c] | $SCN^-$ | KSCN |
| Cyanide ion | $CN^-$ | KCN |
| Hydroxide ion | $OH^-$ | NaOH |
| Nitrite ion | $No_2^-$ | $NaNO_2$ |
| Nitrate ion | $No_3^-$ | $NaNO_3$ |
| Oxalate ion | $C_2O_4^{2-}$ | $CaC_2O_4$ |
| Permanganate ion | $MnO_4^-$ | $KMnO_4$ |
| Phosphate ion | $PO_4^{3-}$ | $Na_3PO_4$ |
| Hydrogen phosphate ion | $HPO_4^{2-}$ | $Na_2HPO_4$ |
| Dihydrogen phosphate ion | $H_2PO_4^-$ | $NaH_2PO_4$ |
| Sulfite ion | $SO_3^{2-}$ | $Na_2SO_3$ |
| Hydrogen sulfite ion (or bisulfite ion)[b] | $HSO_3^-$ | $NaHSO_3$ |
| Sulfate ion | $SO_4^{2-}$ | $Na_2SO_4$ |
| Hydrogen sulfate ion (or bisulfate ion)[b] | $HSO_4^-$ | $NaHSO_4$ |
| Thiosulfate ion[c] | $S_2O_3^{2-}$ | $Na_2S_2O_3$ |

[a] The acetate ion is also represented as $CH_3COO^-$.
[b] The prefix "bi-" means that the ion contains a replaceable H atom. This should not be confused with the prefix "di-," which means two (usually used to represent a doubling of a simpler unit).
[c] The prefix "thio-" means that a sulfur atom has replaced an oxygen atom.

**FIGURE 3.4.** Some Common Polyatomic Ions.

present so to balance the charge on the anion, it must carry a charge of +2. To indicate the metal and its charge the its name should be copper (II).

***copper (II) carbonate***

To review, ionic compounds are formed when a cation and electron form a strong electrostatic interaction. Binary ionic compounds form between one type of metal and one type of nonmetal. Ionic compounds can also form with polyatomic ions acting as cations or anions. Cations and anions interact in a ratio that will balance out positive and negative charges so that the ionic compound is neutral and carries no charge. Finally, when naming ionic compound it is first necessary to identify the type of metal present (if any) and identify any polyatomic ions present.

# 3.3 Covalent Bonding

## Covalent Bond Formation

Molecular compounds (also called covalent compounds) are another category of chemical compounds. In this case, as discussed in the first chapter, atoms share electrons to form a covalent bond. Why do molecular compounds share electrons and not transfer them as occurs with ionic compounds? For electrons to be shared between atoms, the atoms forming the bond must have similar electronegativities. Seeing as the two atoms have similar electronegativies the pull they have on the electrons involved in the bond is similar so one atom will not pull the electron away from the other.

Whereas in ionic compounds electrons are transferred to complete the octet of electrons, in molecular compounds, *sharing* electrons allows the atoms involved in the bond to complete their octet of electrons. Let us consider the bond formed between two fluorine atoms. Each atom needs an additional electron to complete their octet. Seeing as the two atoms involved in the bond are the same, they have an equal pull on the electrons involved in the bond. This is considered a ***nonpolar covalent bond***.

Covalent compounds are usually formed by a chemical reaction between to nonmetals. As discussed earlier, most ionic compounds form solid crystal structures and are not comprised of individual units. Molecular compounds usually occur as individual molecules and can exist as solids, liquids or gases.

## Bond Polarity

While atoms connected by a covalent bond share electrons, they do not always due so equally. In certain cases the difference in the electronegativities result in an unequal sharing of the electrons in the bond. A covalent bond with an unequal distribution of electrons is called a ***polar covalent bond***. The atom with the greater pull of the electrons will be the more negative end and the atom with the weaker pull on the electrons will be the more positive end. While the differences in electronegativities leads to an unequal distribution of electrons the difference is not great enough to result in a complete transfer of the electron from one atom to the other which would lead to the formation of an ionic bond.

A symbol (┼►) is used to indicate a bond is polar. The arrow head points towards the end of the bond with the greater electron distribution. The symbols $\delta^+$ (partial positive) and $\delta^-$ (partial negative) are used to label the poles of the bond.

## Naming Binary Covalent Compounds

Naming covalent compounds follows different rules than ionic compounds. Names must include an indicator of the number of each type of atom present as two of the same type of atom can combine in different ways. For example, nitrogen can combine with 1, 2 or 3 oxygen atoms so the names of compounds containing these two atoms must include information that reveals the number of nitrogen and oxygen atoms present.

1. Elements are named in the order they appear in the formula.
2. A prefix is used to indicate the number of atoms of each element present in the compound.
3. The ending for the last element in the compound is changed to -ide.
4. Mono is usually understood for the first element in a compound so it is not included in the name.
5. The final vowel in a prefix are eliminated when an element's name starts with a vowel.

### Example: Name the Compound $N_2O$

Nitrogen is first in the formula followed by oxygen. There are two nitrogen atoms (di-) (mono-) and one oxygen atom. The last "O" in mono is dropped because oxygen starts with a vowel. The ending of oxygen is change to -ide.

***dinitrogen monoxide***

Some compounds are known by their common names. Water ($H_2O$), ammonia ($NH_3$) and methane ($CH_4$) are a few examples.

## Lewis Structures

Earlier we discussed drawing Lewis symbols. These symbols help in predicting how many electrons an atom would need to gain, or lose to achieve an octet of electrons. At this time, we can take those same symbols and use them to predict the way atoms my come together in chemical compounds. Lewis structures are used to demonstrate the bonds formed between atoms in a particular compound. Later we will use these structures to determine the three-dimensional shape or structure of a compound. Lewis structures reveal what atoms are connected by covalent bonds in a compound and will also reveal if a double or triple bond is present in a structure. The following is a step by step process can be followed to draw Lewis structures.

1. **Determine the total number of valence electrons present in the compound.** To do this you should determine the number of valence electrons present in each atom type by referring to its periodic table position. If an more than one atom of a certain element is present multiply its number of valence electrons by the atoms of that atom that is present. It the compound is an anion, add and electron to the total number of valence electrons. If the compound is a cation, subtract one electron from the number of valence electrons.
2. **Next, you should write the symbol of each element present in the compound.** If there is more than one atom of a particular element present, that symbol should

be written the number of times it is present in the compound. For example, in $NO_2$, the symbol for oxygen should be written twice. Often times, the chemical formula is written in the order that the atoms are connected. In other cases the first atom written will be the central atom. Another method for determining the order of the atoms is to place the least electronegative atom in the middle of structure with the other atoms. The exception here is hydrogen which should never be placed in the middle.

3. **Connect the atoms together with a single line in the order you have followed for second step.** Refer back to the total number of valence electrons in the compound determined in the first step. For each bond drawn, subtract two electrons from this total. Remember, each atom shares one electron in a covalent bond so each bond, represents two electrons.
4. **Next, complete the octets of the atoms surrounding the central atom starting with the outermost atoms (most electronegative) first, moving inward towards the central atom.** Remember to represent the electrons using dots as we did with the Lewis structures. Start with single electrons on the sides not participating in bond formation then double up electrons to form what are called lone pairs. Add the dots until the octet of the atom is complete, being sure to account for the two electrons in any bond the atom is forming. Reduce the number of electrons used to complete the octets from the total number of valence electrons in the compound.
5. **Complete the octet of the central atom with any remaining valence electrons.** Follow the same pattern as practiced with the atoms in the fourth step.
6. **If, after using all of the valence electrons present in the compound, the central atom does not have a complete octet try forming double or triple bonds to the surrounding atoms until its octet is complete.** To do this, use lone pairs from surrounding atoms.

### Example: Draw the Lewis Structure for Nitrogen Trifluoride

1. First, determine the formula based on its name. $\mathbf{NF_3}$
2. Next, determine the number of valence electrons

$$N - \boxed{?}$$
$$F - 7 \times 3 = 21$$

26 valence electrons in the compound

3. Next write the symbols for each atom with the least electronegative atom in the center.

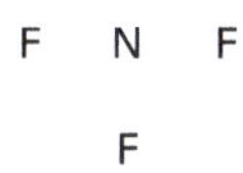


**CHEMICAL STRUCTURE 3.1A.** Lewis Structure for Nitrogen Trifluoride—Step One

4. Connect the F atoms to N using a line to represent a covalent bond. This will use 6 electrons leaving 20 valence electrons remaining.

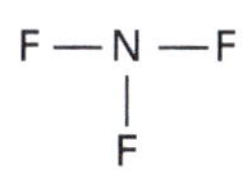


**CHEMICAL STRUCTURE 3.1B.** Lewis Structure for Nitrogen Trifluoride—Step Two

5. Add electrons to complete the octet atoms starting with the most electron negative (outward) atoms. This uses 18 electrons leaving 2 valence electrons remaining.

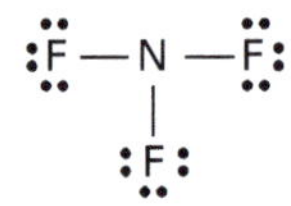

**CHEMICAL STRUCTURE 3.1C.** Lewis Structure for Nitrogen Trifluoride—Step Three

6. Complete the octet of N by adding valence electrons. This uses the remaining 2 electrons.

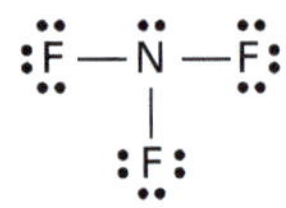

**CHEMICAL STRUCTURE 3.1D.**

7. Check to be sure all valence electrons have been used and confirm octets of atoms are all complete. The Lewis structure above uses all 26 valence electrons and completes the octet of all atoms in the structure.

## Exceptions to the Octet Rule

There are some atoms that do not always follow the octet rule. We have already learned that hydrogen and helium do not follow the octet rule and only need 2 electrons to be stable. There are other times when an atom or atoms may not follow the octet rule. Beryllium, Boron and Aluminum also break the octet rule when they have fewer than eight electrons in certain compounds. Take for example boron trifluoride, in this compound, boron only has 6 valence electrons.

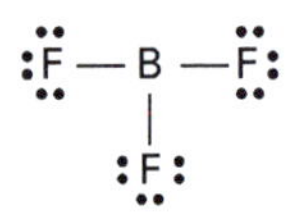

**CHEMICAL STRUCTURE 3.2.** Boron Trifluoride Compound.

Other atoms can have an expanded octet or more than eight electrons in their valence shell. This applies to elements in the third period of the period table or below. In this case, empty *d* orbitals can be used to store extra electrons. Phosphorous is an example of an atom that will have an expanded octet in compounds it forms like phosphorous pentafluoride.

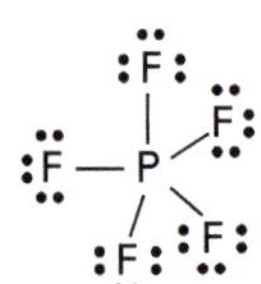

**CHEMICAL STRUCTURE 3.3.** Phosphorous Compound (?)

Finally, the octet rule is also broken when compounds have an odd number of valence electrons. In this case, by having an odd number of valence electrons means that an atom in the compound will have to break the octet rule. Nitrogen monoxide (or nitric oxide) is a compound that breaks the octet rule in this manner. Nitrogen provides 5 valence electrons and oxygen provides 6 valence electrons leaving an odd total of 11 valence electrons. In this case, nitrogen will be one shy of eight electrons in the structure.

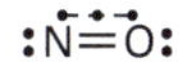

**CHEMICAL STRUCTURE 3.4.** Nitric Oxide Lewis Symbol.

## 3.4 Molecular Geometry—VESPR

Knowing the shape of a molecule is critical to predicting its behavior and properties. The shape of a molecule dictates whether or not it is soluble in certain solvents. It also plays a factor in its boiling points and freezing points. Later in this textbook we will discuss intermolecular forces and knowing the structure of a compound is essential to understanding these interactions.

The easiest shape to predict is the shape of a molecule with only two atoms. In this case, the shape will always be linear. For compounds involving more than two atoms, predicting molecular geometry is a bit more involved. A valuable skill is being able to visualize molecules in 3D. This is an acquired skill.

Shapes of molecules are predominantly dictated by the electrons present in atoms, bonds and on lone pairs surrounding the central atom. Remember that electrons are negatively charged and repel each other. In three dimensional spaces, electrons in molecules try to remain as far apart as possible. It is this phenomenon that dictates the shape of a molecule. By knowing the Lewis structure of a compound, one can then predict what the molecular geometry of a molecule will be using ***valence-shell electron pair repulsion theory or VESPR*** (Figure 3.5).

To predict the geometry of a molecule, the number of lone pairs and bonds around the central atom are counted. The VESPR table provided can be used to predict the geometry of a compound by referring to its Lewis structure (Figure 3.5). Working through some examples we will determine the shapes of different compounds and why these shapes occur.

### $CO_2$

In the Lewis structure for $CO_2$, carbon is the central atom with two double bonds, one to each of the oxygens. This means carbon has no valence electrons and two bonds (single, double and triple bonds all count as one bond). Remember, bonds are formed by sharing electrons so each of those bonds are actually electrons shared between the carbon and oxygen atoms. Again, electrons repel each other so these two double bonds want to be as far apart as possible. The greatest distance between the two would be $180^0$ this creates linear geometry for the molecule:

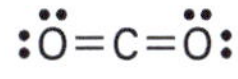

**CHEMICAL STRUCTURE 3.5.** Carbon Dioxide.

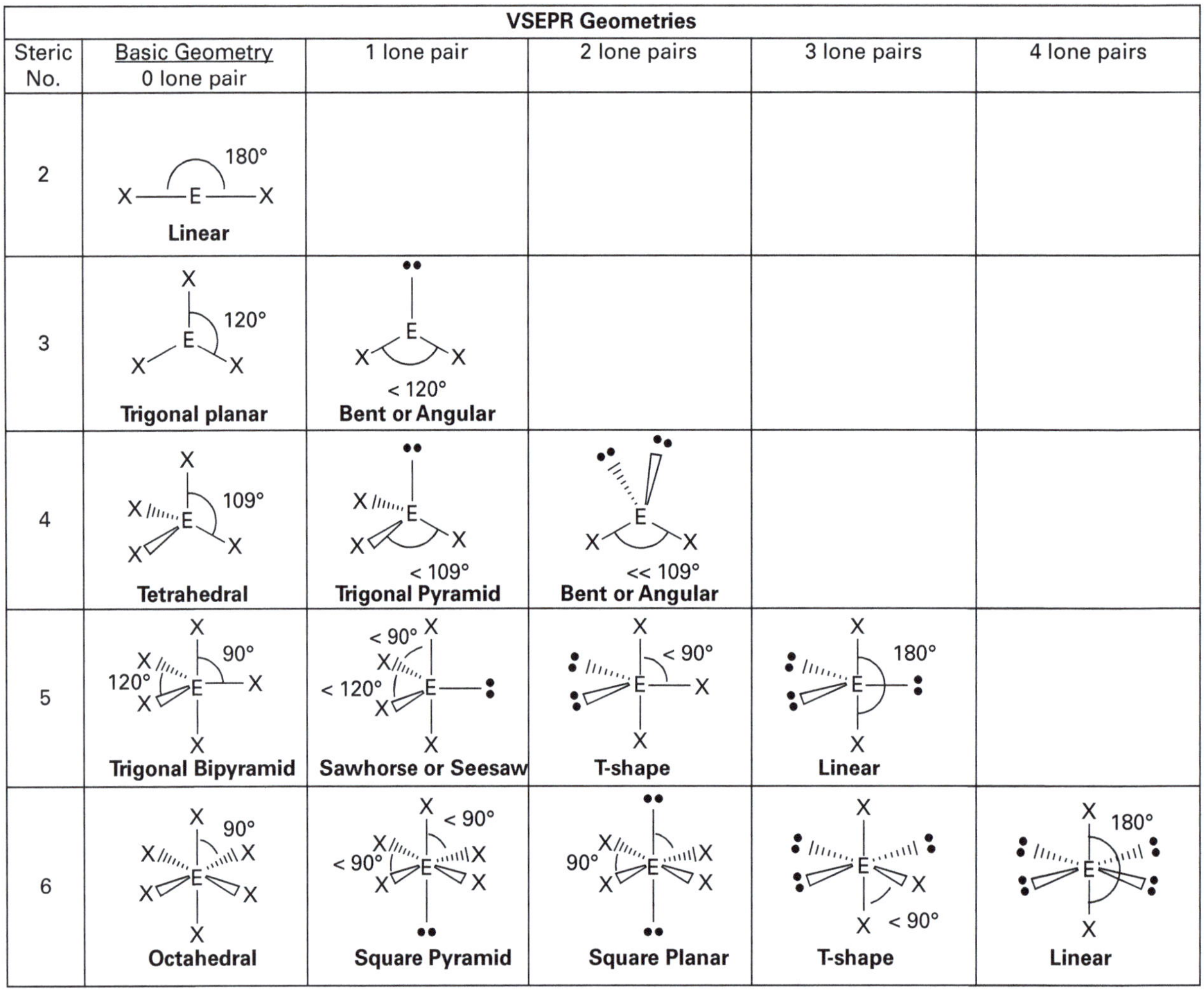


FIGURE 3.5. VESPR Table.

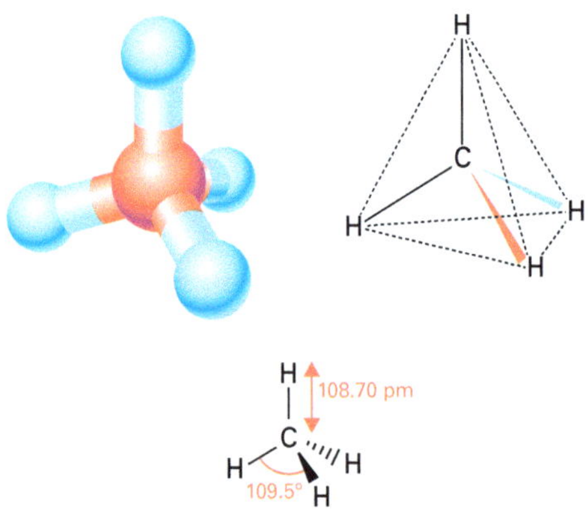


FIGURE 3.6. Methane.

## $CH_4$

As previously mentioned, hydrogen will always be the outermost atom in a compound. In this case, carbon will be the central atom surrounded by four hydrogen atoms. Each hydrogen forms a single bond to carbon. This completes the octet for carbon and leaves the hydrogens each with two electrons. Remember, two electrons completes the valence shell for hydrogen. This means carbon has 4 bonds and no lone pairs. To optimize the distance between each of the bonds between the carbon and the hydrogens, methane will assume a tetrahedral shape. A tetrahedral shape leaves an angle of ~109.5° between the atoms and optimizes the distance between any electrons (Figure 3.6).

## $NH_3$

In the structure of ammonia, nitrogen is connected to three hydrogen atoms through single bonds and also has one lone pair of electrons. As with methane, there are four valence electron components around the central atom. Like the electrons in bonds, lone pairs of electrons also repel each other. To optimize the distance between these electron components, they will distribute themselves around the central atom in a tetrahedral configuration. Again giving a bond angle of ~109.5°. While the *electrons* assume the tetrahedral shape, this does not reveal the geometry of the compound. The geometry of the compound is related to the atoms in the compound. To predict this geometry, the "nonbonding" electrons are used to arrive at the electron pair or basic geometry for the compound but the molecular geometry is determined by considering only the bonding electrons—those electrons involved in bond formation. Taking this into consideration, ammonia has 3 bonding electron regions and one nonbonding electron regions. This will result in a trigonal pyramid molecular geometry for ammonia (Figure 3.7).

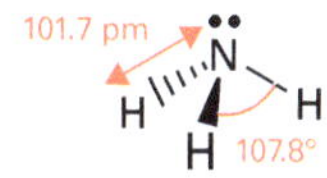


**FIGURE 3.7.** Ammonia.

## $H_2O$

Looking at the formula for water, it might appear as though water has a linear geometry like carbon dioxide but this is not the case. When we consider the Lewis structure for water, oxygen is bound to two hydrogen atoms and also has two lone pairs. The geometry for water will be the geometry that optimizes the greatest distance between all of these electron regions. With four electron regions around the oxygen atom, the basic geometry for this molecule will be tetrahedral. Considering only the two bonding region, the molecular geometry will be angular or bent (Figure 3.8).

The table provided in this chapter can be used to determine the shape a compound will assume based on the number of bonds and lone pairs on the central atom. Refer to this table to predict the structure of various compounds. Two use this table, first follow the rules previously outlined in the chapter to determine the Lewis structure. Then use the Lewis structure to determine the molecular geometry. This skill will become of great value when discussing intermolecular forces in Chapter 6.

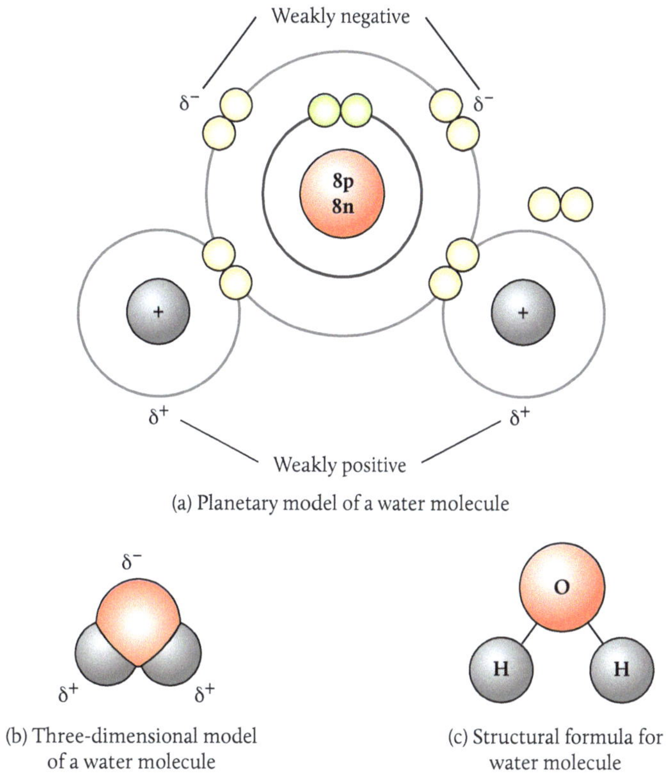


**FIGURE 3.8.** Water.

# Chapter 3 Summary

*Lewis symbols* provide a representation of an atom using the atomic symbol for the element and the *valence* electrons are represented by dots surrounding the symbol.

The *octet rule* states that atoms are most stable when they contain 8 valence electrons like the very stable and mostly inert Noble Gases. This can be achieved by transferring electrons between atoms or sharing electrons between atoms.

A *chemical bond* is a connection formed between two atoms. The two most common types of chemical bonds between atoms are ionic and covalent.

*Chemical nomenclature* is the process by which compounds are given names. There are rules followed for naming compounds that vary depending on the type of compound.

*Lewis structures* are diagrams that show the bonds between atoms of a compound. Also represented around atoms in a lewis structures are lone pairs of electrons that are not involved in bonds between the atoms.

*Lone pairs* are valence electrons of an atom that are not shared with another atom to form a bond.

*Type I metals* are metal atoms that only form ions of one specific charge. These include alkali metals, alkali earth metals, aluminum, zinc and silver.

*Type II metals* are metal atoms capable of forming ions with different or variable charges. These are better known as transition metals—all metals that are not type I.

*Covalent bonds* are formed when two atoms share electrons. These bonds are formed between two nonmetals.

*Nonpolar covalent bonds* are formed when the electrons are shared equally between the atoms forming the bond. These bonds are formed when the electronegativity difference between two atoms is very small (<0.5).

*Polar covalent bonds* are formed when the electrons are not shared equally between the atoms forming the covalent bond. This occurs with large electronegativity differences.

*Polyatomic ions* are covalent compounds that carry a charge and are thus ions. They behave as a single ion and form ionic bonds with other ions to form ionic compounds.

*Valence shell electron pair repulsion (VSEPR)* is a theory used to predict the distribution of electrons around a compound. This theory is used to determine if a compound is nonpolar or polar.

## CHAPTER 3 EXERCISES

1. What is the difference between a polar covalent bond and a nonpolar covalent bond?
2. When making a lewis structure what determines the atom that should be placed in the center of the molecule?
3. To which atoms do you add lone pairs of electrons first when creating a lewis structure?
4. When should double and triple bonds be used in a lewis structure?

5. Explain what will happen when calcium and sulfur combine to form a compound. What will the formula be? What is the name of the compound?
6. What clue can help you easily determine whether a compound is ionic or molecular?
7. Why are noble gases unreactive or inert?
8. What will be the charge on each of the following atoms when they form an ion?

   a. Sr

   b. Br

   c. S

   d. Al

   e. K
9. Give the lewis symbols for each of the following atoms:

   a. B

   b. Ar

   c. I

   d. C

   e. N
10. Name or give the formula for the following polyatomic ions:

    a. Hydroxide

    b. $CO_3^{2-}$

    c. Phosphate

    d. $NO_3^-$

    e. Sulfate

    f. $NH_4^+$
11. Give the formula and name for each of the pairings below. If a compound will not form write N/A.

    a. aluminum with phosphorous

    b. manganese (II) with magnesium

    c. sulfate with rubidium

    d. tin with barium

    e. chromium (IV) with chlorine
12. For each of the following classify as type I ionic, type II ionic or molecular then give the name.

    a. CuO

    b. $Pb_3(PO_4)_2$

c. $N_2O_3$

d. $Ba(NO_3)_2$

e. $SO_3$

13. For each of the following classify as type I ionic, type II ionic or molecular then give the formula.

a. dihydrogen monoxide

b. calcium phosphide

c. lead (IV) carbonate

d. silver nitride

e. tetraphosphorous decoxide

14. For each of the following compounds, give the total number of valence electrons then give the lewis structure:

a. $PO_3^{3-}$

b. $BF_3$

c. $CH_4$

d. $PF_5$

e. $CO_2$

15. For the compounds in question 13, use VSEPR and give the geometric structure for the compound.

16. Use VSEPR to determine the geometrical structure of the following compounds. Also identify each as a polar or nonpolar compound:

a. $CO_2$

b. $H_2O$

c. $CH_4$

d. $NH_3$

e. $CH_3F$

17. For the bonds formed between each of the following atoms, indicate whether the bonds will be polar covalent, nonpolar covalent or ionic.

a. CH

b. LiCl

c. OH

d. OO

e. CF

18. For each atom listed explain what would need to occur for the atom to achieve the octet rule (gain an electron, lose and electron or nothing).

    a. calcium

    b. bromine

    c. radon

    d. carbon

    e. oxygen

19. For each of the polar bonds given, denoted which atom would have a partial positive charge and which would have a partial negative charge.

    a. OH

    b. PF

    c. CO

    d. NF

    e. CCl

20. Complete the following table:

| Compound | Type 1 or 2 | Cation | Anion |
|---|---|---|---|
| $CoCO_3$ | | | |
| cadmium (IV) sulfide | | | |
| ammonium phosphide | | | |
| silver carbonate | | | |
| $AlPO_4$ | | | |

# Credits

Fig. 3.1a: Copyright © 2014 by Daviewales, (CC BY-SA 4.0) at https://commons.wikimedia.org/wiki/File%3ACarbon_Lewis_Structure_PNG.png.

Fig. 3.1b: Source: https://commons.wikimedia.org/wiki/File%3ALewis_dot_Cl.svg.

Fig. 3.1c: Source: https://commons.wikimedia.org/wiki/File%3ALewis_dot_Na.svg.

Fig. 3.1d: Source: https://commons.wikimedia.org/wiki/File%3A%E7%AA%92%E7%B4%A0%E3%81%AE%E4%BE%A1%E9%9B%BB%E5%AD%90.svg.

Fig. 3.2: Copyright © 2013 by OpenStax College, (CC BY 3.0) at https://commons.wikimedia.org/wiki/File%3A207_Ionic_Bonding-01.jpg.

Fig. 3.3: Source: https://commons.wikimedia.org/wiki/File%3ALattice-enthalpy-NaCl-3D-ionic.png.

Fig. 3.4: Source: http://wps.prenhall.com/wps/media/objects/1053/1078773/tools/HPMP.table.2.4.gif.

Chemical Structure 3.4: Source: https://commons.wikimedia.org/wiki/File:Nitric_oxide.svg.

Fig. 3.5: Source: https://commons.wikimedia.org/wiki/File:VSEPR_geometries.PNG.

Chemical Structure 3.5: Source: https://d2gne97vdumgn3.cloudfront.net/api/file/nShwFYC3QaqLaOwrhova.

Fig. 3.6: Adapted from: http://andromeda.rutgers.edu/~huskey/images/methane_w.png.

Fig. 3.7: Source: https://commons.wikimedia.org/wiki/File:Ammonia-dimensions-from-Greenwood%26Earnshaw-2D.png.

Fig. 3.8: Copyright © 2013 by OpenStax College, (CC BY 3.0) at https://commons.wikimedia.org/wiki/File:209_Polar_Covalent_Bonds_in_a_Water_Molecule.jpg.

CHAPTER 4

# Chemical Reactions and Equations

One of the most fundamental topics in chemistry is the study and understanding of chemical reactions. A chemical reaction is probably the first thing that comes to the mind of someone thinking about chemistry. They are a description of a chemical change that occurs. In this chapter you will learn about the components of a chemical reaction, how to write a chemical reaction that is correct, how to use chemical reactions to predict the amounts of products produced or reactants needed and finally common classes of chemical reactions (Figure 4.1).

## 4.1 Components of a Chemical Equation

### Reactants and Products

***Chemical reactions*** are processes during which one or more substances undergo a chemical change to form one or more new substances. During reactions, atoms in a compound or compounds are rearranged to form different compound(s) with entirely new properties. The compounds present at the beginning of the reaction are known as the ***reactants*** (or reactant as the case may be) and the compounds that are formed are called the ***products*** (or product).

Chemical reactions are represented using chemical equations where chemical formulas are used to denote the reactants and products participating in the reaction. In a chemical equation reactants are always written on the left side and products are written on the right. These components are separated by an arrow.

**Reactants** $\longrightarrow$ **Products**

### Arrows

The arrow used in the chemical equation indicates whether a reaction can only go in one direction or if the reaction can go in both directions. An ***irreversible*** reaction is only able to go in one direction so an arrow that points only towards the product is used in chemical equations for such reactions. An example of an irreversible reaction is the burning of coal. In this reaction coal or, more specifically carbon, reacts with oxygen to form carbon dioxide (Figure 4.2).

$$C + O_2 \longrightarrow CO_2$$

In some cases, a chemical reaction can proceed in both directions or is ***reversible***. You will learn a lot more about this category of reactions later in this textbook. For now, it is important to understand that some products can convert back to reactants. In this case, two chemical reactions will occur at the same time. These reactions are very dynamic and never actually stop. As soon as enough product is formed, a portion of the product can be converted back to reactants. This back and forth conversion between reactants and products will continue like a ping pong ball in a game of table tennis. The formation of ammonia from hydrogen and nitrogen is an example of a reaction that occurs reversibly. There are different types of arrows used to depict a reversible reaction, and two examples are listed below. Notice that the products and reactants are separated by arrows that point in both directions, thus indicating that the reaction is reversible. The key to depicting these reactions is the use of an arrow that points towards both the reactants and products.

$$N_2 + 3H_2 \rightleftharpoons 2NH_3$$
$$N_2 + 3H_2 \leftrightarrow 2NH_3$$

FIGURE 4.1. Evidence of a Chemical Reaction.

## Coefficients

You will notice in the equations above for the formation of ammonia $NH_3$), there are numbers listed in front of each of the compounds involved in the reaction (Figure 4.2). These are called coefficients and they reveal how many molecules of each compound is needed for the reaction to take place. This is similar to a recipe where one might need to mix 2 eggs, 2 cups of flour and a cup of milk to make 5 pancakes.

**2 eggs + 2 cups flour + 1 cup milk ⟶ 5 pancakes**

In the case of the formation of ammonia, one molecule of nitrogen reacts with three molecules of hydrogen to form two molecules of ammonia.

$$N_2 + 3H_2 \leftrightarrow 2NH_3$$

Coefficients are a critical component of a chemical reaction. Without the appropriate coefficients, a chemical equation is incomplete and incorrect. Now let's consider the equation for the burning of coal presented earlier.

FIGURE 4.2. Pure Carbon.

$$C + O_2 \longrightarrow CO_2$$

In this equation, there are no coefficients listed for any of the reactants or products. You may think that this means the equation is incorrect, but it is not. In this case one atom of carbon reacts with one molecule of oxygen to produce one molecule of carbon dioxide. When only one molecule of a particular substance is required in a reaction the one is "understood". This means that a "1" is not placed in front of an atom or molecule in a chemical equation. When you see a substance in a correct and complete chemical equation without a coefficient this means that only 1 molecule, or atom as the case may be, of that substance is needed for the reaction to occur. Let's consider the generic equation listed below.

$$2\,A \quad + \quad B \quad \longrightarrow \quad 3\,C$$

Let's translate the meaning of this equation. Looking at the equation, there is a 2 in front of the A, no coefficient for B and a 3 in front of the C. This means that in this reaction, 2 molecules of A react will react with one molecule of B to produce 3 molecules of C. Furthermore, we know that this reaction will proceed irreversibly or only in the direction of the products. How do we know this? Because of the arrow used in the equation which only points in the direction of the products.

## States of Matter

Another important component of a chemical equation indicates the state of matter for each component. It is important to include this information as a substance can have very different properties when it is in the gas, liquid or solid state. Take for consideration water. It has a different appearance, feel as well as differences in other physical properties depending upon if it occurs as steam, a liquid or ice. Each state of matter gets a unique symbol in a chemical equation.

| State | Symbol |
|---|---|
| solid | s |
| liquid | *l* |
| gas | *g* |
| aqueous | *aq* |

These symbols are placed in parenthesis on the right side of the substance in the chemical equation. Let's modify the equations for burning coal and formation of ammonia so that they reflect the states of matter for all of the reactants and products.

$$C(s) \quad + \quad O_2(g) \quad \longrightarrow \quad CO_2(g)$$
$$N_2(g) \quad + \quad 3\,H_2(g) \quad \rightleftharpoons \quad 2\,NH_3(g) \quad \rightleftharpoons \quad 2\,NH_3(l)$$

When translating the equation for the burning of coal, we know now that *solid* coal reacts with *gaseous* oxygen producing *gaseous* carbon dioxide. The equation for ammonia now reveals that *gaseous* nitrogen reacts with *gaseous* hydrogen to produce *gaseous* ammonia followed by a physical change of the latter to the *liquid* ammonia state. Including the states of matter makes these two equation complete and more informative.

It is important to understand the states of matter will depend on the temperature at which the reaction takes place. So if you see equations for the same reaction written with different states of matter, it does not mean that one equation is correct and the other is wrong. This could be an indication that the reactions took place at different temperatures. This is another reason why inclusion of the states of matter makes the equation more informative and is an important component of a chemical equation.

# 4.2 Balancing Chemical Equations

We already learned that the law of conservation of mass states that matter cannot be created or destroyed. This is important when considering chemical reactions and the equations that describe them. In a chemical reaction, whatever atoms start out as reactants end up as products. The only difference when they change from reactants to products is the manner in which the atoms are arranged or what bonds they form to other atoms. This is where those coefficients we discussed earlier play an important role.

When a chemical equation is written it is only correct if the number of each type of atom on the reactant side is equal to the number of atoms of that type on the products side. If matter cannot be created or destroyed than atoms cannot be created or destroyed. When reviewing an equation, if there is a difference between the numbers of atoms on the products side and the reactants the equations needs to be *balanced*.

Before we delve into balancing a chemical equation, it is also important to think of the coefficients as multipliers. When there is a coefficient is in front of a molecule that means that every component of that molecule is increased or multiplied by that coefficient. Let's once again consider the equation for the formation of ammonia.

$$N_2(g) \quad + \quad 3\,H_2(g) \quad \rightleftharpoons \quad 2\,NH_3(g)$$

The lack of a coefficient in front of nitrogen means only one molecule of nitrogen reacts. In that molecule of nitrogen there are two nitrogen atoms so the nitrogen molecule provides two *atoms* of nitrogen. For hydrogen, the coefficient 3 means three hydrogen molecules will react providing two hydrogen atoms for a total of 6 hydrogen atoms. Finally, the 2 in front of the ammonia means two molecules of ammonia will be produced. Those two molecules each have one atom of nitrogen and 3 atoms of hydrogen. So, two molecules of nitrogen need 2 atoms of nitrogen and 6 atoms of hydrogen.

Notice in the example we just reviewed, while the identity of the molecules changed when the reactants were converted to the products, the numbers of each type of atom did not change. The reactants have 2 atoms of nitrogen and 6 atoms of hydrogen and the products have 2 atoms of nitrogen and 6 atoms of hydrogen. This is because this equation is ***balanced***. In this properly balanced equation, the number of atoms of each element remains the same when changing from reactants to products. The only difference is the manner in which those atoms are arranged and connected to each other. This means that the equation is correct and in accordance with the law of conservation of mass.

It is also important to understand the difference between a coefficient in the equation and the subscripts on the molecules. Much like changing the number of protons in an atom changes the identity of the atom, if you change the subscripts for a molecule, you change the identity of that molecule. So, when we balance a chemical equation, you *never change the subscripts* on the molecules. If you change the subscripts you will change the molecules and thus change the reaction the equation is describing. Equations are balanced by using coefficients.

Let's imagine we are working at a bicycle factory. Each bike requires two wheels (W) and one frame (F). Let's create an equation to represent making a bike.

$$F \quad + \quad W \quad \rightleftharpoons \quad FW_2$$

This equation is not balanced. We have 1 frame and 1 wheel on the reactants side but 1 frame and 2 wheels on the product side. The frames are balanced but not the wheels. Let's try and balance by changing the subscripts.

$$F \quad + \quad W \quad \rightleftharpoons \quad FW$$

Now this equation is balanced with 1 frame on both sides of the equation and 1 wheel on both sides of the equation. However, we have changed the identity of the product. This is either describing the formation of a unicycle or a very non-functional and difficult to ride bicycle. In either case, this equation does not represent making a functional bicycle. So, while this equation is balanced, it is incorrect.

Let's now try balancing the same equation using coefficients and leaving the subscripts alone. There is one frame on both sides of the equation so no adjustment is needed here. There is 1 wheel on the reactants side and 2 wheels on the product side. We can balance the wheels by placing a 2 in front of the "W" on the reactants side. Now we are left with 1 frame on both sides, 2 wheels on both sides and we form a proper, functional and easy to ride bicycle on the products side. This equation is balanced and a correct description of the formation of a bicycle.

$$F + 2W \rightleftharpoons FW_2$$

This same logic can be applied to chemical equations. When a chemical equation is complete and accurate, the number of atoms of each element must be equal on both sides of the equation. If this is not the case, then the equation must be *balanced*. Balancing chemical equations involves adjusting coefficients until all of the atoms are equal on both sides of the equation. This concept can be simple to understand but sometimes it can be a challenge to balance an equation. Following the steps below can help make the process easier. One very helpful aspect of balancing equations is that it is easy to check your work by simply verifying whether the number of atoms on either side of the equation is the same. Once they are, you have likely received at the correct answer, so long as you did not break any rules along the way.

1. Write a skeletal equation by writing the formula of each reactant and product.
2. Count and record the number of atoms of each element on each side of the equation.
3. Pick an element to balance starting with an element that appears in only one compound on both sides of the equation. Save elements that are free elements (not combined with other elements) on one side of the equation until ***last***.
4. Starting with the element selected in step 3, determine the least common multiple for the amounts of that element on both sides of the equation. Determine the factor needed on each side of the equation to bring the number of atoms of that element equal to the least common multiple.
5. The factor determined in step 4, becomes a coefficient in the equation for the molecules containing the element. If there is already a coefficient there, multiply it by the factor. *Coefficients are placed in front of entire molecules, not between atoms within a molecule.*
6. Repeat this process for all atoms in the equation.

   Check the final equation but making sure the number of atoms of each type are equal on both sides of the equation.

These rules can be applied to the equation for the formation of water from hydrogen and oxygen. Let's do this step by step. First, we need to write down the skeletal equation. For simplicity, the states of matter have been left out of the equation.

$$H_2 + O_2 \longrightarrow H_2O$$

| $H_2 + O_2$ | $\rightarrow$ $H_2O$ |
| --- | --- |
| 2 H | 2 H |
| 2 O | 1 O |

IMAGE 4.1.

Next, the number of oxygen atoms and hydrogen atoms on both sides of the equation need to be counted.

In this case, the number of hydrogen atoms is the same on both sides, so the obvious place to start is with oxygen. Seeing as there are 2 oxygen atoms on the reactants side and 1 on the products side, the least common multiple is 2. This means we need to multiply the $O_2$ by 1 and the water molecule by 2. Let's add the appropriate coefficients to the equation then count the number of atoms on each side again.

Now here is where balancing equations can get messy. Through the process of balancing the oxygen atoms, the hydrogen atoms are now no longer balanced. It is now necessary to balance the hydrogen atoms. There are two hydrogen atoms on the reactant side and 4 on the product side. The least common multiple of 2 and 4 is 4. So now we must add coefficients to both sides to bring the total number of hydrogen atoms to 4. We can leave the products side alone as we already have 4. On the reactants side, we will add a coefficient of 2 in front of the hydrogen molecule.

$$H_2 + O_2 \rightarrow 2\,H_2O$$

| | |
|---|---|
| 2 H | 4 H |
| 2 O | 2 O |

IMAGE 4.2.

This modification to the equation leaves us with 4 hydrogen atoms on both sides of the equation and 2 oxygen atoms on both sides of the equation. So the complete, balanced equation for the formation of water is:

$$2\,H_2(g) \quad + \quad O_2(g) \quad \longrightarrow \quad 2\,H_2O(g)$$

When balancing equations it is important to always end with a balanced equation in its simplest form. For example, refer to the two equations below.

$$4\,H_2(g) \quad + \quad 2\,O_2(g) \quad \longrightarrow \quad 4\,H_2O(g) \quad \textbf{incorrect, not the simplest form}$$
$$2\,H_2(g) \quad + \quad O_2(g) \quad \longrightarrow \quad 2\,H_2O(g) \quad \textbf{correct, simplest form}$$

While the first version of the equation for the formation of water is balanced with regard to the atoms, it is incorrect. This is because it can be simplified. Each of the coefficients has a common multiple of 2 and therefore should be divided by that value. The second version is the correct version of the balanced equation. When balancing equations always check that the equation is in a simplest form once balanced. If not, make the appropriate adjustments.

Properly balanced equations must have whole number coefficients. A balanced equation cannot have fractions for coefficients. While the equation may have the same number of atoms on both sides of the equation, it will be incorrect if there is a fraction included. It is not possible to add a "part" of a molecule, only a whole molecule can be used or produced in a reaction. The simple correction for this is to multiply all of the coefficients in an equation by a factor that will remove the fraction and yield an equation in its simplest, balanced form. Below is an example of the burning of methanol first balanced with a fraction and then corrected to reflect a proper, balanced equation.

$$CH_3OH(l) \quad + \quad 1.5\,O_2(g) \quad \longrightarrow \quad CO_2\,(g) \quad + \quad 2\,H_2O(g)$$
$$\{CH_3OH(l) \quad + \quad 1.5\,O_2(g) \quad \longrightarrow \quad CO_2\,(g) \quad + \quad 2\,H_2O(g)\,)\} \times 2$$
$$2\,CH_3OH(l) \quad + \quad 3\,O_2(g) \quad \longrightarrow \quad 2\,CO_2\,(g) \quad + \quad 4\,H_2O(g)$$

### A helpful hint

Sometimes it is helpful to balance an equation ***temporarily*** with a fraction. The equation can later be multiplied to change the fraction to a whole number. This is especially helpful with reactions involving oxygen ($O_2$) like combustion reactions. Consider the reaction of oxygen with ammonia presented below.

$$2\,NH_3(g) \quad + \quad 2.5\,O_2(g) \quad \longrightarrow \quad 2\,NO(g) \quad + \quad 3\,H_2O(g)$$

It is balanced as presented above with equal numbers of atoms on both sides of the equation. But it is impossible to add half a molecule of $O_2$ so this equation, while balanced, is wrong. To correct this, the coefficients for the entire equation should be multiplied by 2. The last equation with whole numbers is also in the most simplified form. This is therefore the correct balanced equation.

$$\{2\ NH_3(g) + 2.5\ O_2(g) \longrightarrow 2\ NO(g) + 3\ H_2O(g)\}\ \times 2$$
$$4\ NH_3(g) + 5\ O_2(g) \longrightarrow 4\ NO(g) + 6\ H_2O(g)$$

## 4.3 Mass Relationships in Chemical Reactions

Now that we are able to balance a chemical equation, we can use these equations to predict the amount of a product that can be produced based on the amount of reactants used. It is also possible to figure out the amount of reactants needed to produce a known amount of products. This is very similar to what one needs to do when they are cooking. Let's imagine we are going to make some chocolate chip cookies. The equation below can be considered the recipe. Now this is not an actual recipe just a hypothetical one, but it will work for our purposes.

**2 cups Flour + 3 eggs + 1 cup chocolate chips ⟶ 12 cookies**

Using this equation, determine how many eggs are needed to make 24 cookies. We are doubling the amount of product made, which means we need $2 \times 3$ or 6 eggs. Additionally, if the chef started with 6 cups of flour (and plenty of all the other ingredients) then $3 \times 12$ or 36 cookies will be produced. Most people reading this likely did this calculation in their head as it is fairly straightforward.

This same logic can be applied to chemical equations. The same types of calculations can be performed though they may seem a bit more complicated because the terminology is not as familiar to you. Before moving on to chemical equations, let's work out the calculations from above on paper as it will help when we look at the more complicated chemical equation.

### Example 1 How many eggs will be needed to make 24 cookies?

We know that 3 eggs produces 12 cookies so we can turn this into a *conversion factor.*

$$\frac{12\ cookies}{3\ eggs}\ \textbf{or}\ \frac{3\ eggs}{12\ cookies}$$

In our case, we are looking to determine how many eggs we need so we will use the second conversion factor $\left(\frac{3\ eggs}{12\ cookies}\right)$. What we want to do is convert the number of cookies to the number of eggs:

$$\textbf{24 cookies} \times \frac{3\ eggs}{12\ cookies} = \frac{24\ \cancel{cookies} \times 3\ eggs}{12\ \cancel{cookies}} = \textbf{6 eggs}$$

Imagining cookies are a unit, when cookies are divided by cookies the unit cancels.

We will use a very similar process to perform calculations using chemical equations. The process of calculating amounts of reactants and products using a balanced chemical equation is known as ***stoichiometry***. Let's start with a simple example of the production of ammonia from hydrogen and oxygen. You will recall the balanced chemical equation.

$$N_2(g) \quad + \quad 3\,H_2(g) \rightleftharpoons 2\,NH_3(g)$$

In this equation, one molecule of nitrogen reacts with 3 molecules of hydrogen to produce 2 molecules of ammonia. Let's determine the number of molecules of ammonia produced with 12 molecules of hydrogen completely react with enough nitrogen. Let's approach this step by step.

1. Use the balanced chemical equations to determine the ratio of hydrogen to ammonia

   3 molecules of $H_2$ needed to produce 2 molecules of $NH_3$

   or

   3 molecules of $H_2$ : 2 molecules of $NH_3$

2. Use the ratio determined in step one to create a conversion factor. In this case, we are starting with hydrogen so this should be in the denominator of the conversion factor.

$$\frac{2 \text{ molecules ammonia}}{3 \text{ molecules of hydrogen}}$$

3. Use the conversion factor to complete the calculation.

$$\textbf{12 molecules of } \boldsymbol{H_2} \times \frac{2 \text{ molecules ammonia}}{3 \text{ molecules of hydrogen}} =$$

$$\frac{12 \text{ molecules of hydrogen} \times 2 \text{ molecules ammonia}}{3 \text{ molecules of hydrogen}} = \frac{12 \times 2}{3}$$

$$= 8 \text{ molecules of } NH_3 \text{ produced}$$

This process demonstrates the basic process of stoichiometry. In this example we used the chemical equation to determine the molecules of products produced based on the number of molecules of reactants. While this demonstrates the process of stoichiometry, molecules are not the best unit to use. It is not possible for a person to count molecules of a certain substance. Instead, chemists most often measure quantity by weighing the materials they will use. In this next section we will discuss the use of mass to measure quantities of materials.

## The Mole

During the 19$^{th}$ century an Italian chemist, Amadeo Avogadro, experimentally determined how many atoms of carbon were present in exactly 12 g of carbon. This value is known as ***Avogadro's number***. Seeing as this number is large and cumbersome, it is set equal to 1 mole of material.

**12 g carbon** $= 6.022 \times 10^{23}$ **atoms of carbon = 1 mole of carbon atoms**

This same value can be used to "count" the atoms of any element.

**1 mole of atoms** $= 6.022 \times 10^{23}$ **atoms**

You can think of this counting unit as being similar to how bakers use a "dozen" of baked goods as a unit.

**12 donuts = 1 dozen donuts**

The mol can be used to count anything—atoms, molecules, particles etc. Again, think of it as being similar to a dozen the only difference is that it represents a different quantity of material.

**1 mol of water molecules** $= 6.022 \times 10^{23}$ **water molecules**
**1 mol of Na atoms** $= 6.022 \times 10^{23}$ **Na atoms**
**1 mol of donuts** $= 6.022 \times 10^{23}$ **donuts**

## Molar Mass

So the mol provides chemists with a much less cumbersome number to deal with, it still does not help their ability to measure amounts of material. A chemist cannot simply count numbers of atoms or molecules as a baker can count donuts. Chemists instead determine masses of materials and use that as a means of measurement. This is not as straightforward as it may sound as the masses of atoms of different elements are not the same.

You may recall in the first chapter of this textbook, we learned about the different elements and how the identity of the element is determined by the composition of the atom. More specifically, we learned that changing the number of protons in an atom changes the identity of the atom and its properties (including its mass). We also learned that the number of neutrons in an atom do not change the type of element but does change the mass of the atom. Taking this into account, we know that the ***atoms of different elements have different masses***.

As such, 10 grams of iron contains a different amount of atoms than 10 grams of carbon atoms. To account for this difference, the mass of 1 mol of atoms in gram is known as the molar mass. The atomic mass reported on the periodic table represents mass of one mol of atoms of a given element in grams. Looking at the periodic table we can determine the molar mass of all of the elements very easily. The unit for molar mass is $\frac{grams}{mol}$.

### Example: What is the molar mass for each of the elements listed below?

Sodium (Na) $22.99 \frac{g}{mol}$

Iron (Fe) $55.85 \frac{g}{mol}$

Carbon (C) $12.01 \frac{g}{mol}$

These molar masses can now be used to perform conversions from mol to grams or grams to mol. Let's review some examples.

### Example: How many mols of sulfur atoms are present in 130 grams of sulfur?

We know that there is 1 mol of sulfur in 32.07 grams and we can use this information to determine how many mols of sulfur is present in 130 grams.

$$130\ \cancel{g\ S} \times \frac{1\ mol\ S}{32\ \cancel{g\ S}} = 4.1 \text{ mol of S}$$

### Example: What is the mass of 5.0 mol of magnesium atoms?

Let's use the periodic table to determine the molar mass of Mg and use that to complete this calculation.

$$5.0\ \cancel{mol\ Mg} \times \frac{24.31\ g\ Mg}{1\ \cancel{mol\ Mg}} = 120 \text{ g of Mg}$$

Determining the molar mass of a molecule is not as simple as determining the molar mass of an atom. We still use the periodic table, however, we need to calculate the molar mass of a molecule based on its composition. When determining the molar mass of molecules, refer to the chemical formula as that reflects the types of atoms and number of each atom present in the molecule.

### Example: What is the molar mass of water ($H_2O$)?

*2 atoms of hydrogen:*

$$2 \times 1.01 \frac{g}{mol} \text{ H} = 2.02 \frac{g}{mol} \text{ of H}$$

*1 atom of oxygen:*

$$1 \times 16.00 \frac{g}{mol}\ O = 16.00 \frac{g}{mol} \text{ of O}$$

*Add masses together to get mass of the molecule:*

$$2.02 \frac{g}{mol} \text{ of H} + 16.00 \frac{g}{mol} \text{ of O} = 18.02 \frac{g}{mol} \text{ of } H_2O$$

When determining the mass of an ionic compound that contains a polyatomic ion, remember to distribute subscripts outside of the parenthesis to all atoms inside. The number outside of the parenthesis is multiplying the entire ion. The 2 outside the parenthesis for nitrate tells us there are two nitrate ions. This means there are 2 nitrogen atoms and 6 oxygen atoms in the compound.

$$Ba(NO_3)_2$$

### Example: Determine the molar mass of calcium phosphate.

The formula for calcium phosphate is:

$$Ca_3(PO_4)_2$$

Using the formula we can determine how many of each type of atom is present.

3 Ca atoms

$$1 \times 2 = 2 \text{ P atoms}$$
$$4 \times 2 = 8 \text{ O atoms}$$

Using the numbers of atoms and molar masses from the periodic table, we can determine the molar mass of the compound.

Ca:

$$3 \times 40.00 \frac{g}{mol} \text{ CA} = 120 \frac{g}{mol} \text{ Ca}$$

P:

$$2 \times 30.97 \frac{g}{mol} \text{P} = 61.94 \frac{g}{mol} \text{ P}$$

O

$$8 \times 16.00 \frac{g}{mol} \text{O} = 128 \frac{g}{mol} \text{ O}$$

Add masses together to determine mass of $Ca_3(PO_4)_2$

$$120 \frac{g}{mol} \text{ Ca} + 61.94 \frac{g}{mol} \text{ Ca} + 128 \frac{g}{mol} \text{O}$$

$$= 309.94 \frac{g}{mol} Ca_3(PO_4)_2$$

We can also use the mol in stoichiometry calculations with balanced equations. We can determine mols of products produced based on the mols of reactants and vice versa.

### Example: How many moles of hydrogen is needed to produce 12 mols of ammonia?

First, start with the balanced equation for the formation of ammonia.

$$N_2(g) \quad + \quad 3\,H_2(g) \quad \rightleftharpoons \quad 2\,NH_3(g)$$

Looking at the equation, we know that 3 molecules of hydrogen produces 2 molecules of ammonia. We can also say the 3 mols of $H_2$ molecules will produce 2 mols of ammonia molecules or more simply put 3 mols of $H_2$ produces 2 mols of $NH_3$. Let's create a conversion factor using this information.

2 mol ammonia : 3 mol hydrogren

$$\frac{3 \text{ mol hydrogen}}{2 \text{ mol ammonia}}$$

Now we can use the conversion factor to determine the mols of hydrogen need for the reaction.

$$12 \text{ mol ammonia} \times \frac{3 \text{ mol hydrogen}}{2 \text{ mol ammonia}} = 18 \text{ mol } H_2$$

### Example: How many grams of hydrogen is needed to produce 12 mols of ammonia?

We already know that 18 mol of $H_2$ is need to produce 12 mol of ammonia (see above example). What we need to do now is convert mol of hydrogen to gram of hydrogen using the molar mass of hydrogen (18 g/mol).

$$18\ \textit{mol}\ H_2 \times \frac{18\ \textit{grams hydrogen}}{1\ \textit{mol}} = 324\ \textit{g}\ \mathbf{H}_2 =$$

**320 g $H_2$ (reflecting appropriate significant figures)**

## Limiting Reactant

During any chemical reaction, the amount of product produced is dependent on the amounts of the reactants present. One of the reactants is considered the ***limiting reagent***. This is the reactant that produces the least amount of product and depends on the amounts of each of the reactants present at the start of the reaction. To better explain this, let's consider making a sandwich.

**2 slices of bread + 3 pieces of meat + 1 slice of cheese ⟶ 1 sandwich**

Now let's return to the cookie example. Imagine you start out with 20 cups of flour, 2 cups of chocolate chips and a dozen eggs. How many cookies can you make? The best approach to this question is to calculate the amount of cookies each reactant amount *could* produce.

$$20\ \textit{cups flour} \times \frac{12\ \textit{cookies}}{2\ \textit{cups flour}} = 120\ \textit{cookies}$$

$$2\ \textit{dozen eggs} \times \frac{12\ \textit{eggs}}{1\ \textit{dozen eggs}} \times \frac{12\ \textit{cookies}}{3\ \textit{eggs}} = 96\ \textit{cookies}$$

$$2\ \textit{cups chocolate chips} \times \frac{12\ \textit{cookies}}{1\ \textit{cup chocolate chip}} = 24\ \textit{cookies}$$

Based on the calculations above, how many cookies do you think you can make? If you answered 24 you are correct. You can only produce the least amount of cookies based on each of the amounts of each ingredient. You might be able to produce 120 cookies with 20 cups of flour but you will run out of chocolate chips well before you make that many cookies. In this case, chocolate chips is the limiting reagent because it limits the number of cookies made. All of the other ingredients are said to be in excess. There will be extra of those ingredients once all of the chocolate chips are used.

We can apply this same logic to chemical reactions. One reactant in a chemical reaction will limit the amount of products produced. Using the amounts of each reactant present, you must first determine the amount of product produced based on the amounts of the reactants. The limiting reactant will be the reactant that products the least amount of product (Figure 4.3).

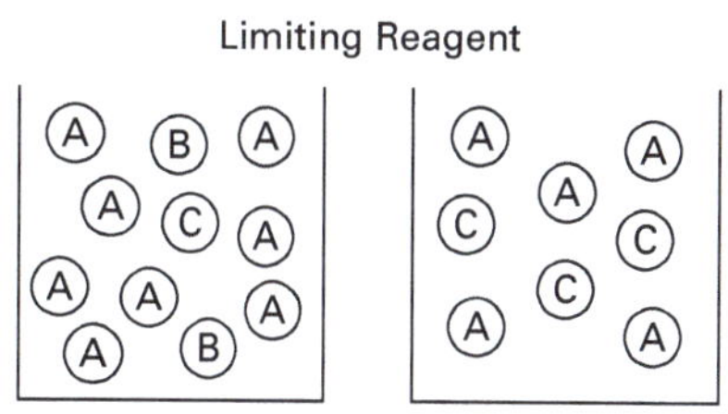


**FIGURE 4.3.** Limiting Reagent.

**Example: How many g of iron will be produced when 55 g of aluminum reacts with 125 g of iron (III) oxide?**

$$2\ Al(s) + Fe_2O_3(s) \longrightarrow Al_2O_3(s) + 2\ Fe(s)$$

Calculate the amount of iron produced from 55 g of aluminum. In this case we need to convert from grams of aluminum to grams of iron but it is not possible to go directly from grams Al to grams Fe. To complete this calculation, we will need to use the following approach:

**grams of Al** ⟶ **mol of Al** ⟶ **mol of Fe** ⟶ **grams Fe**

Let's use conversion factors and the approach from above to complete the calculation.

$$55\ g\ Al \times 55\ \textit{g Al} \times \frac{1\ \textit{mol Al}}{26.98\ \textit{g Al}} \times \frac{2\ \textit{mol Fe}}{2\ \textit{mol Al}} \times \frac{55.85\ \textit{g Fe}}{1\ \textit{mol Fe}} \approx 110\ g\ Fe$$

Now we can complete a similar calculation for amount of Fe produced from $Fe_2O_3$.

$$125\ g\ Fe2O3 \times 125\ g\ Fe_2O_3 \times \frac{1\ \textit{mol iron oxide}}{159.69\ \textit{g iron oxide}} \times \frac{1\ \textit{mol iron oxide}}{2\ \textit{mol Fe}} \times \frac{55.85\ \textit{g Fe}}{1\ \textit{mol Fe}} \approx$$

21.9 g Fe

So now, to answer the question as to the amount of Fe produced. Remember, the reactant that produces will be the limiting reactant and will dictate the amount of product produced.

**21.9 g Fe will be produced**
**$Fe_2O_3$ is the limiting reagent**

## Percent Yield

The calculations we have been working on up until this point predict the amount of product we expect to get based on the limiting reagent. This is the amount of product one can produce if everything goes perfectly. Now think back to the cookie example. Sometimes when cookies are made, the expected amount is not always made. In some cases, a cookie might burn, batter might fall on the floor. When calculated the amount of product produced based on the amount of limiting reagent determines the ***theoretical yield*** of product. This is the amount of product one should obtain in theory is the reaction goes absolutely perfect with no errors along the way. The theoretical yield is the maximum amount of product one should expect to obtain.

Chemists often determine the percent yield they obtain once they complete a chemical reaction. They take the product they obtain and measure its mass to determine the ***actual yield***. They then use this data to determine compare the amount they obtain with the theoretical yield.

$$\text{Percent Yield} = \frac{\textit{Actual Yield}}{\textit{Theoretical Yield}} \times 100$$

Percent yield is important to chemists as it tells them how well they performed the experiment and whether they need to make adjustments. It is also important to optimize experiments and achieve a favorable yield to prevent waste of materials and money when trying to produce chemical compounds.

**Example: A chemist reacts 65 g of nitrogen with excess hydrogen to produce ammonia. If they obtain 25 g of $NH_3$, what was the percent yield?**

$N_2(g) + 3\,H_2(g) \rightleftharpoons 2\,NH_3(g)$

**Theoretical Yield:**

$$65 \quad g \quad \times \quad 65\,g \times \frac{1\ mol\ nitrogen\ gas}{28\ g\ nitrogen\ gas} \times \frac{2\ mol\ ammonia}{1\ mol\ nitrogen\ gas} \times \frac{17\ g\ ammonia}{1\ mol\ ammonia}$$

$= 34\ g\ NH_3$

**34 g $NH_3$ is the theoretical yield**

**Percent Yield:**

$$\text{Percent Yield} = \frac{Actual\ Yield}{Theoretical\ Yield} \times 100$$

$$\text{Percent Yield} = \frac{25\ g\ ammonia}{34\ g\ ammonia} \times 100$$

$\text{Percent Yield} = 74\ \%$

The more efficient the experiment is, the higher the percent yield. It is important to point out that the percent yield cannot be more than 100%. The theoretical yield reports the maximum amount of product that can be produced based on the amounts of reactants used. If a percent yield is calculated to be greater than 100% there is either an error in the measurement of the final product produced or there is an error in the calculation.

## 4.4 Common Reactivity Patterns

There are a countless number of different chemicals reactions that can occur naturally or can occur in a laboratory under certain conditions. While it would be impossible, and probably quite boring, to cover every reaction in this textbook, it is helpful to place reactions into common categories of reaction types. In this section we will discuss these patterns in reactivity. Looking at a chemical equation will then allow you to determine whether the reaction falls into one of these categories.

### Displacement Reactions

Displacement reactions are common in reactions involving ionic compounds. As the name implies, in these reactions, one atom or ion in the reaction is *displaced* by another. In some cases these reactions are called replacement reactions (as one ion replaces the other) and these reactions are the same.

These reactions can be a single-displacement or a double-displacement reaction. First let's consider a single-displacement reaction. In this pattern of reactivity a single atom or ion replaces another in a molecule. The generic equation for this type of reaction is shown below.

$A + BC \longrightarrow AC + B$

In this reaction, A *displaces* B in the molecule BC to form the molecule AC. This is a single displacement because only B is displaced. This is the type of reaction that can occur when certain metals are placed in salt solutions. For instance, when copper metal is placed in a solution of silver nitrate. In this case, copper displaces the silver ion in silver nitrate, producing silver metal and copper (II) nitrate.

$$Cu(s) + 2AgNO_3(aq) \longrightarrow 2Ag(s) + Cu(NO_3)_2(aq)$$

Double-displacement reactions when two ions or atoms displace each other from their compounds to form two new compounds.

$$AB + CD \longrightarrow AD + CB$$

This pattern of reactivity is observed when two ionic compounds are mixed with each other and form a solid product or *precipitate*.

$$NaCl(aq) + AgNO_3(aq) \longrightarrow AgCl(s) + NaNO_3(aq)$$

In the double displacement reaction above, sodium displaces silver and silver displaces sodium making this a double displacement reaction. These two cations exchange "partners" with each other forming new compounds. The silver chloride that is formed is not soluble in water and will be a *precipitate* in the reaction.

## Combination Reactions

In combination reactions, two or more substances react or combine to form a single product. This can occur between two or more elements, between elements and compounds or between two or more compounds. The key to identifying this pattern of reactivity is to look for the multiple reactants forming a single product as in the generic reaction below.

$$A + B \longrightarrow C$$

In this reaction, A and B combine during the reaction to form C. There are many examples of combination reactions. We have already reviewed a few in this chapter, including the formation of ammonia from hydrogen and nitrogen.

$$N_2(g) + 3\,H_2(g) \rightleftharpoons 2\,NH_3(g)$$

In this case, two molecules react to form a single molecule. When magnesium burns it reacts with oxygen in the air to form magnesium oxide. This reaction involves 2 atoms of magnesium reacting with a molecule of oxygen to form two molecules magnesium oxide.

$$2Mg(s) + O_2(g) \longrightarrow 2MgO(s)$$

## Decomposition Reactions

Decomposition reactions are reactions where a single reactant produces two or products. The products can be either elements or simpler compounds. Here is the generic equation for decomposition reactions:

$$AB \longrightarrow A + B$$

Decomposition reactions will require some form of energy in order for the reaction to occur. This can come in the form of light energy (photolysis), electrical energy (electrolysis) or heat energy (thermal decomposition). Water undergoes electrolysis to form hydrogen and oxygen. With adequate heat energy, calcium carbonate can undergo thermal decomposition. These are both great examples of decomposition reactions.

$$2\ H_2O(g) \longrightarrow 2\ H_2(g) + O_2(g)$$
$$CaCO_3(s) \longrightarrow CaO(s) + CO_2(g)$$

## Combustion Reactions

Combustion reactions have oxygen as a reactant. The most common combustion reactions involve hydrogen and carbon reacting with oxygen. When the reactions involve hydrocarbons, the products are usually water and carbon dioxide. Here is the generic equation for the combustion of a hydrocarbon.

$$C_xH_y + O_2 \longrightarrow CO_2 + H_2O$$
$$C_xH_yO_z + O_2 \longrightarrow CO_2 + H_2O$$

We can also further categorize some combination reactions a combustion if they involve oxygen. This includes the reaction of magnesium with oxygen we considered earlier.

$$2Mg(s) + O_2(g) \longrightarrow 2MgO(s)$$

## Precipitation Reactions

Earlier we discussed double-displacement reactions. Some of these reactions produce a solid product or precipitate. These reactions are known as precipitation reaction. Precipitates form in some cases when two ionic compounds are mixed together. Not all combinations of ionic reactions will produce a precipitate but many will.

A list of rules is listed below. These are known as the solubility rules and can be used to determine if a precipitate if two ionic compounds are mixed with each other. First we know that when ionic compounds react a double-displacement reaction will occur, so we can predict what the products of the reaction. Then once we know what the products will be and we can determine whether these products will form a soluble product or precipitate. Let's consider mixing sodium carbonate with two different ionic compounds.

### Solubility Rules

1. All compounds containing column 1 metals, $NH_4^+$, $NO_3^-$ and $C_2H_3O^-$ are always *soluble*.
2. All compounds containing Cl–, Br–, I– are always soluble except when paired with $Ag^+$, $Pb^{2+}$ or $Hg_2^{2+}$
3. All compounds containing $SO_4^{2-}$ except when paired with $Ba^{2+}$, $Pb^{2+}$, $Sr^{2+}$, or $Ca^{2+}$.
4. All compounds containing $CO_3^{2-}$ and $PO_4^{3-}$ are insoluble except when paired with the ions listed in rule number 1 as listed above.

5. All compounds containing OH– are insoluble except when paired with those ions listed in rule number 1 and $Ba^{2+}$ or $Ca^{2+}$.
6. All $S^{2-}$ compounds are insoluble except when paired with those ions listed in rule number 1 and the column 2 metals.

Now consider the reaction of sodium carbonate with lead (II) nitrate. When the double-displacement occurs, lead (II) carbonate and sodium nitrate will form. Reviewing the solubility rules, sodium nitrate is soluble (all nitrates are soluble) but lead (II) carbonate is not soluble in water so it will form a precipitate (carbonates are not soluble and lead is not an exception). When writing the equation for this reaction, an "s" is used to indicate lead (II) carbonate forms a precipitate.

$$Na_2CO_3(aq) + Pb(NO_3)_2(aq) \longrightarrow PbCO_3(s) + NaNO_3(aq)$$

Now let's consider what happens when sodium carbonate is mixed with potassium nitrate. In this case, the products will be potassium carbonate and sodium nitrate. As discussed earlier all nitrates are soluble so, sodium nitrate will not form a precipitate. Looking at the table, we see that carbonates are insoluble but potassium is an exception meaning that potassium carbonate is soluble and will not form a precipitate. This means that neither of the products formed will create a precipitate. Because neither of the products will form a precipitate, no visible reaction will occur. When two ionic compounds do not form a precipitate, the symbol "NR" or "No Reaction" is written on the products side of the reaction.

Let's consider the reaction between sodium carbonate and potassium nitrate. In this case, the two possible products are both soluble in water so as the result is "no reaction" occurs.

$$Na_2CO_3(aq) + KNO_3(aq) \longrightarrow NR$$

When we consider the reaction between sodium chloride and silver nitrate, a reaction does occur and a precipitate does form. In this case, the complete reaction is written as given below.

$$NaCl(aq) + AgNO_3(aq) \longrightarrow AgCl(s) + NaNO_3(aq)$$

Precipitation reactions are often represented using a series of written equations meant to depict what occurs in solution. The types of equations used are the ***molecular equation***, the ***complete ionic equation*** and the ***net ionic equation***.

The first equation is the molecular equation. This equation is similar to the chemical equations we have encountered already in the text. All of the reactants and all of the products are represented in the equation with their states of matter. Let's use the reaction between sodium chloride and silver nitrate we just discussed. The molecular equation will have the reactants and products listed as compounds:

## Molecular Equation

$$NaCl(aq) + AgNO_3(aq) \longrightarrow AgCl(s) + NaNO_3(aq)$$

The next equation used is the complete ionic equation. In this case, the equation is a representation of what physically occurs in solution as the reaction takes place. All of the soluble ionic compounds are separated into their ions. Insoluble compounds (solids) remain written as compounds and are not separated into ions (as they do not form ions in solution).

### Complete Ionic Equation

$$Na^+(aq) + Cl^-(aq) + Ag^+(aq) + NO_3^-(aq) \longrightarrow AgCl(s) + Na^+(aq) + NO_3^-(aq)$$

The last of the equations used is the net ionic equation. In this case those ions that do not form a solid product are removed. These are the formulas that are the same on both sides of the equation. In our example, $Na^+$ and $NO^{3-}$ are removed from the equation.

### Net Ionic Equation

$$Ag^+(aq) + Cl^-(aq) \longrightarrow AgCl(s)$$

In this example, $Na^+$ and $NO^{3-}$ ions in this reaction are called ***spectator ions*** as they do not "contribute" to the precipitation reaction that occurs. Their name is fitting as the are similar to spectators at a sporting event - they are important to the dynamics of the event but they do not contribute to the actual game. Any ions that are removed from the net ionic equation are known as spectator ions. .

## Chapter 4 Summary

*Chemical reactions* are processes during which one or more substances undergo a chemical change to form one or more new substances.

*Chemical equations* are written representations of chemical reactions.

*Reactants* are present at the beginning of a chemical reaction and are converted into ***products*** as the reaction proceeds.

*Irreversible reactions* can only proceed in one direction and equations for these reactions use a one directional arrow ($\longrightarrow$). ***Reversible*** reactions can proceed from reactants to products but are also capable of converting products back to reactants and equations for these reactions use a two directional arrow ($\leftrightarrow$, $\rightleftarrows$, or $\leftrightarrows$).

*Coefficients* are found in front of molecules in a chemical equation. These tell how many molecules of each compound (or element in certain cases) are required for the reaction to proceed.

Letters are used to denote the states of matter of compounds and elements in a chemical reaction. The letters used are s for solid, l for liquid, aq for dissolved in water (aqueous) and g for gas.

Equations must be *balanced* such that the number of each atom type is the same on both the reactants and products side of the equation.

*Stoichiometry* is the process by which quantities of products produced or reactants needed or a chemical reaction to occur are determined using the amount of a single reactant or product and a balanced chemical equation.

The *mole* is an experimentally determined quantity. (1 mole = $6.022 \times 10^{23}$ things)

*Molar Mass* is the mass of one mole of an atom or compound. The atomic weights listed on the periodic table is also equivalent to the molar mass of that atom.

The *limiting reactant* of a chemical reaction is the reactant that is used up first in the production of the products.

*Theoretical yield* is the predicted yield possible to produce based on the amount limiting reactant. It is how much would be produced in a perfectly executed reaction.

The *actual yield* is the amount of product that is actually produced and is always less than the theoretical yield.

*Percent yield* is a measure of the success of a chemical reaction. It is calculated by taking the ratio of the actual yield to the theoretical and multiplying by 100.

*Decomposition reactions* are reactions that start with a single reactant that breaks down to form two or more products ($AB \rightarrow A + B$).

*Combination reactions* occur when two or more reactants combine to produce a single product ($A + B \rightarrow C$).

*Precipitation reactions* are chemical reactions that form a solid product (precipitate).

*Displacement reactions* are also known as metathesis or exchange reactions. In these reactions cations in ionic compounds exchange "partners" or more accurately the anion with which they are paired ($A + BC \rightarrow AC + B$).

*Combustion reactions* occur when a hydrocarbon reacts with oxygen to produce carbon dioxide and water ($C_xH_y + O_2 \rightarrow CO_2 + H_2O$).

*Molecular equations* are used to represent a precipitation reaction and are written with reactants and products as ionic compounds and include all states of matter.

*Complete ionic equations* are used to demonstrate what actually happens in solution for a precipitation reaction. All soluble ionic compounds are written as separated ions with charges instead of paired molecules.

*Net ionic equations* are equations for precipitation reactions that only include the reactants and products that contribute to the precipitation reaction.

*Spectator ions* are ions in a precipitation reaction that remain soluble and do not contribute to the formation of a precipitate.

## CHAPTER 4 EXERCISES

1. When writing a chemical equation, on what side of the arrow do you place the reactants?
2. Write a chemical equation for a reaction where A and B are reactants and C and D are products. The reaction is reversible.
3. What do coefficients in a chemical equation tell you?
4. List and define each of the states of matter and the symbols used for chemical equations.

5. Can you change the subscripts on the compounds of a chemical equation to balance it? Explain.

6. Answer the following regarding the reaction:

   $NaCl(aq) + AgNO_3(aq) \longrightarrow AgCl(s) + NaNO_3(aq)$

   a. What are the reactants?

   b. What state of matter is AgCl (the word not the symbol)?

   c. What state of matter is NaCl (the word not the symbol)?

   d. What are the products of the reaction?

7. Write a chemical equation for the reaction that occurs when 2 moles of liquid isopropanol ($C_3H_8O$) burns in 9 moles of oxygen producing 6 moles of carbon dioxide and 8 moles of water.

8. How many moles of carbon are in 225 g of $Al_2(CO_3)_3$?

9. What is the mass, in grams, of 2.65 mol of LiBr?

10. What mass, in grams, of $(NH_4)_3PO_4$ contains $2.67 \times 10^{24}$ hydrogen atoms?

11. Answer the following questions regarding $C_{14}H_{12}O_6S$

    a. What is the molar mass of this compound?

    b. How many molecules are present in a 535g sample of this compound?

    c. How many grams of this compound are present if there are $7.92 \times 10^{31}$ carbon atoms are present in the sample?

    d. If there are $8.25 \times 10^{25}$ sulfur atoms present in a sample, how many oxygen atoms would be in the sample?

12. Balance each of the following chemical reactions:

    a. $C_8H_{18}\,(l) + \quad O_2\,(g) \longrightarrow CO_2\,(g) + \quad H_2O\,(g)$

    b. $Al(NO_3)_3(aq) + Na2S\,(aq) \longrightarrow Al_2S_3\,(s) + NaNO_3\,(aq)$

13. Balance the following reactions (answer any 3):

    a. $C_5H_6O(l) + O_2(g) \longrightarrow CO_2(g) + H_2O(g)$¬

    b. $Mn + HI \longrightarrow H_2 + MnI_3$

    c. $Al_2O_3 + Na_2CO_3 \longrightarrow NaAlO_2 + CO_2$

    d. $Al_2O_3 + Fe \longrightarrow Fe_3O_4 + Al$

14. Classify the following reactions as decomposition, combustion, or combination.

    a. $Ca(s) + O_2(g) \longrightarrow 2\,CaO(s)$

    b. $2\,NaN_3(s) \longrightarrow 2\,Na(s) + 3\,N_2(g)$

    c. $CH_4(l) + 2\,O_2(g) \longrightarrow CO_2(g) + 2\,H_2O(l)$

15. Cisplatin, $Pt(NH_3)_2Cl_2$, is a cancer drug prepared by the following reaction.

    $K_2PtCl_4 + 2\,NH_3 \longrightarrow Pt(NH_3)_2Cl_2 + 2\,KCl$

a. If 3.42 g of $K_2PtCl_4$ and 1.61 g of $NH_3$ were used in the above reaction, what is the limiting reactant?

b. What is the theoretical yield of cisplatin, $Pt(NH_3)_2Cl_2$?

c. If you obtained 2.08 g of cisplatin, $Pt(NH_3)_2Cl_2$. What was your percent yield?

16. Determine the products then balance the following reactions:

a. $C_2H_4$ (g) + $O_2$(g) $\longrightarrow$ combustion

b. Mg(s) + $O_2$(g) $\longrightarrow$ combination

17. When aqueous solutions of $Na_3PO_4$ and $Cu(NO_3)_2$ are mixed a blue-green precipitate forms. In a particular experiment, 0.525 L of a 0.325 M solution of $Na_3PO_4$ is mixed with 525 mL of a 0.325 M solution of $Cu(NO_3)_2$.

a. Write the balanced equation for the reaction that occurs.

b. Calculate the moles of the solid product formed by each reactant.

c. Based on your answer to b, what is the limiting reactant?

d. Based on your answer to C, calculate the mass in g the solid product formed.

18. For the balanced equation shown below, how many grams of $NO_2$ will be produced by $1.17 \times 10^{25}$ $O_2$ molecules?

$$4\ C_5H_5N + 29\ O_2 \longrightarrow 20\ CO_2 + 10\ H_2O + 4NO_2$$

19. Identify the following ionic compounds as soluble or insoluble:

a. $(NH_4)_3PO_4$ c. $CaCO_3$ e. $Rb_2SO_4$

b. $Sr_2(C_2H_3O_2)_2$ d. $Co_2(CO_3)_4$ f. $Al(NO_3)_3$

20. Answer each of the following questions regarding the reaction between $(NH_4)_3PO_4$ and $Co(NO_3)_4$.

a. What will be the products of the reaction?

b. Which product will form a precipitate?

c. What is the molecular equation?

d. What is the complete ionic equation?

e. What is the net ionic equation?

f. What are the spectator ions?

21. What is the net ionic equation for formation of solid precipitate via the mixing of $Al(NO_3)_3$ with $Na_2SO_4$? What are the spectator ions?

# Credits

# Thermochemistry and Thermodynamics

## The world runs on energy...

***Thermodynamics*** is the study of thermal (heat) energy and its transformation and relationship to other forms of energy. The word itself contains the essence of the topic. Thermo- or heat, and dynamics or motion. This chapter deals primarily with a subset of thermodynamics called thermochemistry. In addition, the material crosses the boundaries with physics, engineering, and biology. It is often difficult to meaningfully discuss topics in these other disciplines without having an understanding of thermochemistry as well.

Energy in all of its different forms is essential to life, and much more. Much like time, energy is a mysterious concept. Both time and energy can be observed and quantified, and yet a basic explanation of the meaning of either is elusive. Much like the periodic table, energy will be a recurring component of many of our subsequent discussions in this book.

## 5.1 The Nature of Energy

### What Is Energy?

As familiar as the term energy is to us, a precise definition is not easily crafted. Often energy is described as the ability to do work, but this is only part of the story. Einstein showed that mass and energy are equivalent at the deepest theoretical levels, but this doesn't change the fact that in practice they can be considered to be separate entities of the utmost importance. A more thorough description of energy might be the following:

***Energy*** is the capability of a system to perform work, exists in many different forms, and may be active or passive in nature.We have yet to define what work is, but certainly energy is needed to perform work of any kind. Its usage and/or transfer can be measured accurately by employing standard procedures, and widely accepted units as will now be described.

#### Units for Measuring Energy as Heat

Heat transfer is the most common form of energy transfer. This is often measured in the form of a temperature change in degrees Celsius (C). There are a number of different kinds of units that have been used historically for measuring energy. In chemistry the two most common units are:

**Joule:** a unit of work or energy equal to the work done by a force of one newton acting through a distance of one meter. Not a particularly accessible definition.

**Calorie:** the amount of heat required at a pressure of one atmosphere to raise the temperature of one gram of water one degree Celsius. Much easier to comprehend, and put into practice.

**Conversion factor:** 1 calorie is equal to 4.19 joules

Both of these units are widely employed in chemistry, but in terms of definition the calorie is far more familiar as it is it is this calorie that shows up on every package of food in terms of nutrition information.

# 5.2 The 1st Law of Thermodynamics

## Law of the Conservation of Energy

Recall from Chapter 1 the ***Law of the Conservation of Mass?*** Its bookend law is the ***Law of Conservation of Energy.*** Simply stated it says:

**Energy cannot be created or destroyed**

This is a good thing. Of course, as we all know, it is readily possible to convert one form of energy into another, and into useful work. We will examine the conversion of forms of energy more closely in the next section.

From a practical standpoint when dealing with energy changes one must account for the total amount of energy of a system (what is being studied), as well as its surroundings (anything outside of the system) before and after energy transfers and conversions occur. These fundamental concepts crossover many different scientific and engineering disciplines.

The ***First Law of Thermodynamics*** is a subset of the Law of Conservation of Energy, but more specific to applied systems. To wit, a possible definition, albeit one that is somewhat narrow in scope:

The first law of thermodynamics is an application of the energy conservation principle to systems involving heat and work. A common formulaic shorthand description of this concept is as follows:

$$\Delta U = Q - W$$

Where: *U = internal energy = total atomic or molecular energy of a system under study*
Q = heat energy
*W = energy associated with the work done by the system under study*

In some cases work is not done by the system. In these instances, the change in energy is equal to the heat or: $\Delta U = Q$.

## Energy and Work of a Given System

In this chapter we are learning about with chemistry as it relates to processes involving work and energy. These topics are intertwined in complex ways, but can be compartmentalized to an extent. This discussion is closely related to physics.

## Kinetic and Potential Energy

***Kinetic Energy*** is the energy associated with motion or activity. Closely related chemical examples are the generation and expansion of a gas resulting from a chemical reaction, and the lighting of a flashlight bulb due to the electrical current produced by a chemical reaction in a battery.

***Potential Energy*** is stored energy, energy waiting to be used. For example the chemical energy of the reactants of in the combustion of gasoline in our vehicles, or the charged battery sitting and waiting in the flashlight.

## Energy, Work, and Heat

Let's look at a couple of examples relating these three entities, but first a couple of straightforward definitions:

***System:*** *A system is defined as the process and components under study.*
***Surroundings:*** *Everything outside of the system being studied.*
For example, take the following hypothetical chemical reaction:

$$Na^{+} \quad + \quad I^{-} \longrightarrow NaI \quad + \quad \text{heat}$$

The reaction is being run in a flask with water as the solvent, and under an atmosphere of pure nitrogen. Here are the conditions which we are defining.

***System:*** By our definition the reaction as noted above is the system of interest.

***Surroundings:*** The rest; water, oxygen and flask are surroundings, again this is our arbitrary definition.

Now, if the reaction generates heat (a likely event) some of this will be transferred to the surroundings the rest will either be used to do work of change the temperature of the system. In accordance with the law of energy conservation, when all is said and done the total amount of energy for the system and surroundings together at the end must be equal to the total that exists at the beginning of the process.

Another example, from physics this time. A rock is dropped down off a sheer cliff to hit the ground at a distance 100 meters below (Figure 5.1). Assume the rock remains intact, and there is no air resistance. How much energy is transferred? What forms of energy are involved in the transfer? Where does the energy go?

*System* (defined by us): The rock, cliff, and ground. Also travel distance, and the force of gravity.

*Surroundings* (our definition too): Everything outside of our system.

Think about what happens here. Since air resistance is being ignored, gravity is the only force acting on the rock. The rock's potential energy before being dropped is simply defined by its height above the ground, 100 meters. The potential energy is first converted to kinetic energy upon being dropped. When the rock hits the ground this kinetic energy (of motion) is translated into mechanical energy in the form of

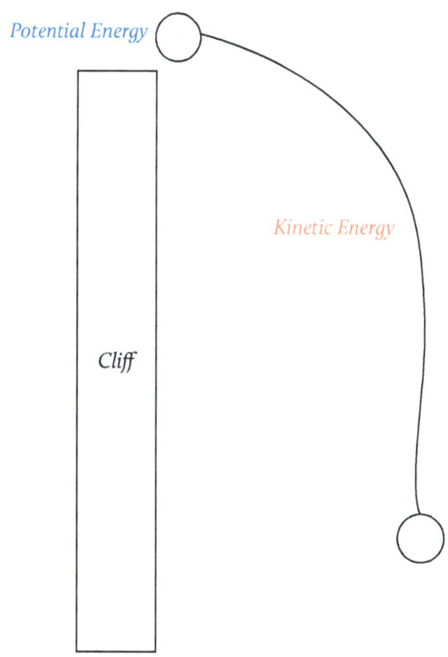


FIGURE 5.1. Potential and Kinetic Energies.

deformation of both the ground and rock surfaces, as well as heat. So, in a qualitative sense we have (E = energy):

**$E_{start}$ (potential energy) = $E_{end}$ (mechanical energy + heat energy)**

For chemical reactions, we are considering (again qualitatively) the following:

**$E_{start}$ (potential energy – in bonds) = $E_{end}$(new bond energy + heat energy)**

# 5.3 Energy: Its Forms and Interconversions

## Forms or Types of Energy

We are all familiar with many different forms of energy. Figure 5.2 summarizes these forms, all of which can be important in chemical processes, some more common than others. If you really think about any of these energy forms you may come to realize how mysterious energy is.

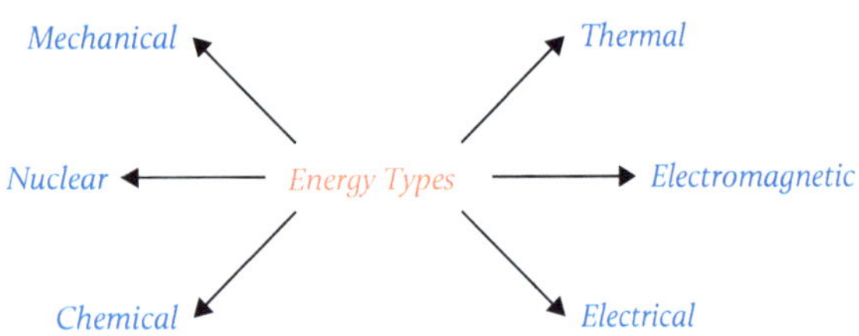


FIGURE 5.2. Forms of Energy.

Potential and kinetic energy are the two broad classes of energy that have already been defined. Considering the subclasses shown in Figure 5.2 there is not one we could easily do without, but most are absolutely essential to survival. The rest of this chapter involves heat energy and its relationship to chemical energy. All chemical reactions rely on this relationship. First, let's look at energy relationships from a somewhat broader viewpoint.

## Interconversions of Different Forms of Energy

Chemical energy is a form of potential energy stored in chemical compounds, more specifically in the atoms and bonds between atoms in molecules. If we ignore nuclear reactions, which we are doing, we can narrow the study to chemical bond energies. Now let's run a familiar reaction, combustion of natural gas in air in the piston-cylinder assembly of a car's engine (Figures 5.2 and 5.3).

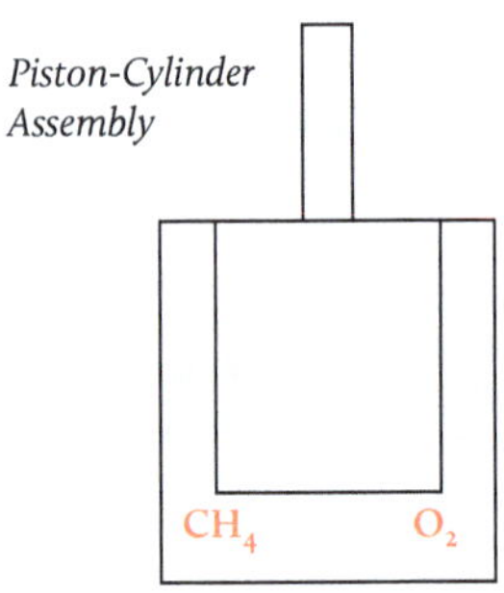


FIGURE 5.3. Simple Piston-Cylinder.

Here is the reaction:

$$CH_4 + 2O_2 \longrightarrow CO_2 + 2H_2O \quad \Delta E = -890\ kJ$$

Natural gas is methane with the formula $CH_4$, and the key component of air here is oxygen, $O_2$. When we burn this gas we break four C – H bonds, and the O–O bond as well. There is a lot going on here, but basically there is a substantial amount of heat produced as well as product gases at high temperature. We have cleverly designed our system to take advantage of the gas produced to drive the piston, and eventually the whole car. From a thermodynamic standpoint what have we done? The process has taken potential chemical energy and converted it into heat energy and kinetic mechanical energy which does useful work.

Note here that in the absence of our designed system the reaction will still occur, but no useful work would be done. Product gases would still be formed, but then just uselessly expand into the atmosphere. However, either pathway will involve heat generation. This is unavoidable and a basic tenet (Carnot Cycle) underlying thermodynamics in general. The relative amount of work obtained is dependent upon biological or engineering design factors. A classic example is the 4 cycle engine employed in the vast majority of cars (Figure 5.4).

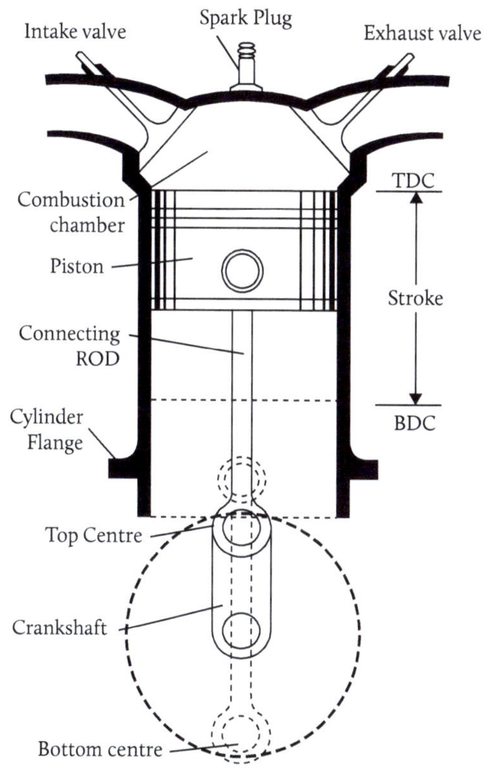


FIGURE 5.4. Piston Cylinder and More: 4-Cycle Engine.

## 5.4 Enthalpy

Crucial biological and inanimate processes rely on energy conversions. For example, photosynthesis is the conversion via chemical reactions of solar energy into potential energy storage in the form of chemical compounds. The subsequent conversion of these compounds into new ones (eg. $CO_2$ and $H_2O$) is accompanied by heat generation, electrical energy generation, and mechanical energy (movement), as well as other energy forms (sound, light, elastic).

### Symbology in Enthalpy

A confusing aspect of this area of chemistry and physics (not uncommon in science) is the the use of symbols (a.k.a. symbology) to represent different types of quantities. The definition of **U** given above is conceptually very important in that it indicates the massive amount of nuclear energy present in an atomic system. However, this value is of no practical significance here whatsoever. We are dealing with *changes* in energy of a system of our choosing, so let's clarify and simplify our symbology as follows for chemical reactions:

$$\Delta \mathbf{E} = \mathbf{E}_{\text{final}} - \mathbf{E}_{\text{initial}}$$

Where:
$\Delta E$ = *change in the total system energy*
$E_{final}$ = *energy at the end of a reaction*
$E_{initial}$ = *energy at the start of a reaction*

ΔE can have either a negative (minus) or positive (plus) sign. An **exothermic** (exo- meaning out) reaction is one that releases energy to the surroundings as heat. By convention such a reaction is given a minus (–) sign. An **endothermic** reaction (endo- meaning in) is one that absorbs heat from the surroundings, and is given a plus (+) sign. We are all very familiar with exothermic reactions such as the burning of natural gas to form heat:

$$CH_4 + 2O_2 \longrightarrow CO_2 + 2H_2O \qquad \Delta E = -890\ kJ$$

Remember here kJ stands for kilojoules = thousands of joules. In this case ΔE is a measurable quantity in the form of heat in units of either joules or calories.

We are all familiar with a reaction that is endothermic, the conversion of $CO_2$ and $H_2O$ from the atmosphere into a sugar and $O_2$:

$$6CO_2 + 6H_2O \longrightarrow C_6H_{12}O_6 + 6O_2 \qquad \Delta E = +5204\ kJ$$

In this case the energy required is supplied by sunlight, and is energetically going way uphill since water and carbon dioxide are such stable molecules. This reaction is basically a summary of photosynthesis, a process that is very complex in the details and actually involves multiple complex steps.

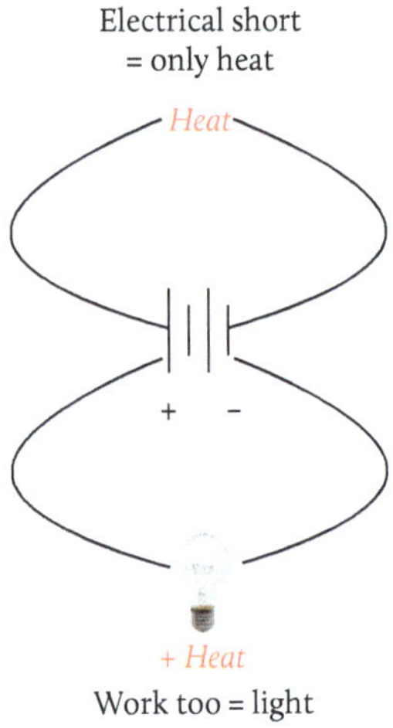


**FIGURE 5.5.** The Conversion of Electrical Energy into Heat or Work + Heat in a Battery.

## Enthalpy

Strictly speaking **enthalpy**, **H**, is defined as the sum of the heat and the product PV, where P and V are the pressure and volume of the system, respectively, representing any work done by gasses that are generated.

$$\Delta H = \Delta E + \Delta PV$$

Most reactions are run at constant pressure so $\Delta PV = 0$ and the PV term above falls out of the equation. So, in this case the enthalpy change is equal to the change in heat, an easily measurable quantity (Figure 5.5). We will use the assumption of constant pressure for the remainder of this chapter. So in examples moving forward: $\Delta H = \Delta E$.

## Intensive and Extensive Properties

Before we proceed with enthalpy involving just heat, a couple of definitions for our chemical systems:

***Intensive Property:*** A property of a substance that is independent of the amount of material present. Examples include melting and boiling point, state of matter, color and density.

***Extensive property:*** A property of a substance that depends upon the amount of material present. Examples include heat released or absorbed upon reaction, mass (weight), volume occupied, and heat capacity.

It is easy to remember which is which. **In**tensive = **In**dependent

# 5.5 Enthalpies of Reaction

We have now reduced enthalpy to heat of reaction $\Delta H = \Delta E$ (recall we are under constant pressure). Also note that enthalpy is an extensive property; doubling the amount of material reacted doubles the amount of heat released. Consider gasoline, you can travel a lot further with a full tank of gas versus a quarter tank of gas. Many years of work in the field of chemistry has resulted in tables of heats associated with chemical reactions. Appendix 4 contains a representative tabulation of enthalpies (heats) associated with the formation of common compounds. A couple of notes:

1. By definition the heats of formation of elements at 298°K and 1 atmosphere of pressure are equal to 0.
2. Transitions between states (gas to liquid to solid) involve heat changes, but no chemical changes, just physical ones.
3. The enthalpy of a reaction is dependent upon the states of the reactants and products
4. A reaction that is reversed involves an enthalpy change equal in magnitude to the forward reaction, but opposite in sign (plus or minus).

The first note is followed by convention.here is done by convention. Here are examples of numbers 2-4 taken from Appendix 4.

$$\mathbf{Br_2\ (liquid) \longrightarrow Br_2\ (gas)} \qquad \Delta H = +30.91\ kJ$$

Our system here is just bromine at 298°K. One mole of liquid bromine is vaporized to gas. The plus sign means that this is an endothermic process (heat is added into the system). Also remember this process is a physical change only. There is no chemistry going on.

$$C\,(s)) + 2S\,(s)) \longrightarrow CS_2\,(g) \quad \Delta H = -+116.7\ kJ$$
$$C\,(s) + 2S\,(s) \longrightarrow CS_2\,(l) \quad \Delta H = -+89.4\ kJ$$

The plus signs means that in both cases the formation of carbon disulfide from the elements requires the input of heat energy. Not surprisingly a substantially larger heat influx is required to go to the gas phase compared to the liquid phase synthesis.

$$Ca + O \longrightarrow CaO \quad \Delta H = -635\ kJ$$
$$CaO \longrightarrow Ca + O \quad \Delta H = +635\ kJ$$

These numbers mean that the combination of elemental calcium and oxygen into calcium oxide, a primary component of lime, is avery exothermic and will release heat to the surround. The reverse process will absorb heat from the surroundings in order for it to occur making it an endothermic process.

## 5.6 Calorimetry

The analytical measurement of heat flow is known as **calorimetry**. This involves monitoring the change of temperature as a function of time of a given mass of material or a system. Although simple in principle, these measurements can be tricky in practice since temperature changes of the surroundings must be carefully measured as well as those of the system being studied.

### Heat Capacity and Specific Heat

These quantities are closely related, and understandably they are often a source of confusion. Let us define heat capacity and specific heat:

***Heat Capacity:*** The amount of heat required to raise the temperature of a given amount of a substance 1 K (°C).

This is a generic definition of qualitative value only since the amount of the substance is not specified. A more useful definition is:

***Specific Heat ($C_s$):*** The amount of heat required to raise the temperature of 1 gram (g) of a substance 1 K (°C).

As you can see specific heat is a subset of heat capacity. The heat capacity can be determined as follows:

$$C_s = q / (m \times \Delta T)$$

where:

$q$ = heat
$m$ = mass in grams
$C_s$ = the specific heat of the the material
$\Delta T = T_{final} - T_{initial}$ ** For this text celsius will be used for enthalpy calculations.

Table 5.1 contains representative specific heats of some familiar substances.

TABLE 5.1 **Specific Heats of Common Materials (J/°C)**

| | |
|---|---|
| **Liquid Water** | **4.18** |
| **Ice** | **2.11** |
| **Water Vapor** | **2.00** |
| **Dry Air** | **1.01** |
| **Granite** | **0.84** |
| **Iron** | **0.79** |
| **Copper** | **0.45** |
| **Lead** | **0.38** |

The range of specific heats is from close to 0 with radon at: 0.094J/g °C) to a high with hydrogen at 14.3 J/g °C. **Water** as a liquid is noteworthy with a very high specific heat of **4.18 J/g** °C. This is very important for our earth's ecosystems, especially the oceans, since they can reversibly absorb and release tremendous amounts of heat.

Heat capacities are somewhat temperature dependent, but changes of phase can have a dramatic effect. **Ice** at 0°C has a specific heat of **2.05 J/g** °C, half that of the liquid phase.

Tables of specific heats are readily available, and the associated calculations using the same are straightforward. By way of example let's take a rock. Rocks vary a great deal in their chemical composition, but not so much in their heat capacities. Taking granite as a representative rock, this has a specific heat of 0.80 J/g °C. So let's assume we have 1000g (mass or m) of granite (about 2.2lbs), and the sample temperature is raised from 15°C to 25°C (ΔT or final temperature minus the initial temperature). Using the specific heat equation above we can calculate the amount of heat (q) absorbed from the sun as follows:

$$C_s = q / (m \times \Delta T) \textbf{ becomes:}$$

$$q = mC\Delta T$$

$$q = (1000\,g)(0.80\,J / g°C)(25 - 15°C)$$

$$q = 8{,}000\,J \; or \; 8\,kJ$$

How are such calculations important? Any number of ways, but let's look at a specific example. The human body must maintain a temperature of ~ 98.6°F (37°C) simply to survive, a ΔT = 15°C above a room temperature of 22°C. The average human adult consumes about 2000 calories a day = 8,372 joules of heat energy. The heat capacity of sugar is about 1.24 J/g°C. So the amount of sugar required to raise the body temperature from room temperature, 22°C to 37°C on sugar alone would be:

$$q = mC\Delta T$$

**can rearrange to:**

$$m = q / (C\Delta T)$$

$$m = (8{,}372\,J)(0.80\,J / g°C)(37 - 22°C)$$

$$m = 450\,g$$

This is equivalent to roughly one pound. So, over 24 hours this amounts to 450g/24hrs or 0.04 lbs/hr or 0.66oz/hr.

## 5.7 Hess's Law

Arbitrarily, let's assume we are trying to synthesize compound **E**. The synthetic route of choice involves two reaction steps as shown below.

**Step 1:** $A + B \longrightarrow C \quad \Delta H1 = +20\ kJ$
**Step 2:** $C + D \longrightarrow E \quad \Delta H2 = -140\ kJ$

Now we can add these steps together, and simplify the result by subtracting out the term common to both sides of the equation, which is **C**. *Also the two enthalpies are additive to give us the energy results for the combined synthetic process*

**Step 3:** $A + B + \cancel{C} + D \longrightarrow \cancel{C} + E \quad \Delta H1 + H2 = +20 - 140kJ = -120$

***Hess's Law*** states that the sum of the enthalpies of the individual steps will be equal to the enthalpy of the combined reaction steps.

As mentioned above the enthalpies of a large number of reactions are known. Hess's Law is a powerful tool for calculating the enthalpy of an unknown reaction by combining the enthalpy information of two or more known reactions. The best way to learn this technique is to run through a couple of practical examples. So here we go.

### Example #1

Liquified natural gas (LNG) is composed primarily methane ($CH_4$), and accounts for about 10% of the total natural gas used worldwide every year; a tremendous quantity of energy. Transport as a liquid rather than a compressed gas is efficient, but hazardous as the LNG is very flammable to the point of being dangerously explosive. Predicting the amount of heat to be released upon combustion is thus important. This process can be analyzed in two steps. The liquid is first vaporized followed by combustion with oxygen. The energies associated with these steps are shown below:

**Step 1:** $CH_4(l) \longrightarrow CH_4(g) \quad \Delta H1 = +8kJ$
**Step 2:** $\cancel{CH_4}(g) + 2O_2(g) \longrightarrow CO_2(g) + 2H_2O(g) \quad \Delta H2 = -802kJ$

---

**Step 3:** $CH_4(l) + \cancel{2O}(g) \longrightarrow CO(g) + 2H\,O(g) \quad \Delta H2 = -794kJ$

This is a simple sequence of steps. Just like mathematical equations the equations are added together and any common entities on both sides of the arrow are subtracted out. Similarly, the enthalpies are added together taking note of the plus and minus signs.

### Example # 2

Coal is basically just carbon that can be upgraded to value added products in a number of ways using a group of reactions called gasifications. These reactions can also be combined in useful ways. Here are a couple of these reactions:

$$\textbf{Step 1:}\quad C(s) + 2H_2 \longrightarrow CH_4(g) \qquad \Delta H1 = -75\ kJ$$

$$\textbf{Step 2:}\quad [C(s) + H_2O(g) \longrightarrow CO(g) + H_2(g) \qquad \Delta H2 = -131 kJ] \times 2$$

$$\textbf{Step 3:}\quad 3C(s) + 2H_2(g) + 2H_2O \longrightarrow 2CO(g) + 2H_2(g) + CH_4 \quad \Delta H2 = 337kJ$$

This sequence of steps is very similar to example #1 except that the complete equation of step 2 is multiplied through by 2. Why? So that when we add the equations together and simplify to yield step 3 the $H_2$ molecules all cancel out. *Note here though too that the value of ΔH is also multiplied by 2!* So the enthalpy change for step 3 is: $\Delta H = -75kJ + (-131kJ \times 2) = -337kJ$.

## 5.8 Spontaneous Reactions

A ***spontaneous reaction*** is thermodynamically defined as a reaction that, *given time*, is favored to occur. The "given time" part of this sentence is key here. For example, if a chemist is given a mixture of hydrogen ($H_2$), and chlorine ($Cl_2$) gasses he/she will instantly surmise the mixture is thermodynamically highly reactive and over time will react spontaneously, possibly explosively, to form the strong acid hydrochloric acid:

$$H_2 + Cl_2 \longrightarrow 2HCl \quad \Delta H = -92.3\ kJ$$

The minus sign tells us the reaction is exothermic and releases heat. However, thermodynamics says nothing about how fast or slow the reaction will occur, only that it will. It could occur in a nanosecond, 10 days or in 100 million years. In reality this mixture of gasses is stable as is until given some sort of assistance, such as raising the temperature or adding a catalyst or a spark.

How can we tell if a proposed reaction will be spontaneous? In the simplest cases the examination of heats of formation will give us an indication; however, oft times Hess's Law (Section 5.6) must be employed to make a prediction. Then there is entropy.

## 5.9 Entropy: the 2nd and 3rd Laws of Thermodynamics

### The Concept of Entropy

There are concepts in science that are not intuitive even if they are very familiar to us in practice. We have touched on two of these already, time and energy. The concept of **entropy** is another.

Simply put **entropy** (by convention given the symbol **S**) is a measure of disorder. The concept is intellectually mystifying, and indeed a strange one; yet very familiar to all of us. How is it mystifying? The entropy, disorder or randomness of the universe is constantly increasing. Does this mean we cannot have ordered systems? Of course we can. Take as an example the human body. This is a very organized and sophisticated system. To construct it involves a decrease in entropy (**–S**) for the system (the body). The system becomes more organized and less disordered which is less favorable hence the negative sign. owever, the entropy of the surroundings of that body increase by a value that is positive and of greater absolute magnitude than –S. That is:

$$\left|S_{surroundings}\right| > \left|-S_{system}\right|$$

A classic example of entropy changes is the scrambled egg. Before scrambling an egg is a pretty well-organized, and highly ordered entity. Scrambling certainly increases the disorder and is associated with a positive entropy (+S). In this case the system becomes more disordered or less organized as reflected in the positive sign. Could we unscramble the egg? In principle yes, but the reverse process would involve another increase in entropy once both the system (the egg) and the surroundings (everything else) are taken into account.

*The entropy of a given system depends only upon the initial and final states, not upon the path taken in between.* This characteristic defines it as being a **state function**. The concept is much like the difference in physics between a distance traveled, and the displacement between two points. The displacement is the shortest route traveled. The distance may be that, or one that is much longer and more torturous. Either way the starting and ending points are the same.

By definition the following is true for an ***isothermal*** and ***reversible*** process involving a system:

$$\Delta S = S_{final} - S_{initial}$$

$S = \frac{q}{T}$, where q is heat and T is the temperature in Kelvin

Let's look at a theoretical example of the entropy change involved with a straightforward reversible and isothermal process.

## Phase Changes and Entropy

Consider the melting-freezing of water at 273K. We can approximate this process as being truly reversible (in a thermodynamic sense) by imposing infinitely small temperature changes up or down. The heat of fusion associated with the melting of one mole of ice to form water at 273K and one atmosphere of pressure is $\Delta H = +6.011$. This is an endothermic change, hence the plus sign. This is q for this process, so the entropy associated is:

$$\Delta S_{freezing} = \frac{q}{T} = \frac{\Delta H}{T} = +\frac{6.01}{273} = +0.022kJ \,?\, K$$

The reverse process, melting has the same magnitude of s, but a negative sign:

$$\Delta S_{melting} = -0.022kJ / K$$

This discussion leads us to the second, and third laws of thermodynamics.

## The 2nd Law of Thermodynamics

The first law of thermodynamics dealt with energy, and its conservation. The second law deals with entropy, a quantity that is *not* conserved. That is, for any system plus its surroundings (essentially the whole universe) the entropy change for any irreversible process is always positive. This is the ***Second Law of Thermodynamics***. Once again, this does not mean that the system or surroundings alone must increase in entropy, rather only that the net entropy change when added together must be positive.

For a reversible process $\Delta S = 0$. Now the question arises concerning if a truly reversible process is possible. This is a strange topic. Consider the water melting-freezing-melting process postulated above to occur by infinitely small steps, and thus be theoretically reversible. Consider the fact too that you observed this process. To be truly reversible the finish and starting points of the system and the surroundings would have to be *exactly* the same. This means your memory would be reversed. So would time. Draw your own conclusions here.

Can we predict the change in entropy in a qualitative manner using our knowledge of chemistry? Yes and no. An example:

Magnesium spontaneous ignites and burns in air via a simple reaction:

$$2Mg + O_2 \longrightarrow 2MgO$$

By experiment we know that this reaction readily runs from reactants to products; that is favors the right hand side of equation (MgO). *Any spontaneous reaction involves an increase in entropy of the system and surroundings combined.* What about just the system, in this case Mg, O and MgO? This is harder to tell because there are two factors now involve, enthalpy and entropy. We know the reaction is exothermic so $\Delta H$ is negative, and substantial at -602kJ/mole. The enthalpy tells us nothing about the change in entropy though, although the fact that a reaction has occurred and new chemical products suggest a change in entropy has likely occurred.

## The 3rd Law of Thermodynamics

This law is mostly of theoretical interest. It states that the entropy of a perfect crystalline substance equals zero ($S = 0$) at absolute zero (0K). In practice this state cannot be reached. There is always molecular motion, and there is no such thing as a perfect crystal, that is, a crystal without imperfections. This law does set a lower limit of zero for absolute entropy values of substances. In practice all materials have an inherent entropy. *Note this applies to S, and not $\Delta S$, which can have a negative value.*

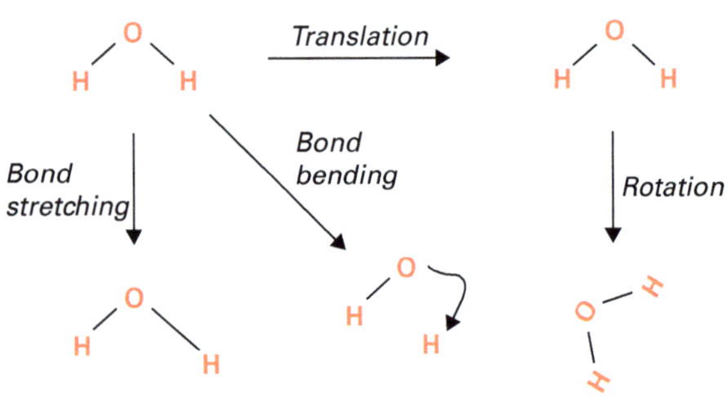


**CHEMICAL STRUCTURE 5.1.** Molecular Motions of a Water Molecule.

## Qualitative Assessment of Entropy

### Molecular Motions & Entropy

Picture a single water molecule as a gas in a complete vacuum at 298K, surely an idealized situation. What types of motion are available to this molecule (Chemical Structure 5.1)?

It has kinetic energy and is moving rapidly from place to place; so it has **translational motion**. It is also free to rotate in three dimensions, well-named as **rotational motion**. More complex are bond **vibrational motions**. There are a lot of possibilities here including bond stretching and bending to greater or lesser degrees. Now take this molecule and arbitrarily freeze it in time and space. This is called a **microstate**. The result would be a molecule in just one of a virtually infinite number of combinations of translational, rotational and vibration orientations. As the number of possibilities goes up so does the number of microstates possible, and so does the value of the entropy!

Now suppose we add three more variables; increases in volume, temperature and number of molecules. Any of these changes significantly increases the associated entropy. From a qualitative standpoint any group of molecules possess a substantial, and variable entropy via a very large number of possible molecular arrangements based upon translational, rotational and vibrational motions. Increases in number of molecules, volume and/or temperature increases the number of possible microstates available.

What about the liquid and solid states of materials relative to the gas phase? Chapter 6 discusses states of matter and the interactions of molecules in detail, but a brief discussion of this topic in the context of entropy makes sense. Figure 5.6 shows the qualitative change in entropy as we change from the solid to the liquid to the gas phase

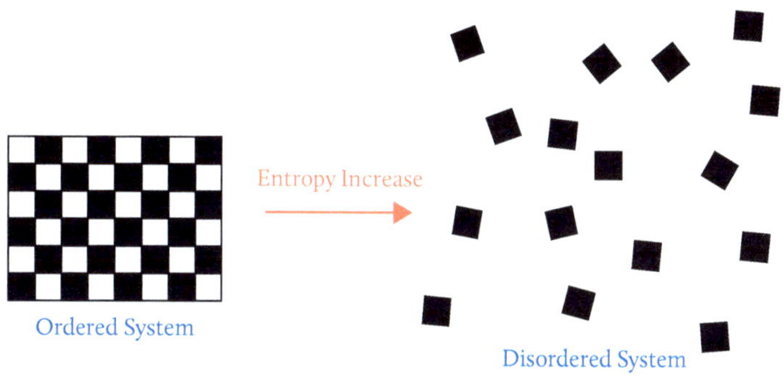


**FIGURE 5.6.** Entropy—Organized to Disorganized.

***Liquid Phase:*** Intuition suggests fewer degrees of freedom of motion here compared to the gas phase. However, in reality there is just one that disappears that of volume. To a first approximation the molar volume of a liquid does not change with temperature. The other motions are still possible; translational, rotational, and vibrational. Molecules in the liquid phase cannot move with complete freedom throughout a container as they still experience intermolecular forces from surrounding molecules.

***Solid Phase:*** Compared to the liquid state solid state molecules are frozen in place, so the translational option disappears as well as volume. Depending upon the molecule the rotational option may also not be available. For the most part, the only movement observed in the solid phase are vibrations in one place.

## Entropy and Chemical Reactions, Including the Surroundings

As mentioned above the third law of thermodynamics sets a minimum value for a substances entropy at 0 at 0K. It is possible to calculate entropies for substances, but the techniques needed are beyond the scope of this text. However, **standard entropies** ($S_o$) are readily available for many elements and compounds. Here are some generalizations about standard entropies:

1. They are all positive and of varying magnitude.
2. They tend to increase with increasing molar mass.
3. They tend to increase as the number of atoms in the molecule increase.
4. Gas phase $S_o$ > liquid phase $S_o$ > solid phase $S_o$

Are these entropies important in a quantitative sense?

Yes, we now combine our $\Delta H$ and $\Delta S$ information into the calculation of free energies.

### Entropy and Microstates

The ***Boltzman equation*** is a simple equation that represents yet another somewhat mysterious concept:

$$S = (k)\ln W$$

Here S is entropy, k is a constant and W is the number of microstates a system can have. What does this latter mean? Consider the $CH_4$ reaction with $O_2$ described above. Basically W is a measure of how many positions and orientations the reactant molecules can attain, a measure of disorder or entropy. Clearly a very large number.

## 5.10 Free Energy: Combining Enthalpy, Entropy, and Temperature

This section combines what we have learned about enthalpy and entropy into a single entity known as **free energy.** Doing so will allow us to predict if a chemical reaction is spontaneous or not. Is it important to be able to do so? Absolutely, yes. For example, say a chemist is asked to synthesize a new wonder drug in an efficient and practical manner. Examining the energetics of the system allows the chemist to predict the best synthetic route to a viable process.

The change in **free energy (G)** is defined by the following equation:

$$\Delta G = \Delta H - T\Delta S$$

So, if for a given reaction we know the enthalpy change (**ΔH**), the temperature (**T**) and the entropy change (**ΔS**) we know the change in free energy (**ΔG**). If **ΔG** is negative, the reaction is favored and will go **spontaneously**. Let's do two qualitative examples.

### Example 1

When the ionic salt ammonium chloride ($NH_4Cl$) is dissolved in water in a flask the flask gets noticeably colder. What is going on here? As experimentalists we observe/conclude a number of facts:

1. The dissolution of the salt readily occurs so the process is spontaneous. Thus the free energy change is negative (–**ΔG**)
2. Heat is absorbed from the surroundings so the process is endothermic. That is the enthalpy change is positive (+**ΔH**).
3. For **ΔG** to be negative the product of **TΔS** must be large and positive, so there is a substantial entropy increase.

The entropy results make sense as the process involves taking a very crystalline salt, that is well organized, and randomly dissolving it in water as a solvent.

### Example 2

Gallium melts at 30°C and has a heat of fusion ($\Delta H$) = 5.59 kJ/mol. When the liquid and solid phase are in equilibrium at 30°C, $\Delta G$ = 0 kJ/mole. What is $\Delta S$ for 100g going from solid to liquid? How about liquid to solid? (Note that S is in the units of J/mol, $\Delta H$ in kJ/mol, and T should be in degrees Kelvin.

This chapter lays the foundations for thermodynamics, a topic that we are very familiar with in everyday experience, but also a topic that is mystifying to all of us in the details.

# Chapter 5 Summary

*Energy:* Thermochemistry is the chemistry part of thermodynamics dealing primarily with heats of reaction and bonding energetics.

*Kinetic Energy:* Is the energy of motion.

*Potential Energy:* Is stored energy

*System:* A system is defined as the process and components under study.

*Surroundings:* Everything outside of the system being studied.

*1st Law of Thermodynamics:* As it applies to chemistry the first law of thermodynamics is an application of the energy conservation principle to systems involving heat and work. A common formulaic shorthand description of this concept is as follows:

$$\Delta \boldsymbol{U} = \boldsymbol{Q} - \boldsymbol{W}$$

Where: $U$ = *internal energy (the total atomic or molecular energy of a system under study)*
$Q$ = *heat energy*
$W$ = *energy associated with the work done by the system under study*

*Energy Forms and Interconversions:* Common forms of energy include chemical, electrical, magnetic, acoustical, thermal, nuclear, photolytic, and gravitational.

Within the context of chemistry the most important types of energy are thermal (heat), potential in the form of compounds, electrical and photolytic wherein these various forms can be interconverted.

*Enthalpy ($\Delta H$):* Enthalpy is the sum of heat (***ΔE***) and PV (pressure-volume) work, but in chemistry it is generally equal to the heat of reaction. Heats of reactions have been measured for many reactions.

*Heat Capacity:* The amount of heat required to raise the temperature of a given amount of a substance 1 K (°C).

*Specific Heat ($C_s$):* The amount of heat required to raise the temperature of 1 gram (g) of a substance 1 K (°C).

*Hess's Law:* states that the sum of the enthalpies of the individual steps will be equal to the enthalpy of the **combined** reaction steps

*Spontaneous Reaction:* A reactions that is predicted to occur based upon thermodynamic considerations; however, this does not infer any information concerning the rate of the reaction which could be very fast of very slow.

*2nd Law of Thermodynamics: Entropy:* Entropy (S) is a measure of a system's order. If the system under study is the universe entropy is always increasing. That is disorder is always getting larger. As opposed to energy, entropy is not conserved, rather it is always increasing.

*3rd Law of Thermodynamics: Entropy:* The entropy of a perfectly crystalline substance is zero at absolute zero.

*Free Energy:* Free energy (G) combines enthalpy and entropy into a single entity defined by the following equation.

$$\Delta G = \Delta H - T\Delta S$$

Where ΔG is the change in free energy, ΔH is the change in enthalpy, ΔS is the change in entropy, and T is the temperature. A negative ΔG indicates that a reaction is spontaneous while a positive ΔG indicates the reaction is not spontaneous.

# CHAPTER 5 EXERCISES

1. What is the difference between energy and work?
2. We all know that a chemical reaction occurs, and heat energy is released when propane ($C_3H_8$) is reacted with oxygen ($O_2$) in a simple combustion:

   $$C_3H_8 + 5O_2 \longrightarrow 3CO_2 + 4H_2O + \text{Heat}$$

   What chemistry is actually occurring in this reaction? Why is heat produced at all?
3. An automobile is a complex device that involves many different kinds of potential and kinetic energies. Consider a car at first just sitting at rest, then being started and driven.

   a. Identify the kinds of potential and kinetic energies that are present when the car is at rest.

   b. Identify the kinds of potential and kinetic energies that are present when the car is at being driven.

   c. Identify what types of energy conversions are taking place during the whole process.
4. What is the difference between an energy change (ΔE) and an enthalpy (ΔH) change?
5. What is the enthalpy change when 221g of Mg is burned in oxygen?

   $$2Mg + O_2 \longrightarrow 2MgO \quad \Delta H = -1200\ kJ$$
6. What is the enthalpy change when 1.23 moles of sodium is reacted with an excess of water?

   $$2Na + 2H_2O \longrightarrow 2NaOH + H_2 \quad \Delta H = -368\ kJ$$

7. What is the enthalpy change during the formation of 44.4g of carbon dioxide in the following decomposition reaction?

$$CaCO_3 \longrightarrow CaO + CO_2 \qquad \Delta H = -178.1\ kJ\ /\ mole$$

8. How many moles of $SO_2$ is formed form the reaction of sulfur with excess oxygen?

$$S + O_2 \longrightarrow SO_2 \qquad \Delta H = -297\ kJ\ /\ mole$$

9. Gaseous or liquid propane is burned exothermically to form a mixture of CO, $CO_2$ and $H_2O$ products at the same temperature. Which releases more net energy in the form of heat?

10. Using the bond energies contained in Appendix ** determine the ΔH for the following reaction:

$$C_3H_8 + 5O_2 \longrightarrow 3CO_2 + 4H_2O \qquad \Delta H = ?$$

11. Using the bond energies contained in Appendix ** determine the ΔH for the following reaction:

$$CH_4 + Cl_2 \longrightarrow CH_3Cl + HCl \qquad \Delta H = ?$$

12. Using the bond energies contained in Appendix ** determine the ΔH for the following reaction:

$$H_2C = CH_2 + Br_2 \longrightarrow BrH_2C = CH_2Br \qquad \Delta H = ?$$

13. By applying the Hess's: Law principle determine the ΔH for this reaction:

$$Ca(s) + 2H_2O \qquad Ca^{2+} + 2OH^- + H_2$$

by combining the following reactions:

$$H^+ + OH^- \qquad H_2O \qquad \Delta H = -55.9\ kJ$$
$$Ca + 2H^+ \qquad Ca^{2+} + H_2 \qquad \Delta H = -543.0\ kJ$$

*Answer:* –431.2 kJ

14. By applying the Hess's: Law principle determine the ΔH for this reaction:

$$P_4O_6 + 2O_2 \longrightarrow P_4O_{10}$$

by combining the following reactions:

$$P_4 + 3O_2 \longrightarrow P_4O_6 \qquad \Delta H = -1640\ kJ$$
$$P_4 + 5O_2 \longrightarrow P_4O_{10} \qquad \Delta H = -2940\ kJ$$

15. How much heat is required to heat 66g of zinc from 22°C to 77°C ? The specific heat of zinc is 0.388 J/(g·°C).

16. Methyl alcohol, $CH_3OH$, is toxic because liver enzymes oxidize it to formaldehyde, HCHO which can coagulate protein via crosslinking reactions. These types of reactions are not an issue with ethanol a molecule that is just one carbon atom bigger at $CH_3CH_2OH$. Calculate ΔH° for the following reaction given the

standard enthalpies of formations of (in kJ/mol): $CH_3OH$ = −245.9; HCHO = −150.2; $H_2O$ = −285.8.

$$2CH_3OH \quad + \quad O_2 \longrightarrow 2HCHO \quad + \quad 2H_2O$$

17. Sulfur, $S_8$, burns in air to produce sulfur dioxide per the following mechanism:

$$S8 \; + \; 8O_2 \longrightarrow 8SO_2 \qquad \Delta H = -2.39 \times 103 \; kJ$$

How much heat is released when a kilogram (~2.2lbs) of sulfur dioxide is produced?

18. The following processes all involve a decrease in entropy. What is meant by this statement?

a. scrambling an egg.

b. subliming dry ice

c. the desalination of water

19. A water filled can is placed in a freezer. After the water freezes the can has split open. Work is clearer being done in this process. Where did the required energy for this effort come from?

20. What is free energy (ΔG) composed of? Be specific.

# Credit

Fig. 5.5a: Copyright © 2004 by KMJ, (CC BY-SA 3.0) at https://commons.wikimedia.org/wiki/File:Gluehlampe_01_KMJ.jpg.

CHAPTER 6

# States of Matter and Intermolecular Forces

## The fundamentals of attraction and repulsion between atoms and molecules.

***Intramolecular Forces*** are attractive and repulsive forces ***within*** molecules.
***Intermolecular Forces*** are attractive and repulsive forces ***between*** molecules

Intermolecular are generally much weaker than intramolecular bonds, that is, those between molecules or ions. Both categories of forces are essentially governed by the attraction between opposite charges (+ or –) and repulsion of the same charges (+/+ or –/). The difference in strength ultimately comes down to the magnitude of the positive and negative charges. Thus it is much easier to melt or boil water wherein intermolecular bonds are broken than to decompose water molecules by breaking the O-H covalent bonds. Consider also the very high melting points of the pure ionic solid NaCl (801°C or 1474°F) as well as iron metal (801°C or 1474°F) as examples of the remarkably robust nature of intermolecular and ionic and metallic attractive forces.

## 6.1 States of Matter—The Basics

There are more than the three states of matter we all take for granted. For example, two other states are liquid crystalline and plasmas. However, from a conceptual standpoint the observer might conclude that the differences between the various states is a continuum rather than being distinct from one another. That said it is certainly appropriate to consider just gases, liquids and solids, and their relative physical and chemical properties. (Figure 6.1)

*Gases:* This state possesses the following traits.

1. Have relatively weak attractive forces between atom, s ions or molecules
2. Assume the shape and volume of their containers
3. Expand to fill the available space

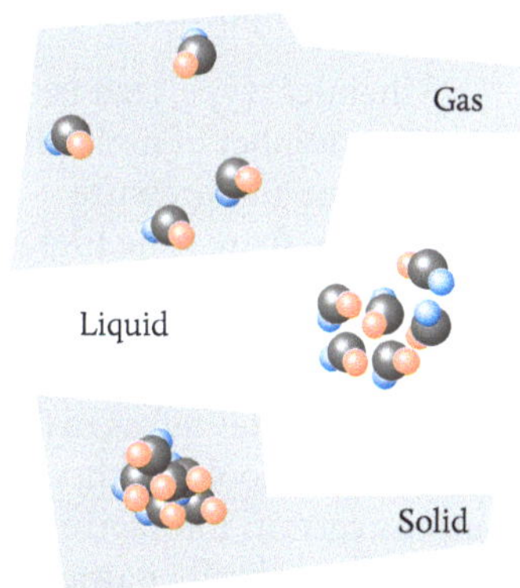


FIGURE 6.1. States of Matter.

4. Are compressible
5. Allow diffusion of other species within the gas to occur rapidly

***Liquids:*** This state possesses the following traits.

1. Have relatively strong attractive forces between atoms, ions or molecules
2. Partially assume the shape and volume of a container
3. Do not expand to fill all available space
4. Are incompressible
5. Flow readily
6. Allow diffusion of other species, but this occurs slowly relative to gases

***Solids:*** This state possesses the following traits.

1. Have strong attractive forces between atoms or molecules
2. Do not assume the shape and volume of a container
3. Do not expand to fill all available space
4. Are incompressible
5. May or may not allow diffusion of other species within them.

Logically liquids and solids are referred to as condensed phases. Even in the gas phase intermolecular and interatomic forces are very significant. The only exception to this rule is the case of say a single molecule; however, this circumstance never occurs.

## 6.2 Intermolecular Forces

Why do certain elements and compounds assume different states of matter under specific temperature and pressure conditions? Why do the states sometimes change when the conditions are changed? The reasons are rooted in chemical bonding forces between elements, molecules, and ions. Intramolecular forces are the first level of interactions that occur. These forces occur between atoms within a molecule. At the next level, are intermolecular forces with occur between atoms or molecules that are separated in three dimensional space. At the next level, this chapter's material, there are the interactions between molecules or atoms in three dimensional space. What variables are important here?

1. Chemical type, that is, the kinds of atoms and how they are bonded together
2. The number of molecules or atoms present
3. Volume and pressure
4. Temperature. The higher the temperature the greater the kinetic energy of the system

Let us consider five very different chemicals in our discussion of intermolecular forces; helium (He), nitrogen ($N_2$), hydrochloric acid (HCl), calcium oxide (CaO), and water ($H_2O$). We will look at other examples in later sections. For each og these compounds or atoms certain interactions occur that will impact their properties. Van der Waals forces is an umbrella name given to the four basic types of bonds that exist between atoms and molecules. These bonding forces are described below.

## Dispersion Forces

Often called London Dispersion Forces, *these attractive forces exist between all atoms and molecules.* The classic example that demonstrates this type of force very well is that of simple helium. At atomic number two it has just two protons and two electrons, and is a very chemically stable inert gas. Recall our discussion of entropy in which we concluded that for even the simplest system, such as helium, there are an almost infinite number of possible microstates possible each containing the nucleus and two electrons at various locations in 3D space, and various energies. The top part of Figure 6.2 shows the two atoms with their electrons frozen into one of these microstates, and equally distributed this leaves the atom electronically neutral.

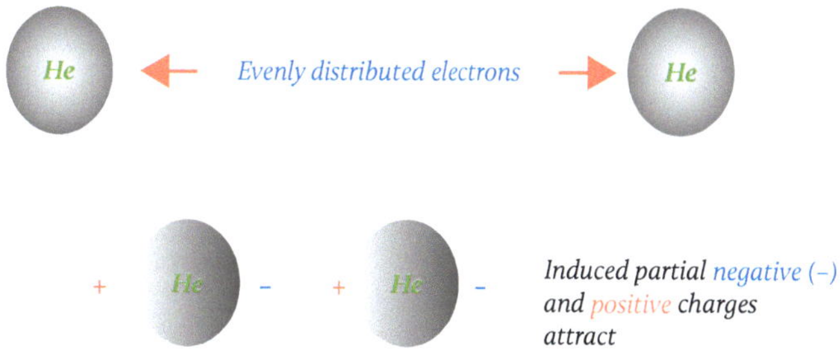


FIGURE 6.2. London Dispersion Force Interactions.

In the bottom half of Figure 6.2, the electrons of the first atom are both near the second atom's electronic cloud. In essence this is an induced positive-negative charge separation. Because the electrons are all negatively charged they repel each other. At that instant in time each the first atom's uneven electron distrubution (due to the random motion of electrons about the atom) *induces* a negative and positive pole on the second atom. As a result, both atoms have positive and negative ends and theopposite charges attract! The net result here is a weak attraction between the helium atoms. We know it is very weak because experimentally helium's boiling point is just 4.2K or almost absolute zero. It does not solidify at all unless put under very high pressure.

Helium is a very simple case in which dispersion forces are the only forces that exist between He and other atoms of any kind. Our other cases above are more complex to envision, but dispersion forces always exist there too.

## Dipole-Dipole Forces

A **dipole** is simply a bond between two atoms wherein the bond has a partial positive electronic charge on one end, and a partial negative charge on the other end. Recall this is due to the unequal sharing of electrons between atoms in a covalent bond. THe move electronegative atom will be at the negative end the less eletronegative atom at the positive end. By convention this is shown as an arrow (the arrowhead being negative) with a plus end:

δ+  δ–

Here also the δ means partial, so δ+ means a partial positive charge at one end while δ– means partial negative at the other.

Not all bonds are dipoles. Let's look at our examples above:

*He:* No bond, no dipole

*$N_2$:* Two nitrogen nuclei joined by a triple bond in between. Since the two nuclei are identical the electronic distribution is even as well, thus no positive or negative end and no dipole exists.

*HCl:* Chlorine is substantially more electronegative than hydrogen thus a dipole exists;

δ+ **H-Cl** δ–

Now suppose two HCl molecules come near to each other, both with dipoles. What happens? Much like the north and south poles of two magnets, like repels like as in A below:

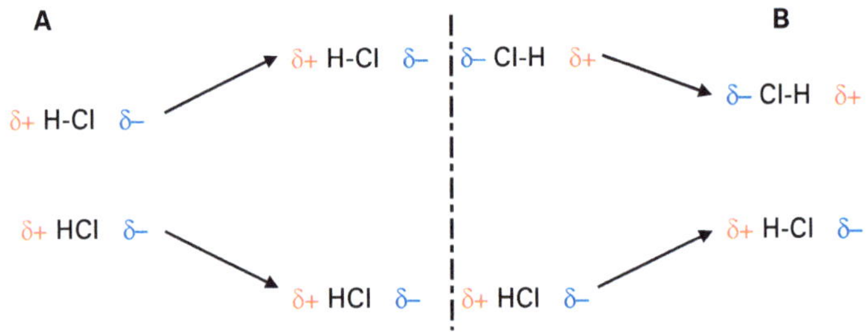


**CHEMICAL STRUCTURE 6.1.** Two HCl Molecule Structure.

Here the red plus ends of the molecules near each other create a repulsion, and the same is true with the blue negative ends. In B one dipole is reversed, and there is a net attraction between the δ+ and the δ– ends of the molecules. These are classic dipole-dipole force attractions (and repulsions).

Water molecules also have substantial dipole-dipole interactions that are somewhat more involved than HCl. Because of the two lone pairs of electrons on the oxygen atom water has a bent structure and an associated dipole as follows:

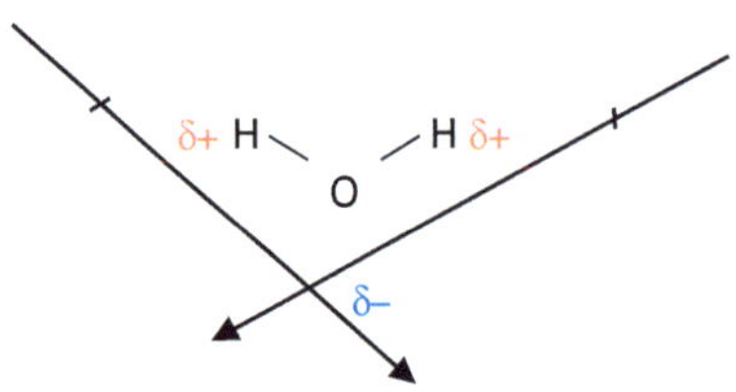


**CHEMICAL STRUCTURE 6.2.** Water Molecule Structure.

In this case both hydrogens have partial positive charges (δ+) while the oxygen is partially negative (δ–) since it is far more electronegative. Now, if two water molecules arrive next to each other the attractive net orientations will favor alternating δ– and δ+ charges as is true for all molecules that experience dipole dipole interactions

## Ion-Dipole Forces

Ion-dipole interactions are sort of a subset of dipole-dipole. In this case the dipole is defined in exactly the same manner. Chemical Structure 6.3 shows how these forces of attraction work for the case of calcium oxide (CaO) dissolved in water.

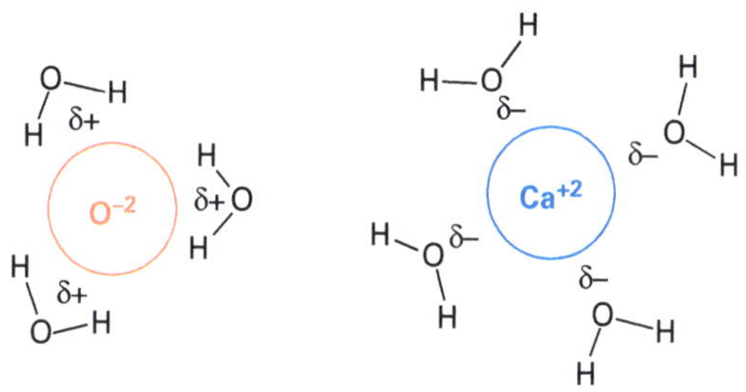


CHEMICAL STRUCTURE 6.3. Ion-Dipole Forces.

This is an ionic compound wherein the divalent calcium $Ca^{+2}$ is solvated (surrounded) by water molecules with the negative ends of their dipoles bonding to the Ca ion. The reverse is true for the $O^{-2}$ ion in solution. These forces are very strong.

## Hydrogen Bonding Forces

Hydrogen bonds are the strongest of the intermolecular bonds; however, they are also the rarest which makes learning where they occur easy. To have a hydrogen bond you obviously must have hydrogen be present. Besides H you must have present one of three other elements, ***nitrogen, oxygen or fluorine***. Simply put, no N, O or F, no hydrogen bond. Below is an example of H bonding with HF. Here the dotted lines represent hydrogen bonds which are much weaker than the covalent bonds between the H and F atoms (solid lines).

H — F ........ H — F ........ H — F

It is also noteworthy that in all three heteroatom cases there are lone pairs of electrons present. An example of hydrogen bonding for water is shown in Figure 6.3.

The relatively strong nature of hydrogen bonds in demonstrated by Figure 6.4 which shows the boiling points of the hydrides of groups 14-17.

The three compounds that can form hydrogen bonds ($H_2O$, $NH_2$, and HF) show strikingly higher boiling points than would be predicted based just upon molecular weights.

**PRACTICE** Draw out the Lewis structures of the following molecules, and predict the kinds of bonding each contains: $H_2$, $CO_3$, $C_2H_6$, $CaSO_4$, Benzene, benzoic acid, urea and formic acid.

A couple of generalizations concerning relative intermolecular forces and estimating the same are in order.

1. When molecules are about the same size their relative dispersion forces will also be similar. In that case dipole-dipole and hydrogen bonding, if present, will tell the tale.

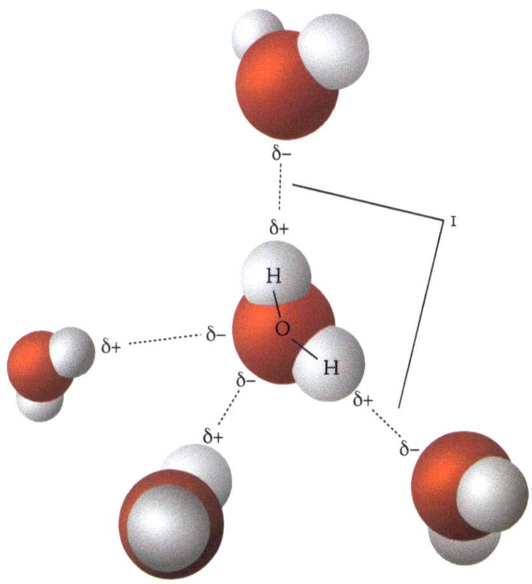


FIGURE 6.3. Hydrogen Bonding in Water.

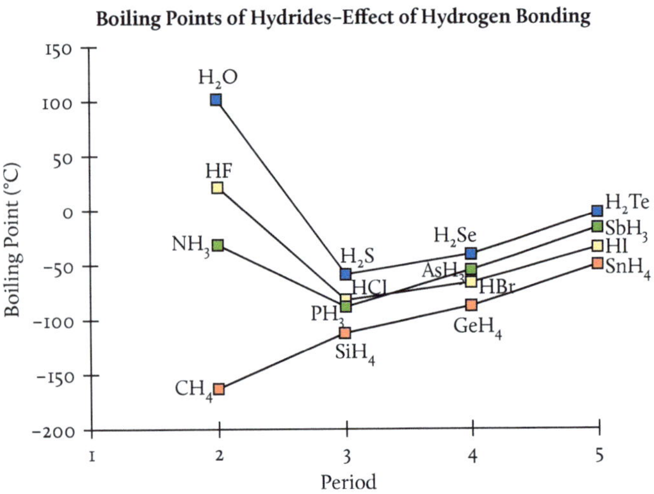


**FIGURE 6.4.** Boiling Points of Group IV-VI Hydrides.

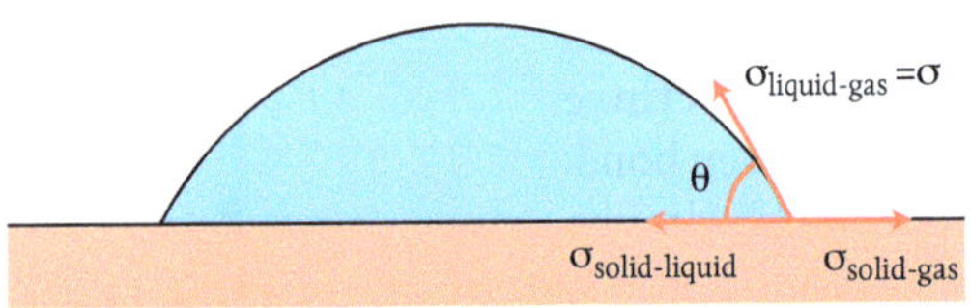


**FIGURE 6.5.** Surface Tension.

2. Dispersion forces tend to increase with increasing molecular weight. As the molecular size gets larger the dispersion forces get larger and larger, and can become more significant that a small number of dipole-dipole interactions.

## 6.3 Liquid Properties

### Surface Tension

The ***surface tension*** of a liquid is its tendency to attain a minimum surface area. Figure 6.5 shows the forces that act upon molecules within the bulk of a liquid as well as those that act on the molecules at the surface.

In the former case the forces surrounding the bulk molecules equal out if the system is at equilibrium. At the surface; however, the molecules are not subject to upward forces. This means that they feel a net pull downward into the bulk. The effect of this net force is to minimize the number of molecules at the surface and lower the surface tension. We have all observed droplets of liquids or bubbles that form perfect spheres in the air. This occurs because a sphere is the shape that possesses the lowest surface tension.

### Viscosity

***Viscosity*** is the resistance of a liquid to flow. The phenomena is driven by intermolecular bonding of molecules as well as the shape of the molecules. ong molecular chains will tend to become physically twisted and entangled much like the entanglement of spaghetti strands in a pot, but on a molecular level. The stronger the intermolecular forces, the greater the viscosity of the substance. Also, the longer the molecules the greater the viscosity of the substance.In general, viscosity is inversely proportional to the temperature that is; as the temperature goes up the viscosity goes down because as temperature increases, more heat energy is available to break up some of the intermolecular forces between the molecules

***Cohesive forces*** are intermolecular forces that bond molecules to each other. *Adhesive forces* are intermolecular forces that bond molecules to a surface. *Capillary Action* is an interesting phenomena that occurs in narrow tubes. The adhesive forces binding the liquid to the surface tend to increase the surface area of the liquid. By climbing the surface walls the surface tension of the liquid tends to be lowered. This process is very important biologically. For example, plants use capillary action to overcome the force of gravity to move water and nutrients up and through the plant.

## 6.4 Phase Changes

We are all very familiar with phase changes, the reversible conversion of solid to liquids to gas; the ice-water-steam conversion being a most common experience.

Most, but by no means all, substances can be made to undergo these transitions which are especially temperature and pressure dependent. Consider the following graph of phase changes (Figure 6.6).

For the time being consider just phase changes for pure substances under constant pressure conditions.

The melting of a solid to form a liquid requires heat. The amount of heat required is dependent upon the physical properties of the substance being melted (specific heat/heat capacity). This is a reversible process, and the amount of heat required is exactly the same each time the process is psignerformed under the same conditions. All that changes upon reversing the process is whether the the heat exchange is endothermic or exothermic.

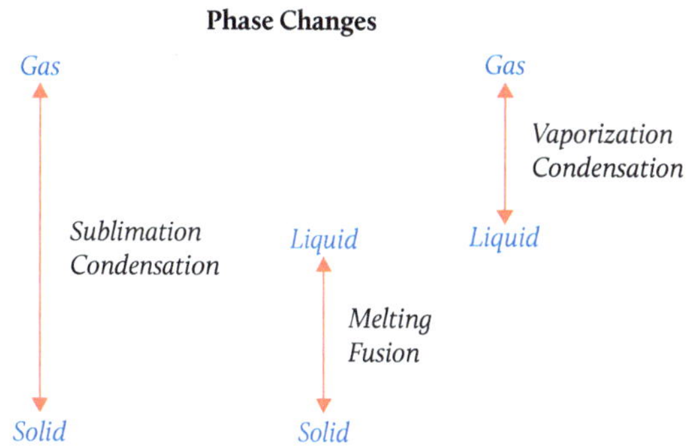


**FIGURE 6.6.** Phase Changes.

Melting is an endothermic process hence a positive sign is associated with the heat gain. The reverse is an exothermic process so a negative sign is used: For water:

$$\Delta H_f = +6.02\ kJ\ /\ mole$$

***1 mole of ice*** $\rightleftarrows$ ***1 mole liquid water***

$\Delta H_f$ is known as the **heat of fusion**.

The vaporization of a liquid to form a gas also requires heat. The amount of heat required is dependent upon the physical properties of the substance being melted. We are considering just a specific system here so the process is considered sreversible, and the amount of heat required is exactly the same each time the process is performed. All that changes upon reversing the process is the sign. Vaporization is an endothermic process hence a positive sign is associated with the heat gain. The reverse is an exothermic process so a negative sign is used:

$\Delta H_v = +40.7\ kJ\ /\ mole$

***1 mole of liquid water*** $\rightleftarrows$ ***1 mole gaseous water***

$\Delta H_c = -40.7\ kJ\ /\ mole$

$\Delta H_v$ is known as the **heat of vaporization**. The heat of vaporization is generally much higher than the heat of fusion. The solid to liquid transition involves breaking some, but not all, of the intermolecular bonds between molecules. The liquid to vapor transition; however, involves the breakage of virtually all such bond forces.

The direct conversion of a solid to form a gas is a rarer occurrence, but also requires heat. Dry ice (solid $CO_2$ to gaseous $CO_2$) is the only example of this phenomena most of us encounter. The amount of heat required is dependent upon the physical properties of the substance being melted. We are considering a system, and not the surroundings so the process is reversible, and the amount of heat required is exactly the same each time the process is performed. All that changes upon reversing the process is the sign.

Vaporization is an endothermic process hence a positive sign is associated with the heat gain. The reverse is an exothermic process so a negative sign is used:

$\Delta H_s = +32.3\ kJ\ /\ mole$

***1 mole of solid*** $CO_2$ $\rightleftarrows$ ***1 mole gaseous*** $CO_2$

$\Delta H_s = -32.3\ kJ\ /\ mole$

$\Delta H_{sf}$ is known as the **heat of sublimation**. The heat of sublimation is almost always greater than the heat of vaporization which is logical since in essence the direct conversion from a solid to a gas involves a greater thermodynamic change even though an intermediate liquid phase is not involved. Compare the value of $\Delta H_v$ for water with its $\Delta H_s$. Note that for water the sum of $\Delta H_v$ and $\Delta H_f$ is about equal the $\Delta H_s$.

$\Delta H_s = +47\ kJ\ /\ mole$

***1 mole of solid*** $H_2O$ $\rightleftarrows$ ***1 mole gaseous*** $H_2O$

$\Delta H_s = -47\ kJ\ /\ mole$

There is another sublimation that we encounter but most do not recognize it as such. This is the formation of *Rime* ice crystals in the winter. Look it up.

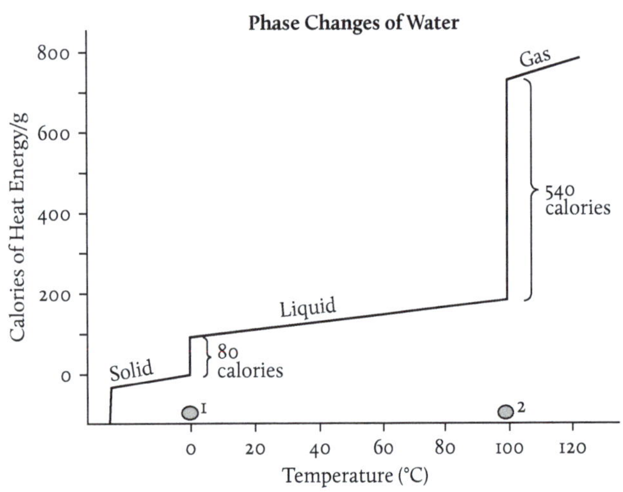


FIGURE 6.7. Phase Changes with Temperature.

Figure 6.7 contains the heating curve for a typical sequence of phase changes for water. The enthalpy unit here is calories.

From an arbitrary starting temperature of −25°C as heat is added the solid absorbs heat until the melting point is reached at O °C. At this point the added heat goes into melting the ice to form liquid water. During this process both ice and water coexist and the temperature remains at O °C. Thereafter the water rises in temperature until 100°C is reached at which point the liquid water is converted to a vapor phase. Once the liquid phase is gone, the gas just heats up.

## Critical Temperature and Pressure

Generally the addition of pressure will result in a gas liquefying. However, at some point with increasing temperature this no longer becomes possible; that is the gas remains a gas regardless of how much additional pressure is applied. These inflection points are called the **critical temperature** and **critical pressure**, respectively.

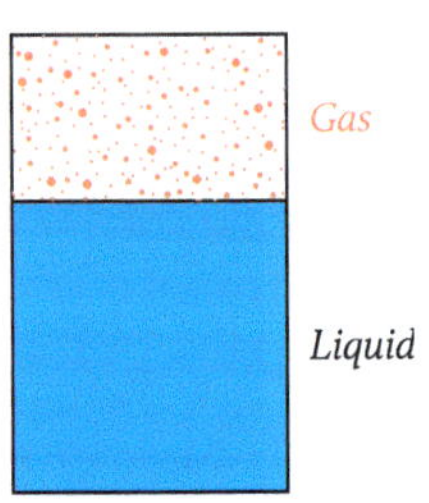


FIGURE 6.8. Gas Vapor-Liquid Equilibrium.

## Vapor Pressures

Common humidity is a result of an equilibrium between water as a liquid and the gas phase (Figure 6.8), in this case a vapor as a component of our atmosphere. Not as familiar, but just as important, is the equilibrium between a solid phase and its vapor phase.

The process of going back and forth between states is a matter of the relative strengths of the intermolecular forces of attraction. The forces involved here include the dispersion, dipole-dipole, hydrogen bonding, and ion-dipole forces as described earlier in this chapter. This is a dynamic process that is very dependent upon the experimental conditions, especially the ambient temperatures. It also occurs primarily at the interface of the vapor phase and the condensed phase.

There is a second factor involved in predicting vapor pressures qualitatively, and that is atomic or molecular weight. Everything else being equal the species with the higher molecular weight will have a lower vapor pressure. This is logical as the more mass you have, the more energy it takes to lift it.

We can use our newfound knowledge of intermolecular bonding to explain why elements and compounds have the vapor pressures they do as well as make qualitative predictions of relative vapor pressures for different substances. For example, all of the noble gases (group 18) exist in the vapor phase under ambient conditions. Why? Take neon as a random example of this group. Neon has 10 protons and ten electrons, the latter split into two shells of $2e^-$ and $8e^-$. The octet rule says this element should be stable and unreactive, and it is. What kinds of bonding occur here? For the Noble Gases, dispersion forces are present since these always exist, but no dipoles are there, and certainly no H bonding (what H?). Thus the expectation is for very weak interatomic attractions leading to a stable gaseous state, which is the reality.

How about say $H_2S$? A quick look at the periodic table reveals sulfur is in group 16 right below oxygen in period 3, and is more electronegative than hydrogen. Like O it also has two lone pairs on the S and a bent geometry similar to $H_2O$:

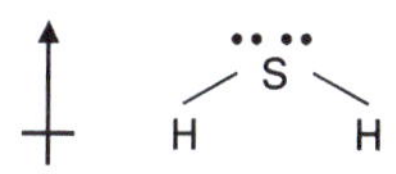


**CHEMICAL STRUCTURE 6.4.** $H_2S$ Structure.

Looking at our intermolecular bonding guidelines this molecule has dispersion force attractions, it also has dipole-dipole bonding; however, there is no hydrogen bonding present since there is no N, O or F present. $H_2O$ is the same, but also has very strong hydrogen bonding. On the basis on the latter the immediate expectation is that water will have the lower vapor pressure since it will be harder to break the intermolecular bonds. Counteracting that thought is the higher molecular weight of the $H_2S$. The only way to determine the truth is to measure the vapor pressures experimentally!

**PRACTICE** Arrange the following elements and compounds in the order of increasing vapor pressure; that is lowest to highest.

a. $F_2$, $I_2$, $Br_2$, $Cl_2$.

b. $AsH_2$, $SbH_2$, $PH_2$, $NH_2$

c. Fe, S, Kr

d. $CH_3OH$, $CH_3CH_2OH$, $CH_3CH(OH)_2$

## Phase Diagrams

***Phase diagrams*** show the state of a matter of a substance as a function of temperature and pressure. In practice these plots are generated from the results of laboratory

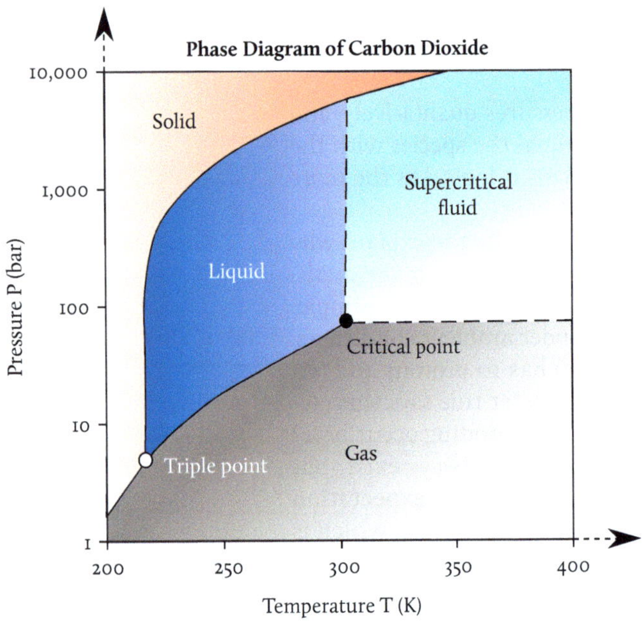


FIGURE 6.9. Phase Diagram of Carbon Dioxide.

experiments. Figure 6.9 shows the phase diagram for $CO_2$ which is not a liquid below a pressure of ~5.5 atmospheres.

As shown low pressures and temperatures favor the solid phase while high temperatures favor liquid and gaseous phases. The lines of interface between phases is where two phases can coexist under very specific pressures and temperatures. The triple point is unique in that this is the only place where all three phases can exist together. As mentioned above the critical point is that temperature and pressure above which only the gas phase is found.

Another practical example is Figure 6.10 for water.

Although the transitional points and interfacial phase separation lines differ substantially, the basic plot form is the same for both

# 6.5 Gases

The interactions of molecules with each other and the surroundings is complex in any phase; gas, liquid, or solid. The gas phase is in many ways the easiest to deal with since intermolecular interactions are overcome, and we can simplify the situation somewhat. We start with some underlying principles of ***kinetic molecular theory***.

## Kinetic Molecular Theory of Gases

This theory makes the following assumptions. Also the term molecules here will encompas atoms such as the noble gases exist as independent, non-bonded entities.

1. Gases consist of large numbers of molecules in continuous motion. Also the velocities of these entities in motion is random.
2. The total volume of the gas molecules is negligible compared to the container volume.
3. Attractive and repulsive forces between molecules do not take place.
4. Although energy can be transferred between molecules, the average kinetic energy of the molecular system is constant as a function of time.
5. The average kinetic energy of the molecular system is proportional to the absolute temperature.

This is quite a set of assumptions, but as we will see now it provides for a gas phase system that can be analyzed and practically applied to the real world, as it is every day! By extension these assumptions lead to the following logical conclusions:

The pressure of a gas is due to very large numbers of gas molecule colliding with the walls of the container vessel (Figure 6.11).

Equal numbers of moles of two gases at the same temperature have the same kinetic energies.

If the temperature of a gas sample is doubled the sample's kinetic energy doubles.

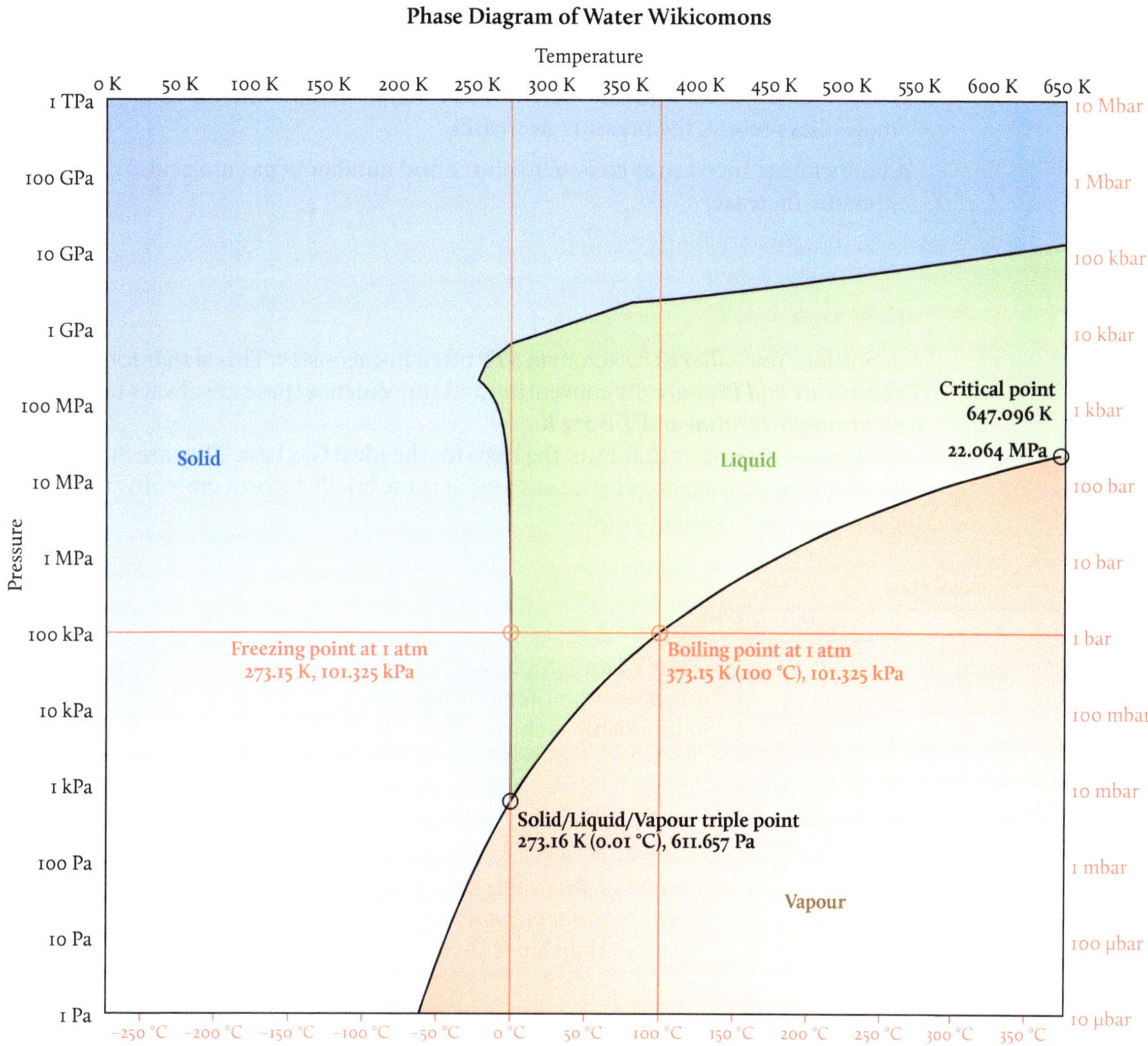


**FIGURE 6.10.** Phase Diagram of Water.

The last point is shown qualitatively by Figure 6.10. At any temperature there is a distribution of molecular energies, but at higher temperatures the average kinetic energy, and thus average molecular speeds, increases. Also the distribution of energies gets broader as the temperature is raised.

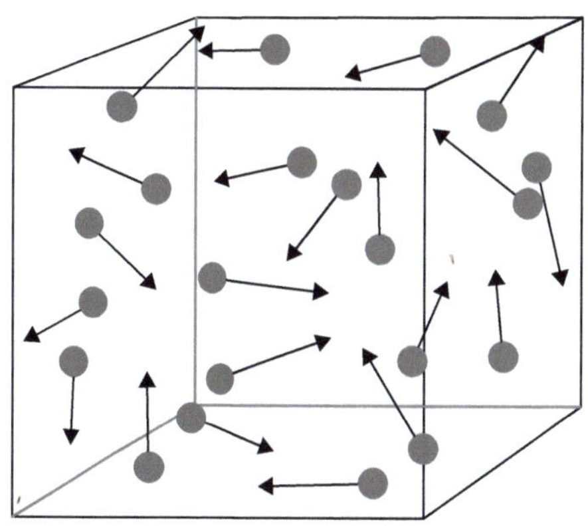

**FIGURE 6.11.** Gas Molecule (Atom) Collisions with a Container Wall.

**PRACTICE** The kinetic energy of a gas molecule is given by the equation $E_k = (1/2)mv^2$ where m = mass and v = velocity. Assume two gas samples have the same kinetic energy, but the mass of gas A is twice that of gas B. Qualitatively what is the velocity of gas A relative to that of gas B?

A couple more observations concerning the kinetic gas theory are probably intuitive, but lead directly to the ideal gas law:

If the volume of a container is increased at constant temperature and number of gas molecules present, the pressure decreases.

A temperature increase at constant volume and number of gas molecules causes a pressure increase.

## Ideal Gas Law

A definition. You will see the acronym ***STP*** often in chemistry. This stands for ***Standard Temperature and Pressure***. By convention and convenience these are always taken to be ***P = 1 atmosphere (atm)*** and ***T = 273 K***.

There are three laws that form the basis for the Ideal Gas Law. They are Boyle's Law, Charles Law and Avogadro's Law. Let's look at these briefly before combining them into the powerful ideal gas equation.

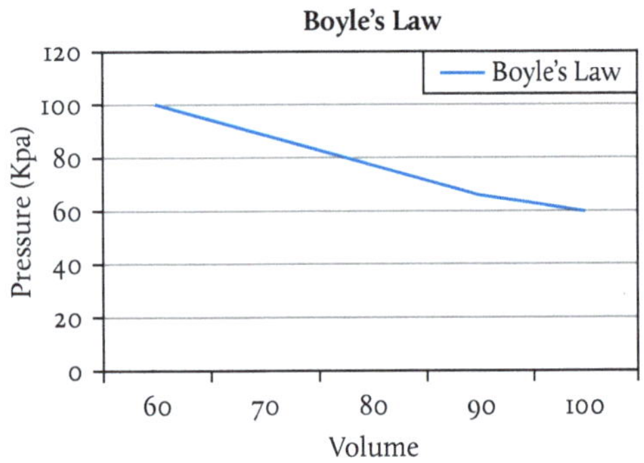


FIGURE 6.12. Boyle's Law: The Relationship of a Gas's Volume to Pressure.

### Boyle's Law

*Boyle's Law* simply states that *the volume of a fixed quantity of a gas at constant temperature varies inversely with the pressure*. This relationship is shown in Figure 6.12.

### Charles's Law

*Charles's Law* simply states that the *volume of a fixed quantity of a gas at constant pressure varies in direct proportion to the temperature*. This relationship is shown in Figure 6.13.

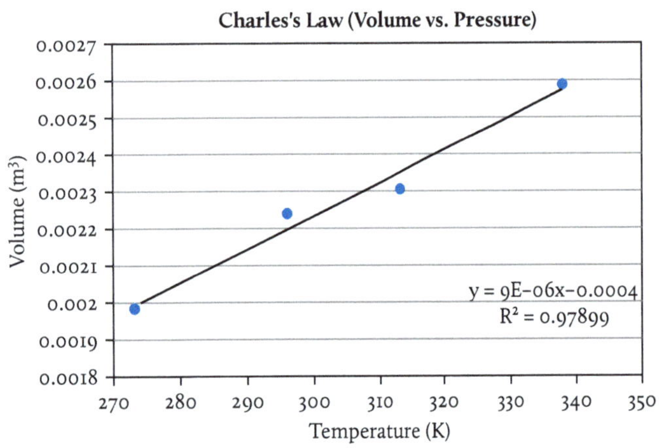


FIGURE 6.13. Charles's Law: The Relationship of a Gas's Volume to Temperature.

### Avogadro's (Gay-Lussac's) Law

*Avagadro's Law* is not as easily accessible, but states that *equal volumes of gases at the same pressure and temperature contain the same number of molecules*. This relationship is not as straightforward as the Boyle's and Charle's Laws, but it is crucial for any scientist to understand.

Table 6.1 shows three different gases under to which is applied Avogadro's law. Everything is the same for these gases except the total masses present. The number of molecules of each is the same at $6.02 \times 10^{23}$ molecules (atoms), Avogadro's Number. In each case this represents one mole of gas, but because their molar masses are different the weights are different.

TABLE 6.1 **Under STP Conditions: Differences in Masses and Densities of Gases**

| Gas | He | $N_2$ | $CO_2$ | $CH_4$ |
|---|---|---|---|---|
| P(atm) | 1 | 1 | 1 | 1 |
| T(K) | 273 | 273 | 273 | 273 |
| V(L) | 22.4 | 22.4 | 22.4 | 22.4 |
| Mol | 1 | 1 | 1 | 1 |
| Wt. (1 mol) | 2 | 28 | 44 | 16 |
| d (g/L) | 0.09 | 1.25 | 1.96 | 0.71 |

This concept is important to understand as it affects the reactivities and properties of different materials as a function of molecular weight.

**PRACTICE** with Avogadro's law**

The three laws above can be distilled into the following relationships for gases:

$V \alpha 1 / P$ (at constant n, T)

$V \alpha T$ (at constant n, P)

$V \alpha n$ (at constant T, P)

These equations can then be combined into a general proportionality:

$V \alpha nT / P$

Introducing a proportionality constant R yields:

$\boldsymbol{V = R(nT / P)}$ **or rearranging the equation**

$PV = nRT$

This is the ***Ideal Gas Equation.*** Here P is the pressure in atmospheres, V is the volume in liters, n is the number of moles of gas, R is the universal gas constant, and T is the temperature in Kelvin. R is a universal constant in the general sense of the universe. The most commonly used value used in chemistry is **0.082 L-atm/mole-K**. Note the units here.

The ***Ideal Gas Law*** is just that, ideal but not fully accurate. Calculations using this law are done under the assumptions of kinetic molecular theory. If the conditions are set correctly it is very accurate. What are ideal conditions? *The ideal gas law only works well under the conditions of low pressure and high temperature*, not so well under either of the reverse conditions. This can be explained due to the increases in strengths and number of attractions between molecules at high pressures (closer together) and low temperatures (not enough energy to overcome interactions).

This equation has four potential variables and one constant. So, if you know or can derive three of the variables you can then calculate the fourth one. As chemistry goes this is pretty logical and straightforward material. Many times these equations will simplify themselves upon perusal of a given problem.

### Solving Ideal Gas Law Problems

Gas law word problems are tailor-made for ready solving if one is logical, and follows a stepwise solution process. Here is a guide to approaching such problems:

1. Write down what you know, that is what is given in the problem. Assess what is needed to solve the problem. Oft times information is given that is not needed for the solution!
2. Convert all numbers into units that are needed. For example, any temperature in oC must be converted to degrees K.
3. Rearrange the equation as necessary to solve for the unknown.
4. Follow through with dimensional analysis! If the units at the end are correct, the numbers will be correct too.

A few examples dealing with the ideal gas law are in order.

*Example 1: Constant n and R. $P_1V_1 / T_1 = P_2V_2 / T_2$
*Example 2: Constant V and P and R. $n_1 / T_1 = n_2 / T_2$
*Example 3: given # of grams of a gas
*Example 4: calculation of a gas density.

## Dalton's Law of Partial Pressures

***Dalton's Law*** is straightforward. It says that the total pressure of a mixture of gases is equal to the sum of the partial pressures of the individual gases. That is for a mixture of four gases, A, B, C, and D with partial pressures $P_A$, $P_B$, $P_C$, and $P_D$, the total pressure, $P_{tot}$, of the mixture would be:

$$P_A + P_A + P_A + P_A = P_{tot}$$

**PRACTICE** A mixture of $O_2$, He, $N_2$ and $CO_2$ has a total pressure $P_{tot}$ = 1.2atm, the temperature is 27 °C, and the volume is 10L. If $PO_2$ = 0.3atm, PHe = 0.05atm and $PN_2$ = 0.25, what is the pressure of $PCO_2$?

## Gas Reactions

More interesting than Dalton's law is that we can relate the partial pressures of each gas back to an amount in moles or grams of each gas. To do this we use the ideal gas law. Let's look at the practice example just given above. In this instance how many grams of He are present in the mixture?

$PV = nRT$[ We know $P_{He} = 0.05\ atm, V = 10L,$ and $T = 27 + 273 = 300K$. We also know $R = 0.082$ L-atm/mole-K. So substituting and solving for moles of He, n:

$n = (0.05\ atm)(10) / (0.082)(300K) = 2 \times 10^{-2}$ moles of He. Assuming He is atomic weight 4g/mole (from the periodic table) then: $(2 \times 10^{-2}\ moles)(4g / moles) = 0.08g$.

The calculations for the other gas components is carried out the same way, and the total number of moles or weight of gases present in the mixture is also easily determined.

# 6.6 Solids

The subject of **solids** covers a lot of territory in the sense of chemical structure, physical properties and chemical reactivity. There are a number of basic classes of solids that are more or less distinct, although certainly there are rich areas of hybridization. The basic areas are as follows:

***Metals:*** This group includes the vast majority of the elements in the periodic table. The properties of metals are defined by their structure which is a rigid framework of nuclei that are surrounded by a mobile sea of electrons. This electron mobility is responsible for the outstanding ability of metals to carry electrical current, as well as heat. In addition the ductility and elongation are due to the electron's fluidity. Most are not brittle.

***Ionic Solids:*** These are held together by the strong electrostatic attraction of cations (usually metallic) and anions. They tend to be brittle and fracture easily. Also they are poor electrical and thermal conductors. Metallic and ionic bonding mechanisms are drastically different.

***Covalent Solids:*** As the title suggests these materials are held together by covalent bonds which can lead to a wide range of physical properties. As examples diamond and rubber bands are defined by covalent bonds, but are clearly very different materials in so many ways. These kinds of solids can be rigid or flexible, and tend to be poor electrical and thermal conductors.

***Semiconductors*** are hybrid materials that are somewhere between conductors and insulators, nonmetals and metals and upon which virtually all of our modern electronics are dependent. This is a fairly small group of materials.

***Molecular Solids:*** These are held together by the intermolecular forces described above. Because these types of bonds are relatively weak, these compounds tend to have poor physical properties and be poor thermal and electrical conductors. That said, they are extraordinarily important as all of biological life is based upon them.

There are other types of solids that are basically subsets of these major classes. Examples include **polymers** and **nanomaterials** among others; however, a discussion of these subclasses is outside the scope of this book.

## Crystalline and Amorphous Solids; Crystal Structures

It is important to understand the difference between crystalline and amorphous solids.

***Crystalline Solid:*** One that is characterized by a very regular, repeating structure as shown in Figure 6.14 for gold.

FIGURE 6.14. Gold Crystals.

There are seven basic crystal structures in nature. The structure any given material takes is dictated by thermodynamics, but tends to favor the structure that yields the most efficient three dimensional packing.

***Amorphous Solid:*** A substance that resembles a liquid in the sense that the atoms, molecules or ions are randomly distributed. However, in the solid phase the molecules lack the freedom of motion typical of the liquid state (Figure 6.15).

FIGURE 6.15. Geode.

## Chapter 6 Summary

*Intramolecular Forces:* attractive and repulsive forces ***within*** molecules

*Intermolecular Forces:* attractive and repulsive forces ***between*** molecules

*States of Matter:* There are three primary states of matter, gas, liquid and solid. They have the following characteristics:

*Gases*

Have relatively weak attractive forces between atoms or molecules
Assume the shape and volume of their containers
Expand to fill the available space
Are compressible
Allow fast diffusion of other species within a gas to occur very rapidly

*Liquids*

Have relatively strong attractive forces between atoms or molecules
Partially assume the shape and volume of a container
Do not expand to fill all available space
Are incompressible
Flow readily
Allow diffusion of other species, but this occurs very slowly compared to the gas phase.

*Solids*

Have strong attractive forces between atoms or molecules
Do not assume the shape and volume of a container
Do not expand to fill all available space
Are incompressible
May or may not allow diffusion of other species within them. If so, the rates of diffusion are almost invariably slower than the corresponding diffusion rates in the liquid phase.

*Intermolecular Forces:* These are the forces of attraction and repulsion that occur between molecules. These forces are determined by the negative and positive charges of the electrons and nucleus, which in turn, determine the three dimensional structures of molecules. They are weaker than the corresponding intramolecular covalent or pure ionic forces. The primary forces involved are:

*London Dispersion (LDF) Forces:* Generally most forces of attraction these types of forces occur for all atoms and molecules.

*Dipole-Dipole Interactive Forces:* Are present in asymmetrical molecules where dipole moments are present. Dipole-dipole forces are stronger in magnitude than are LDF forces. A subset of this category is ion-dipole interactions.

*Hydrogen Bonding Forces:* Can occur in molecules that contain hydrogen and oxygen, nitrogen and/or fluorine atoms, these are stronger than dipole-dipole forces.

*Properties of Liquids:* The properties of a specific liquid are to a large extent determined by the intramolecular bonding forces just described.

*Surface Tension:* a liquid's tendency to attain a minimum surface area, meaning a minimum number of molecules at the surface. Basically the net forces on the surface molecules sum to pull the molecules back into the liquid.

*Viscosity* is the resistance of a liquid to flow. In general the greater the intramolecular bonding is, the greater will be the viscosity.

### *Phase Changes*

The three common types of phase changes are:

*Vaporization/Condensation:* The reversible conversion of a liquid into a gas. The energy associated with this change is known as the ***Heat of Vaporization***, $\Delta H_v$.

*Freezing/Melting:* The reversible conversion of a liquid into a solid. The energy associated with this change is known as the ***Heat of Fusion***, $\Delta H_f$.

*Solidification/Vaporization:* The reversible conversion of a solid directly into a gas. The energy associated with this change is known as the ***Heat of Sublimation***, $\Delta H_s$.

*Phase Diagrams:* show the state of a matter of a substance as a function of temperature and pressure.

### *Gases*

*Kinetic Molecular Theory of Gases:* This theory makes the following assumptions:

1. Gases consist of large numbers of molecules in continuous motion. Also the velocities of these entities in motion is random.
2. The total volume of the gas molecules is negligible compared to the container volume.
3. Attractive and repulsive forces between molecules do not take place.
4. Although energy can be transferred between molecules, the average kinetic energy of the molecular system is constant as a function of time.
5. The average kinetic energy of the molecular system is proportional to the absolute temperature.

### *Ideal Gases*

Equations based upon the above assumptions.

*Boyle's Law:* That the volume of a gas is inversely proportional to the pressure at constant temperature.

*Charles's Law:* That the volume of a gas is inversely proportional to the temperature at constant pressure.

*Avogadro's/Gay Lussac's Law:* The volume of a gas is directly proportional the amount of gas present (number of moles) at constant temperature and pressure.

*Dalton's Law* teaches that the total pressure of a mixture of gases is equal to the sum of the partial pressures of the individual gases.

### *Solids*

The basic classes of solids include:

*Metals:* The properties of metals are defined by their structure which is a rigid framework of nuclei that are surrounded by a mobile sea of mobile electrons.

*Ionic Solids:* These are held together by the strong electrostatic attraction of cations (usually metallic) and anions.

*Covalent Solids:* These materials are held together by covalent bonds which can lead to a wide range of physical properties.

*Molecular Solids:* These are held together by the intermolecular forces described above. Because these types of bonds are relatively weak compared with covalently and ionically bonded materials.

# CHAPTER 6 EXERCISES

1. As the ice is heated it progressively changes phases from solid to liquid to gas. Note that this process involves physical changes, not chemical ones. From a molecular standpoint what is going on?
2. What molecular or structural variables determine what state a given compound type will be in at a given temperature?
3. Which class of bonding is typically stronger intramolecular (within a molecule) or intermolecular (between molecules)?
4. What types of intermolecular forces must be overcome to convert the following molecules from the liquid state to a gas?

   a. CO

   b. $PH_3$

   c. $CH_3OH$

   d. CCl4
5. Besides molecular bonding forces what is the other critical molecular proprety that determines the state of a substance?
6. Explain the difference in boiling points for the following pairs of substances:

   a. HF (20 °C) and HCl (−85 °C)

   b. $CHCl_3$ (61 °C) and $CHBr_3$ (150 °C)

c. $Br_2$ (59 °C) and ICl (97 °C)

d. $H_2O$ (100 °C), $H_2S$ (– 60 °C), $H_2Se$ (–41 °C), $H_2Te$ (–2.2 °C),

7. Explain why ammonia's ($NH_3$) melting point of –78 is so much higher than that of phosphine ($PH_3$) at –133. Based just upon molecular weight the opposite trend would be expected.
8. What types of intermolecular bonding forces can be expected for ethylene glycol and ethanol? Which molecule has stronger bonding strengths and why?
9. Give a molecular bonding explanation for why surface tension occurs at all.
10. What information does a phase diagram possess?
11. Draw an intramolecular bonding structure for two molecules of ethylene glycol, $HOCH_2CH_2CH_2OH$. For simplicity use line drawings.
12. The ideal gas law is most accurate at what pressures and temperatures? Why?
13. Determine the densities of the following gases at 1.00 atm and 298 K: $SO_2$, HBr, and $CO_2$. Explain the trend.
14. Calculate the density of sulfur hexafluoride at 1.1 atmospheres and 300 °C.
15. Calcium hydride, $CaH_2$, reacts with water to form hydrogen gas per the following reaction:

$$CaH_2(s) + 2H_2O(l) \longrightarrow Ca(OH)_2(aq) + 2H_2(g)$$

How many grams of $CaH_2$ are needed to generate 660 L of $H_2$ gas if the pressure of $H_2$ is 3300 torr at 21°C?

16. $SiH_4$ has a higher molecular weight than $CH_4$ and yet the average kinetic energies of equal quantities of these two compounds at the same temperature is the same. How can this be?
17. Air bags can be inflated via the decomposition of sodium azide, $NaN_3$.

$$2NaN_3 \longrightarrow 2\,Na + 3N_2$$

What mass of the azide would be required to inflate a bag to 25L at a pressure of 1.3 atm.?

18. A mixture containing 0.765 mol He(g), 0.330 mol Ne(g), and 0.110 mol Ar(g) is confined in a 10.00-L vessel at 25 °C (a) What is the partial pressure of each of the gases in the mixture.
19. In terms of molecular structures, describe the differences between a crystalline and an amorphous solid.
20. Chlorine is a commonly used disinfectant. Assume the volume of a particular sample of $Cl_2$ gas is 10 L at 996 torr and 27 °C.

    a. How many grams of $Cl_2$ are in the sample?

    b. What volume will the $Cl_2$ occupy at STP?

21. How does a metallic solid differ from a non-metallic solid?

# Credits

Fig. 6.1: Copyright © 2007 by Yupi666, (CC BY-SA 3.0) at https://commons.wikimedia.org/wiki/File:Solid_liquid_gas.jpg.

Fig. 6.3: Copyright © 2011 by Qwerter, (CC BY-SA 3.0) at https://commons.wikimedia.org/wiki/File:3D_model_hydrogen_bonds_in_water.svg.

Fig. 6.4: Source: http://www.vias.org/genchem/img/boilpts_hydrides.png.

Fig. 6.5: Copyright © 2008 by Pmarmottant, (CC BY-SA 3.0) at https://commons.wikimedia.org/wiki/File:DropletOnSurface.svg.

Fig. 6.7: Copyright © 2013 by Provenzano15, (CC BY-SA 3.0) at https://commons.wikimedia.org/wiki/File:Water_Phase_Changes.svg.

Fig. 6.9: Source: https://commons.wikimedia.org/wiki/File:Carbon_dioxide_pressure-temperature_phase_diagram.svg.

Fig. 6.10: Copyright © 2016 by cmglee, (CC BY-SA 3.0) at https://commons.wikimedia.org/wiki/File:Phase_diagram_of_water_simplified.svg.

Fig. 6.11: Adapted from: https://web.archive.org/web/20170425042045/http://hendrix2.uoregon.edu/~imamura/102/section3/chapter15.html.

Fig. 6.12: Copyright © 2010 by Arman Cagle, (CC BY-SA 3.0) at https://commons.wikimedia.org/wiki/File:Boyles_law_00.PNG.

Fig. 6.13: Copyright © 2013 by Reginaprice2013, (CC BY-SA 3.0) at https://commons.wikimedia.org/wiki/File:Charles%27_Law_Graph.png.

Fig. 6.14: Copyright © 2009 by Alchemist-hp, (CC BY-SA 3.0) at https://commons.wikimedia.org/wiki/File:Gold-crystals.jpg.

Fig. 6.15: Copyright © 2010 by Rob Lavinsky, (CC BY-SA 3.0) at https://commons.wikimedia.org/wiki/File:Allophane-329818.jpg.

# Solutions

When most people think of chemistry, they likely envision a chemist mixing to materials together, creating an explosion, a dramatic release of smoke or what appears to be a magical color change. While these observations can be made for some chemical reactions, most are not as dramatic. An important starting point for chemical reactions is the creation of a solution. In fact, the average person witnesses a form of chemistry known as solution chemistry when they enjoy their morning coffee, make their own salad dressing or sprinkling salt on a driveway before a snow storm. When these solutions are made, observations can be made. Sugar will dissolve in the hot coffee easily but the oil and water in the salad dressing will not mix very well. On the portion of the driveway where salt was added, no ice formed but on areas where salt was not added, ice can be observed. This chapter will review the chemistry involved in solution formation. Once the chemistry is reviewed, an explanation will be made for each of the observations discussed.

## 7.1 Energetics of Solution Formation

The most common type of solution formed is when a solid is mixed in a liquid, most likely water. In this example, water would be the ***solvent*** and the solid would be the ***solute***. The solvent is the component of the solution in the larger amount, which is often water, and the solute is the component in the lesser amount. The most basic observation one can make about solution formation is that sometimes when two compounds are mixed, they form one **homogenous** mixture. In other cases, however, when two compounds are mixed, they will not form a homogenous mixture (Figure 7.1 A and B) and, instead, they will remain separated. Sugar and salt both readily *dissolve* or mix with water while oil does not.

### The Enthalpy of Solution Formation

What determines whether a homogeneous mixture will form are the energetics of solution formation for the two combined compounds. Many factors play a role in solution formation. Enthalpy (H) and Entropy (S), discussed in Chapter 5, are two of these factors. Consider what must happen in order for a solution to form. First, intermolecular

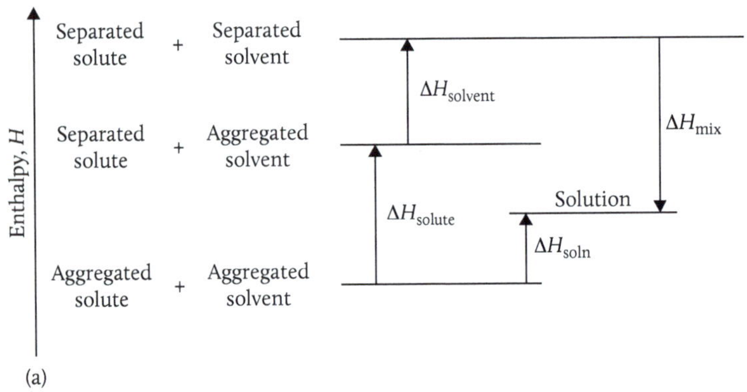


FIGURE 7.1A. Endothermic Solution Formation.

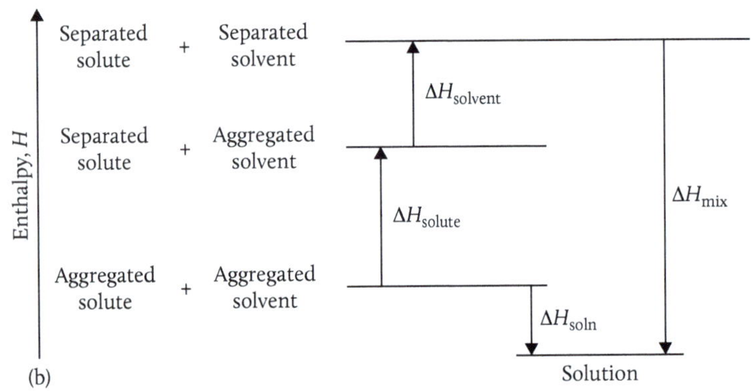


FIGURE 7.1B. Exothermic Solution Formation.

forces must be broken. There are intermolecular forces between the solute molecules as well as between the solvent molecules. These forces must be broken in order for a solution to form. Breaking these bonds will require energy. Some of this energy will be recovered when new interactions are formed between the solute and solvent molecules (Figure 7.1 A and B). As depicted in Figure 7.1 A and B, there will be enthalpy changes for the separation and formation of the intermolecular forces. The separation of the solute-solute and solvent-solvent interactions are endothermic processes while the formation of interactions between solvent and solute molecules is an exothermic process.

You may have noticed that when some solutions are made they will release heat to the surroundings but other solutions will absorb heat from the surroundings. The overall enthalpy change for solution formation can be calculated as follows:

$$\Delta H_{solution} = \Delta H_{mix} + \Delta H_{solvent} + \Delta H_{solute}$$

If the overall enthalpy change of the solution is positive, the solution formation is an endothermic process, absorbing heat from the surroundings and feeling cold as it forms. If the overall enthalpy change is negative, energy is released to the surroundings and the solution will feel warm to the touch as it forms. In the case of instant cold packs, often used by sports teams, when the ammonium nitrate present in the pack mixes with water it absorbs heat from the surroundings creating its cooling effect.

## Spontaneity

You may think it is impossible for solution formation to be endothermic. How can something have a positive enthalpy change and form a solution? The answer is simple—entropy! So long as the change in entropy is large enough, it can compensate for the positive change in enthalpy and make endothermic solution formation favorable.

Chemists use the term ***spontaneous*** to describe a process that occurs without outside intervention. For example, solutions are spontaneous if they will form (given enough time) without adding heat, pressure, etc. to force the solution to form. In the case of solution formation, a solution process that is exothermic will be spontaneous because of the negative enthalpy for the solution. It is possible for endothermic processes to be spontaneous if a favorable disorder, entropy, accompanies the solution formation. In the case of ammonium nitrate found in ice packs, its ions are in a very ordered arrangement in the solid state. Upon mixing with water, the ions are now allowed more freedom to move and are thus more disordered making the solution formation spontaneous and endothermic.

# 7.2 Solubility

***Solubility*** is the ability of a solute to dissolve in a given substance like water. It is best thought of as the concentration of a solute that dissolves in a given substance. Chemists will often describe a solute as being "soluble" in a solvent when a large amount of the solute is able to dissolve and insoluble when a miniscule amount of the solute can dissolve in the solvent.

There is usually a limit to the amount a solute can dissolve in a solvent. Think of adding sugar to water. If you add a small amount, let's say a pinch, it will readily dissolve in the water. If you were to take that same amount of water and added a heaping tablespoon of sugar, some of it would dissolve but some would also fall to the bottom of the container and never dissolve. The most accurate definition of solubility is the concentration at which the rate of dissolution of the solid is in equilibrium with the rate of crystallization of the solid.

***Solute + Solvent ⇄ Solution***

When a solution is formed, the amount of solute added to the solution can create an unsaturated, saturated or a supersaturated solution. If the amount of solute added to the solvent is less than the solubility the solution is ***unsaturated.*** A solution is ***saturated*** when the concentration of the solute is equal to the solubility.

A ***supersaturated*** solution is a solution that has solute concentration greater than the solubility. For most solids, solubility increases at increasing temperature (Figure 7.2). This can be explained by considering intermolecular forces. At a higher temperature, a greater percentage of molecules will be capable of overcoming intermolecular forces and separating to form solvent-solute interactions. This means more of the solute particles can dissolve in the solvent. Exploiting this property of solids, a solution can be made at an elevated temperature and slowly cooled. The cooled solution will then have more solid dissolved than the solubility at this lower temperature, forming a supersaturated solution. Such a solution is unstable and simply agitating the solution or adding a single crystal of the solid to the solution can cause any excess solid dissolved above the saturation point to precipitate out of solution.

FIGURE 7.2. Supersaturated Solution.

# 7.3 Solids

## Ionic Solids

You will recall from previous discussions, ionic compounds are formed when negatively charged anions are attracted to positively charged cations through electrostatic interactions. When ionic compounds dissolve in water, the electrostatic interactions between the ions are separated and new interactions occur between the water and the

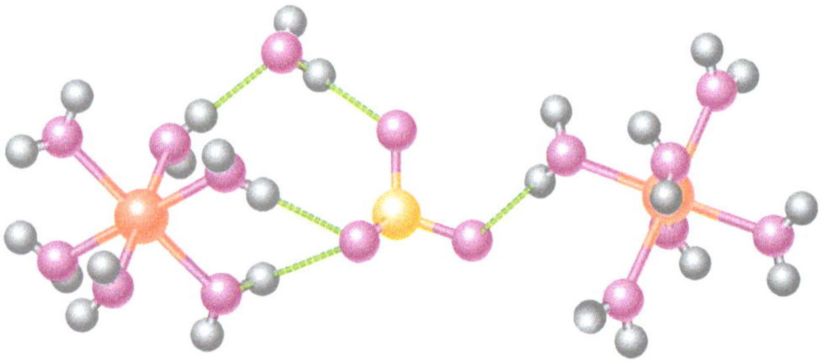

FIGURE 7.3. Waters of Hydration.

ions. These interactions are known as ion-dipole interactions.

The water molecules surround or *hydrate* each ion (Figure 7.3). The positive end of the dipole for the water molecules will be oriented towards the negatively charged anions. The negative end of the dipole will be oriented towards the positively charged cations. The number of water molecules surrounding each ion depends of the charge of the ion and the size of the ion. As discussed earlier, the ions from the ionic compound will now have a greater freedom of movement so the disorder increases. Conversely, the solvent molecules can experience a decrease or increase in disorder depending on the ions dissolved in solution. Overall, the combination of enthalpy changes and entropy changes dictates whether the solution process is spontaneous and the extent to which different ionic compounds dissolve in water.

Due to the many factors involved in dissolving ionic compounds, it is difficult to predict which ionic compounds are soluble in water and which are not. This is why the solubility table was provided in Chapter 4. Some factors affecting solubility of ionic compounds include the charge of the ion and its size. In terms of charge, a charge that is large, for example, +4, will require more energy to separate it from its anions in the solid. It will also take more water molecules to hydrate and shield the large positive charge. These two factors make it more difficult for large charges to dissolve in water. First, it will require more energy to separate their intermolecular forces and second the solution will "run out" of water molecules faster than when ions with lesser charge mix with water. Large ions are also at risk of running out of molecules during mixing as their size will require more water molecules to hydrate than smaller ions. While not perfect at predicting solubility of ionic compounds, in most cases large sized ions and/or ions with a large charge will be less likely to dissolve in water.

## Molecular Solids

Similar to what occurs with ionic compounds, when molecular solids mix with water, the interactions within the solid must be separated and new ones formed with water molecules. Nonpolar molecular compounds interact with each other via London-dispersion forces. Depending on their elemental composition and three-dimensional arrangement, polar compounds can form dipole-dipole interactions and hydrogen bonds in addition to London-dispersion forces.

When it comes to making solutions with molecular compounds, chemists use the phrase "like dissolves like". This refers to the fact that nonpolar compounds will form solutions with other nonpolar compounds. Similarly, polar compounds will form solutions with other polar compounds. This is best explained by considering what must happen once the solvent-solvent and solute-solute interactions are separated. All of the molecules in the solution much form new interactions. The new interactions that form must compensate for those that were broken. This is most likely when the new interactions that form are similar to those that were previously broken. Keeping this in mind the optimal interactions to form are dipole-dipole interactions with polar solutes which will occur with polar solvents but not with nonpolar solvents. Furthermore, compounds capable of hydrogen bonding would mix best with solvents also capable of hydrogen bonding. Nonpolar solutes will best mix with nonpolar solvents as London dispersion forces will best form between them.

## 7.4 Concentration Units

An important descriptor for solutions is their ***concentration.*** This value expresses the amount of a solute dissolved in a solvent. There are many different ways to express concentration and thus many different units that can be used. In general, a unit is chosen that best suits the application for the concentration of a solution. In this section, several different methods for giving a concentration value will be described.

### Molarity (M)

Molarity is the most common unit a chemist will use to express a concentration for a solution. As the name implies, it uses the mole in the expression for concentration. Molarity is defined as the number of moles of a solute dissolved in 1 liter of *solution.* The unit used for molarity is **M**. Keep in mind that the volume of the solution will be the combined volume of the solute and the solvent.

$$\textbf{Molarity} = \frac{moles\ of\ solute}{Liter\ of\ solution}$$

### Molality (m)

As you will recall from the chapter on gases, as the temperature changes, volume changes. For this reason, a unit related to molarity is often used in cases where the calculation may involve a temperature change. With molarity, if the temperature changes, than the expression for molarity would no longer be valid as the volume of the solution would be different. To address this issue, the expression for molality involves a mass unit instead of volume. Molality is defined as the number of moles of a solute dissolved in 1 kilogram of *solvent.* The unit becomes $\frac{mol}{kg}$ or ***m.*** You should be very careful not to confuse molality and molarity. While these to expressions sound similar they are different. Also recall that the calculation for molarity involves the volume of the solution whereas the molality involves the weight of just the solvent.

$$\textbf{Molality} = \frac{\boldsymbol{moles\ of\ solute}}{\boldsymbol{kg\ of\ solvent}}$$

### Mole Fraction ($\chi$)

The mole faction is a calculation. The name here implies how this can be calculated. The mole fraction of a solution is the moles of a given component in a solution divided by the total number of moles of all of the components present. For example, the mole fraction of component A would be calculated as follows:

$$\chi_A = \frac{\boldsymbol{moles\ of\ A}}{\boldsymbol{moles\ of\ A + moles\ of\ B + moles\ of\ C \ldots}}$$

This calculation can be used in a similar manner to express the mole fraction for any of the components of a solution. A more generic version of the expression is offered below. This expression is useful for gases, particularly when dealing with partial pressure calculations.

$$\chi_{\text{component}} = \frac{\textit{moles of component}}{\textit{total moles of solution}}$$

## Mass Percent (%)

Another expression for the concentration of a solution is its mass percent composition. This is another easy calculation as the approach is the same as any type of percentage calculation. In this case the mass of the solute is compared to the total mass of the solution. It ultimately gives a value of the solute per 100 grams of solution.

$$\textit{Mass percent of solute} = \frac{\textit{grams of solute}}{\textit{total grams of solution}} \times 100\%$$

It is worth noting that this calculation can be made using any mass unit so long as the unit used for the solute mass is the same unit used for the total mass of solution. Chemists do not often use this unit, however it is often used in pharmacies and many products sold to the average consumer. A couple of examples include antifreeze, hydrogen peroxide and vinegar.

## Parts Per Million, Parts Per Billion and Parts Per Trillion

Closely related to the mass percent calculation is the calculations for parts per million (ppm), parts per billion (ppb) and parts per trillion (ppt). In each case, the mass fraction portion of the calculation is the same but the multiplier changes. For mass percent, the multiplier is 100 (or parts per 100). For ppm the multiplier is 1 million ($10^6$), for ppb its 1 billion ($10^9$) and so on. These expressions are used when the levels of solute in a solution is very small. These units are very common in environmental applications.

$$\textit{ppm} = \frac{\textit{grams of solute}}{\textit{total grams of solution}} \times 10^6$$

$$\textit{ppb} = \frac{\textit{grams of solute}}{\textit{total grams of solution}} \times 10^9$$

$$\textit{ppt} = \frac{\textit{grams of solute}}{\textit{total grams of solution}} \times 10^{12}$$

# 7.5 Colligative Properties

Once a solution is formed, the properties of both the solute and the solvent are altered. One set of properties, known as ***colligative properties,*** are those properties that change based on the amount of solute particles present in solution and not the chemical identity of the solute. In this section, we will explore different colligative properties. It is important to keep in mind that anion and cation components of ionic compounds separate upon dissolution in water. *This is important because it impacts the concentration of the particles of solute in the solution.* For example, for every on formula unit of sodium chloride that dissolves in water, 2 ions dissociate into solution. This will become important later, when we discuss calculations related to the colligative properties.

## Vapor Pressure Lowering

The vapor pressure above a solvent is altered once a solution is formed. You should recall that vapor pressure is the pressure above a solvent when the transition between the liquid and its vapor (gas phase) is in equilibrium. The vapor pressure above a solution made with a nonvolatile solute (one with a very low vapor pressure itself) will be lower than that of a pure solvent. How can this be explained? There are a couple of factors that come into play. The first is that the surface of the solution will have fewer solvent molecules than the surface of the pure solvent in the same container (Figure 7.4). This is because some of the solvent molecules that were at the surface of the pure solvent are now replaced by solute molecules. The next factor to explain this is the new intermolecular forces between the solute and the solvent. In order to transition to the gas phase, solvent molecules would have to break intermolecular forces holding the solute in solution. This would not be very favorable and thus makes forming a vapor more difficult. ***Raoult's Law*** is used to calculate the effect a solute has on the vapor pressure of a solution. In the expression, the vapor pressure of the solvent above the solution ($P_{solv}$) is equal to the vapor pressure of the pure solvent ($P°_{solv}$) multiplied by the mole fraction of the solvent in the solution ($\chi_{solv}$).

$$P_{solv} = (\chi_{solv})(P^{o}_{solv})$$

In the case of mixtures involving both a volute solvent and a volatile solute, the vapor pressure above the solution will be equal to the vapor pressure of the solvent combined with the vapor pressure of the solute. The individual vapor pressures can be determined using Raoult's Law then added together.

$$P_{solute} = \chi_{solute} \times P^{o}_{solute}$$

$$P_{solvent} = \chi_{solvent} \times P^{o}_{solvent}$$

$$P_{solution} = \chi_{solvent} \times P^{o}_{solute}$$

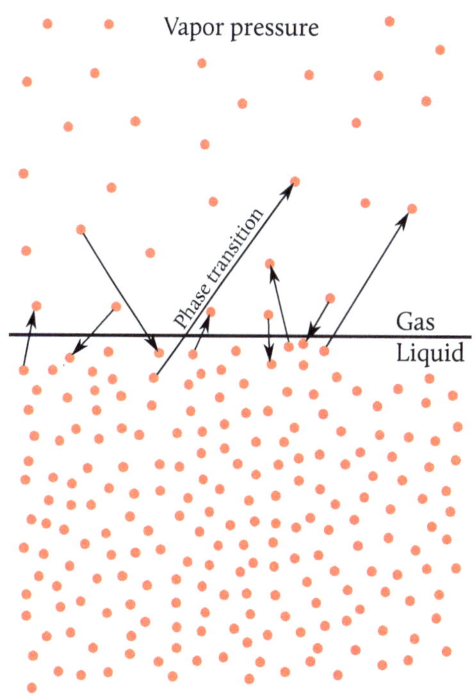

FIGURE 7.4. Vapor Pressure Lowering.

## Boiling-Point Elevation

Boiling-point elevation is the result of the change in the vapor pressure of the solution compared to the pure solvent (Figure 7.5). As discussed earlier, the boiling point of a substance is the temperature at which the vapor pressure of the substance is equal to the atmospheric pressure. We just learned that solutions have a lower vapor pressure than the pure solvent. This being the case, for a solution at the same atmospheric pressure as the pure solvent, the solution will require more heat for the vapor pressure of the solution to reach the atmospheric pressure. Thus a solution will boil at a higher temperature than the pure solvent. Often cooks will add

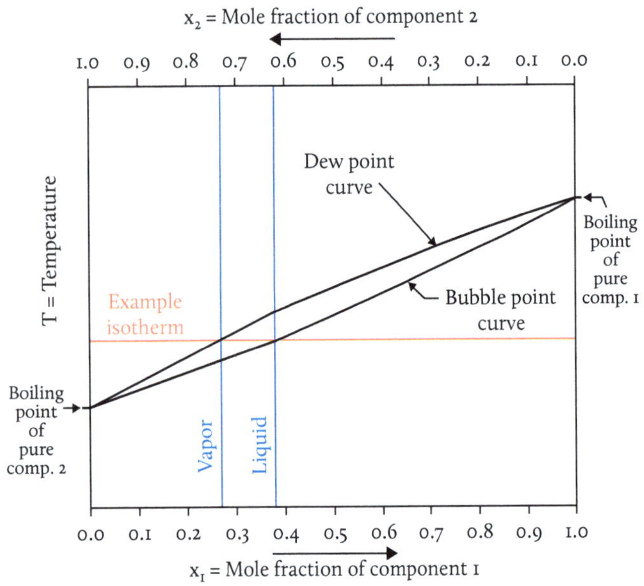


**FIGURE 7.5.** Boiling Point Elevation of a Solution.

salt to water. Many wrongly believe this will result in the water "boiling faster". The reality is that the new solution will boil at a high temperature. As a result food will cook faster in the water solution than pure water because it is being cooked at a higher temperature.

As with the lowering of the vapor pressure, the extent to which the boiling point is elevated depends upon the concentration of the solute in the solution. To determine the change in the boiling point ($\Delta T_b$), the following equation can be used:

$$\Delta T_b = mK_b$$

In this equation, $_m$ is the molality of the solution and $K_b$ is the boiling point elevation constant. The $K_b$ value is a property of a given solvent at particular pressure. The change in the boiling point of the solution must be added to the boiling point of the pure solvent to determine the elevated boiling point of the solution. Table 7.1 provides the constant for several solutions at 1 atm along with the boiling points for the pure solvents.

TABLE 7.1 **Colligative Property Constants for Some Solvents**

| Solvent | Boiling Point (°C) | Kb (°C/m) | Freezing Point (°C) | Kf (°C/m) |
|---|---|---|---|---|
| acetic acid | 117.90 | 3.22 | 16.64 | 3.63 |
| benzene | 80.09 | 2.64 | 5.49 | 5.07 |
| carbon disulfide | 46.2 | 2.42 | −112.1 | 3.74 |
| carbon tetrachloride | 76.8 | 5.26 | −22.62 | 31.4 |
| chloroform | 61.17 | 3.80 | −63.41 | 4.60 |
| nitrobenzene | 210.8 | 5.24 | 5.70 | 6.87 |
| **water** | **100.00** | **0.51** | **0.00** | **1.86** |

## Freezing-Point Depression

Another result of the lowered vapor pressure of a solution is the lowering of the freezing point temperature. This particular colligative property is exploited during winter months when salt is applied to roadways and walkways to prevent the formation of ice as temperatures drop below freezing. Due to the change in vapor pressure of solutions compared to pure solvents, the triple point of solutions changes which then impacts the freezing point of the solution.

Thinking about how molecules in solution behave versus a pure solvent can help explain this change as well. Seeing as the solution now has solute particles mixed in with the solvent molecules, it will be harder for the solvent molecules to optimize their interactions in the process of forming the solid phase. Given this, it is understandable that the freezing point of a solution will be lower. The lowered freezing point of a solution is also dependent on the concentration of the solution. The following equation can be used to calculate the change freezing point of the solution ($\Delta T_f$).

$$\Delta T_f = mK_f$$

In this equation, *m* is the molality of the solution and $K_f$ is the freezing point depression constant. As with $K_b$, the $K_f$ value is a property of a given solvent at particular pressure. The change in the freezing point of the solution must be subtracted from the freezing point of the pure solvent to determine the elevated boiling point of the solution.

## Osmotic Pressure

The last colligative property to discuss is osmotic pressure. Semipermeable membranes are thin layers of a specialized material that only allows small molecules to pass across. You can think of these as acting as very specific filters. In general, water is able to pass across a semipermeable membrane but solutes, and other molecules cannot. Such membranes can be used to purify water. Bladders, leaves and contact lenses are all examples of semipermeable membranes. The passive diffusion of a fluid, like water, across a membrane is known as ***osmosis.*** This process is critical to the function of many cells and organisms.

When two solutions in water of with different solute concentrations are on either side of a semipermeable membrane, the net movement of water will be to the side with the higher solute concentration. As water moves across the membrane, the solution with the higher solute concentration will be diluted. Water will continue to flow until the concentration on both sides is equal.

It is pertinent to define different solution types as they pertain to osmosis and two solutions on either side of a semi-permeable membrane. ***Isotonic*** solutions are two solutions where the solute concentration for the two solutions are equal. In such a case, there will be not new flow of water across the membrane. A solution is said to be ***hypotonic*** when its solute concentration is lower than the solute concentration of the solution on the other side of the membrane. Water will flow from the hypotonic solution to the other side of the membrane. A solution is considered hypertonic when its solute concentration is higher than the solute concentration of the solution on the opposite side of the membrane. Water will flow into the ***hypertonic*** solution.

Osmotic pressure is the pressure required to stop the flow of solvent molecules across a membrane separating a solution from the pure solvent (Figure 7.6). As with the other colligative properties, this property is dependent on the concentration of solute particles in the solution. It can be calculated using a variant of the ideal-gas law (**PV** = ***nRT***):

$$P = \frac{n}{V}RT$$

or

| Osmotic Pressure Equation: $\pi = MRT$ |
|---|

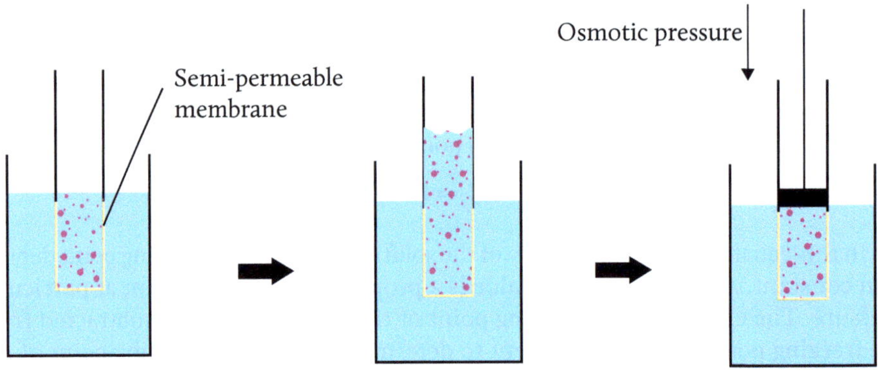


**FIGURE 7.6.** Osmotic Pressure.

In a somewhat odd twist the pi symbol, $\pi$, is used to denote osmostic pressure. This has nothing to do with the symbol used to represent the common constant number 3.14. Pressure (P) in this case is the osmotic pressure ($\pi$), the concentration used will be molarity where $M = \frac{moles}{liter}$ or in this case $\frac{n}{V}$. R is the ideal gas constant ($0.0821 \frac{L \cdot atm}{mol \cdot K}$) and T is the temperature in Kelvin.

## The van't Hoff Factor and Electrolyte Solutions

To this point you should notice that in each of the calculations related to colligative properties the only factor that changed the particular property was the concentration of solute. The colligative properties of the solution differ from the pure solvent based on a factor related to this concentration.

Special consideration needs to be taken when considering the concentration of electrolyte solutions. Why? As you will recall, when electrolytes, like salts acids and bases, dissolve in solution, ions are separated. This is different from molecules dissolving in solution as molecules remain as intact, single units. The reason this matters is that for every undissolved electrolyte molecule or formula unit, multiple ions will be dispersed into solution once the electrolyte dissolves. Here are some dissociations for electrolytes in water:

$$NaCl(s) \longrightarrow Na^{+}(aq) + Cl^{-}(aq)$$
$$HCl(l) \longrightarrow H^{+}(aq) + Cl^{-}(aq)$$
$$CaCl_2(s) \longrightarrow Ca^{2+}(aq) + 2Cl^{-}(aq)$$

For both sodium chloride and hydrochloric acid two ions are formed upon dissolution in water. For calcium chloride, 3 ions form upon dissolution—one calcium ion and two chloride ions. This means for each of these substances, the total concentration of particles in solution is higher than what would be calculated for the intact compound because they dissociate. Seeing as colligative properties are dependent on the total concentration of particles in solution, these dissociation events must be considered when determining changes in colligative properties of a solution.

The ***van't Hoff factor (symbol i)*** accounts for these dissociation events and can be added to the equations for the different colligative properties. The value for the colligative property will then be properly calculated and reflective of the effect of the true concentration of particles in solution.

$$\Delta T_f = iK_f m$$

$$\Delta T_b = iK_b m$$

$$\pi = iMRT$$

As ions freely move about in solution, every now and then they will be attracted to each other. The frequency with which these attraction events occur will increase as the concentration of the solute increases because there are more of the ions in solution. The van't Hoff factor accounts for difference in the measured value for a colligative property versus the calculated value. Table 7.2 lists the van't Hoff factors for different substances at varying concentrations.

TABLE 7.2 van't Hoff Factors (i)

| Compound | i | Comments |
|---|---|---|
| nothing | 0 | Pure solvent |
| NaCl (table, sea salt) | 2 | 2 ions |
| $CaCl_2$ | 3 | 3 ions |
| $K_3PO_4$ | 4 | $PO_4^{-3}$ is one ion |
| $CH_3OH$ | 1 | Non-dissociated |
| sugar | 1 | Non-dissociated |
| $(NH_4)_2SO_4$ | 3 | 2 $NH_4^+$; 1 $SO_4^{-2}$ |

Below is an equation that can be used to determine the van't Hoff factor for a solution.

$$i = \frac{\textit{moles of particles in solution}}{\textit{moles of formula units dissolved}}$$

The van't Hoff factor can be determined experimentally by measuring a colligative property in the laboratory and then comparing the value obtained to the value calculated. The calculated value should reflect full dissociation of the electrolyte.

$$i = \frac{\textit{measured colligatice property value}}{\textit{calculated colligative property value}}$$

# Chapter 7 Summary

A *solution* is mixture of two or more substances that do not chemically react upon mixing.

The *solvent* is the component of a solution present in the greater amount.

The *solute* in a solution is the component present in the lesser amount.

A *saturated solution* is a solution where the solute is present at a level equal to its solubility in the solvent.

An ***unsaturated solution*** is a solution where the solute is present at an amount below the solute's solubility in the solvent.

A ***supersaturated solution*** is a solution that has solute dissolve in the solvent at an amount above the solubility for that solute. This is an unstable solution and will crystalize if the solution is disrupted by motion or introduction of a small additional amount of solute.

***Concentration*** is a measure of the amount of solute present in a solvent.

***Molarity (M)*** measures concentration in terms of moles of solute per liter of solution.

***Molality (m)*** measures concentration in terms of moles of solute per kg of solvent.

***Mole fraction (χ)*** measures concentration in terms of moles of any component in solution per total moles of all solution components.

***Mass percent*** measures concentration in terms of grams of solute per total grams of solution multiplied by 100.

***Parts per million (ppm)*** measures concentration in terms of grams of solute per total grams of solution multiplied by ($10^6$).

***Parts per billion (ppb)*** measures concentration in terms of grams of solute per total grams of solution multiplied by ($10^9$).

***Parts per million (ppm)*** measures concentration in terms of grams of solute per total grams of solution multiplied by ($10^6$).

***Parts per trillion (ppt)*** measures concentration in terms of grams of solute per total grams of solution multiplied by ($10^{12}$).

***Colligative properties*** are those properties of solutions that change based on the amount of solute particles present in solution and not the chemical identity of the solute. The impact of these properties is based on the total number of molecules (for molecular compounds) or ions (for ionic compounds) present in the solution.

***Vapor pressure lowering*** is a colligative property whereby the vapor pressure of a solution is lower than the vapor pressure of the pure solvent.

***Boiling point elevation*** is a colligative property whereby the boiling point of the solution is higher than that of the pure solvent.

***Freezing point depression*** is a colligative property whereby the freezing point of the solution is lower than that of the pure solvent.

***Osmosis*** is a process of passive diffusion where water molecules move across a semi-permeable membrane. Water flows from a location of low solute concentration to the side with high solute concentration. The bigger the concentration difference the greater the flow of water.

***Osmotic pressure*** is the amount of pressure to stop the flow of water across the membrane (i.e. stop osmosis). This colligative property increases as the solute concentration increases.

***Isotonic*** solutions are two solutions where the solute concentration for the two solutions are equal. In such a case, there will be not new flow of water across the membrane.

A solution is said to be ***hypotonic*** when its solute concentration is lower than the solute concentration of the solution on the other side of the membrane.

A solution is considered ***hypertonic*** when its solute concentration is higher than the solute concentration of the solution on the opposite side of the membrane.

The ***van't Hoff factor (symbol i)*** accounts for the extent to which ionic compounds dissociate in solution and must be accounted for as it relates to colligative properties.

# CHAPTER 7 EXERCISES

1. Explain how a solution can form spontaneously even though the process is endothermic.
2. Explain the energetics involved in the formation of a solution. Give an equation for this process.
3. Draw an energy diagram showing the formation of a solution that is an exothermic process.
4. What does it mean if the formation of a solution to be a "spontaneous" process?
5. Define unsaturated, supersaturated and saturated solutions.
6. Explain to how you would go about preparing a supersaturated solution.
7. A sample of $KNO_3$ (60.0 g) is dissolved in 100 g of water at 100°C, with precautions taken to avoid evaporation of any water. The solution is cooled to 30.0°C and no precipitate is observed (solubility of $KNO_3$ at this temperature is 45 g).

   a. What type of solution is formed after cooling to 30.0°C with no precipitate?

   b. If a single crystal is added to the 30.0°C solution, how much precipitate would you expect to form? Explain your reasoning.
8. What is the ***percent by mass*** of a solution made by dissolving 0.906 mol of glucose (180 g/mol) in 0.755 kg of water?
9. What is the concentration in ppb of Polychlorobenzenes (PCB) in a chemical spill, if there is 0.060 mg in 4,600 Kg of soil?
10. What is the concentration (M) of a NaCl solution prepared by dissolving 9.3 g of NaCl in sufficient water to give 350 mL of solution?
11. What are the respective concentrations (M) of $K^+$ and $CO_3^{2-}$ afforded by dissolving 0.530 mol $K_2CO_3$ in water and diluting to 1.50 L?
12. A solution is made by mixing 12.8 g of $NH_4NO_3$ and 0.567 mol of $K_3PO_4$ in a total volume of 575 mL.

   a. What is the concentration (M) of $NH_4NO_3$?

   b. What is the concentration (M) of $K_3PO_4$?

   c. What is the ***total ion*** (as in ALL of the ions) concentration (M) in the solution?

13. There are 25 mg of lead in a 175 mL total sample volume. What is this concentration in ppb? (Assume a density of 1 g/mL)

14. A stream contains 0.010 μg of dioxin per 1.0 mL sample (assume a density of 1 g/mL).

    a. What is this concentration in % mass?

    b. What is this concentration in ppm?

15. A solution is prepared by dissolving 4.8 grams of NaCl and 7.2 grams of KBr in 76 g of $H_2O$.

    a. Calculate the mole fraction of KBr in the solution.

    b. Calculate the molality of NaCl in solution.

16. Convert 0.006% to ppm

17. Calculate the mass % and molality of $Na_2SO_4$ in a solution containing 10.6 g $Na_2SO_4$ in 483 g of water.

18. 10.0 g of glucose ($C_6H_{12}O_6$) is added to 1 L of water (density = 1.0 g/ml). What would the boiling point and freezing point of the solution be? ($K_b$ = 0.51 °C/*m*, $K_f$ = 1.86 °C/*m*)

19. Calculate the vapor pressure at 25 °C of a solution made by adding 50.0 mL of glycerin ($C_3H_8O_3$, non-electrolyte, 1.26 g/mL @ 25°C) to 500.0 mL of water. Vapor pressure of pure water is 23.8 torr @ 25 °C, and its density is 1.00 g/mL.

20. Describe what will happen to a red blood cell if it is placed in each of the following solutions. Will the cell stay the same size, swell and/or burst, or will the cell shrink. Explain your reasoning.

    a. Hypotonic

    b. Isotonic

    c. Hypertonic

# Credits

# Kinetics and Equilibria

This chapter merges two closely related topics that are usually discussed separately in general chemistry textbooks:

***Reaction Rates:*** How fast or slow a reaction proceeds; that is, how does the concentration of reactants and/or products change as a function of time?

***Equilibria:*** Within the context of reversible chemical reactions this means the state where the concentrations of reactants and products are unchanging. However, this does not mean that reactions have ceased altogether, rather just that the rates of the forward and reverse reactions are equal.

## Part A: Kinetics—Chemical Reaction Rates

How fast or slow a chemical reaction occurs is usually important, often critically so. We present here an overview of this topic with some relevant examples of reaction rates on both ends of the time spectrum.

### 8.1A A Typical Chemical Reaction Profile

Clearly a chemical reaction is a time driven phenomena regardless of how fast or slow it may be. Figure 8.1A and 8.1B show generic profiles of two chemical reactions, one exothermic and one endothermic. The reaction is a simple displacement of A by C; basically a breaking of the bond between A and B and a new bond forming between B and C with A now physically removed.

There is a lot going on here, and although the reaction shown shows the same chemistry, C displacing A, the energetics are very different. Despite the similarities in these two profiles note that the exothermic reaction will yield a negative heat of reaction ($-\Delta H$), while the endothermic reaction is a $+\Delta H$. Both reaction sequences go through a what is well named as **the transition state**, basically the chemistry involved, a transition from reactants AB + C to A + BC. This requires the addition of an energy, the **activation energy, Ea**, the magnitude of which is characteristic of the specific reaction being studied. This magnitude of this energy has a significant impact on reaction rates.

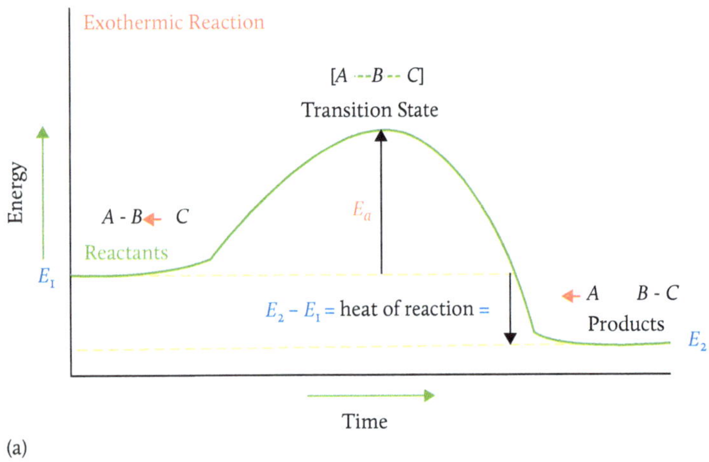


**FIGURE 8.1A.** Exothermic and Endothermic Reaction Energy Profiles.

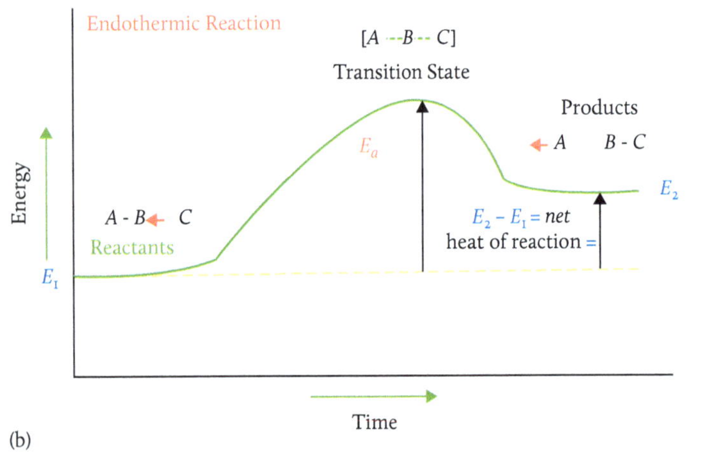


**FIGURE 8.1B.** Endo.

In general transition states are short-lived phenomena that involve unstable intermediates. Remember too that the above discussion ignores the entropy change, and that in the final analysis it is ΔG that determines spontaneity.

## 8.2A Factors Affecting Reaction Kinetics

It should be apparent that for a reaction to occur the reactants must get together. This is often not a trivial process. By way of example most batteries are heterogeneous constructs wherein the actual reactions occur on the surfaces of the solid state electrodes. This process has a lot of prerequisites to occur. Ions must travel through the liquid electrolyte, adsorb onto the surface or absorb into the interior, reaction occurs and electrons are transferred from hither to yon to complete a circuit. All of this has to occur a in a timely manner that is not too fast or slow depending upon the application. Other factors that affect the reaction are:

***Temperature*** is acritical variable in almost any chemical process. In general reactions proceed faster at higher temperatures, but there are upper and lower limits too. Generally speaking, increasing temperature increases the rate at which molecules move and at higher temperatures more energy is present to to break bonds all of which increase reaction rates. For example enzymes tend to denature (unravel and deactivate) at modest temperatures. Batteries tend to perform poorly below −20°C and above 60°C. On the other hand the synthesis of ammonia runs at an optimal temperature of 450°C.

***The higher the concentration of the reactants***, the more likely the are to molecular collide with one another. Increasing the number of collisions will lead to an increased rate of reaction.

***Pressure*** is subset of the concentration category. It is an important variable in gas phase reactions, but not solids and liquids. Increasing the pressure of a system increases the concentration of the gases which, in turn, increases the likelihood of collisions and increases the rate of reaction. As discussed in Chapter 6 entropy is a driving force in the gas phase.

***Reactant orientation*** is also very important to reaction rates.For a reaction to occur the reactants to not only collide with each other, but also be oriented correctly in 3D space. For example consider the two possible collision geometries of Figure 8.2 The reaction is a simple displacement of a bromide anion from sulfur dibromide by a chloride anion:

In the top half image the chloride ($Cl^-$) approaches to the electron-rich bromine atom and the electronic clouds make the reactants repel each other (dashed lines). In the bottom image the chloride approaches to the electron poor, positively charged sulfur, and reaction proceeds with $Br^-$ as the leaving group.

***Catalysts*** can help speed up a reaction. Section 8.4A discusses catalysts in detail and they can be very simple or very complex. They are not consumed in the reaction they simply help the reaction proceed as it would if given more time.

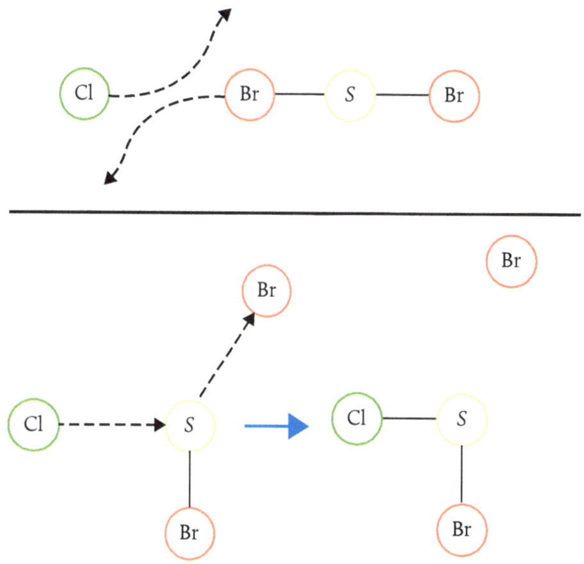


FIGURE 8.2. Reactant Orientation Effect on Kinetics.

## 8.3A Rate Law Expressions—Reaction Order

Let's look at a simple decomposition reaction, the decomposition of hydrogen peroxide ($H_2O_2$) at say 25°C. This is the reaction that slowly occurs in the disinfectant solutions one buys in the drug store.

**8.1A** $$2H_2O_2 \rightleftharpoons 2H_2O + O_2$$

We are ignoring here the state of matter of the reactant and products as well as the energetics involved, but it is important that the equation be balanced. Also in practice we must define the temperature of the reaction. So, let's say 100°C. As a generalization the rates of chemical reactions double with each 10°C increase in temperature. The decomposition rate of $H_2O_2$ at 100°C is modest, but the concentration does change as a function of time as the decomposition takes place. The rate of decomposition also changes as a function of time as the reactant concentration declines. This rate, in this case measured as a change in concentration as a function of time, can be represented mathematically as follows:

$$\text{Rate} = k[H_2O_2]^a$$

This equation is known as the **rate-law expression** where **k** is the **rate law constant**. Both k and the rate are specific to the reaction being studied, and are determined experimentally. The concentrations are usually, but not always, measured in mol/L (Molarity, M).

The mysterious superscript "a" in the represents what is known as the **order** of the reaction, that is, it defines mathematically how the $[H_2O_2]$ changes as a function of time. *"a" may or may not relate to the coefficients in the reaction equation, and is determined expermentally. Also "a" does not have to be a whole number.*

There are really two reaction orders to discuss. One is the order in each reactant, while the other is the total order for the reaction which is a sum of the orders of the reactants. Let's look at a couple of examples.

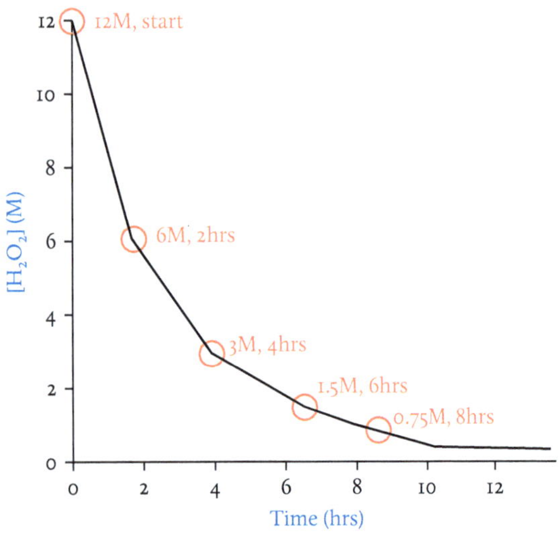


**FIGURE 8.3.** First Order Decomposition of Hydrogen Peroxide.

Figure 8.3 shows an example of a **first order** decomposition plot for $H_2O_2$.

The ordinate contains the concentration of hydrogen peroxide in mol/L (M) as a function of time in hours which is shown by the abscissa. The numbers are arbitrary, but demonstrate a first order reaction behavior. Here the concentration of $H_2O_2$ starts at 12M and after 2 hours is cut in half to 6M. Upon passage of another two hours the concentration is cut in half again to 3M, and so forth. This type of rate behavior is typical for a first order reaction. The time involved in each step is the same and is known as the **half-life (t1/2)**, the time it takes or half of the remaining material to react.

**Second order** reactions are also well known, but not as common as first order. Here are a couple of examples. Remember that in all cases the rate order must be determined experimentally.

**8.2 A3NO $\longrightarrow$ $N_2O$ + $NO_2$ rate = k[NO]2**

This is a reaction type we have not seen before, called a **disproportionation.** A disproportionation is a reaction where the same starting material results in the formation of more than one product.

Here the experimentally measured exponent of the reactant's concentration dependence is 2, thus the reaction order is also 2. Note that the coefficient of NO in the balanced equation is 3, however.

Here is a reaction with three reactants:

**8.3A** $H_2O_2$ + $2H^+$ + $3I^-$ $\longrightarrow$ 2H2O + I3$^-$

This reaction is the basis of the iodine clock, a widely used magic trick in chemistry. There are three reactants here, but it turns out the rate of the reaction is only dependent upon the concentration of two of these. Here's the rate equation:

**Rate = $k[H_2O_2][I^-]$**

Thus the reaction is first order in $H_2O_2$, first order in $I^-$ and zero order in $H^+$. The total order of the reaction is then 1 + 1 = 2.

**Zero order** reactions are not concentration dependent at all. Third and higher order reactions are pretty rare. Which brings us to the concept of the rate limiting step.

## Rate Limiting Steps

Multiple reaction steps are often needed to synthesize a desired compound. This is especially true in the field of pharmaceutical chemistry where the value added nature of the product justifies the substantial additional cost of following multiple synthetic steps. However, it is almost sure that the reaction steps will not proceed at the same rate. That is, one step will be the slowest. The slow step is the one that will determine the overall rate of the synthesis. This step is called the **rate-limiting step**. The concept of a limiting reagent was already discussed in an earlier chapter, and the current topic is an extension of this earlier concept to how fast a sequence of reaction occurs.

Let's run through a hypothetical example; the formation of acid rain. Sulfuric acid ($H_2SO_4$) is produced via the combustion of sulfur-laced coal to make power (and lots of heat) in a power plant. Sulfur is present mined bituminous coal at a level of 0.7-4% by weight.

Here one possible sequence of reactions, there are many others:

| | | | |
|---|---|---|---|
| **8.4A** | $S + O_2$ | $\longrightarrow$ | $SO_2$ |
| **8.5A** | $SO_2 + 1/2O_2$ | $\rightleftarrows$ | $SO_3$ |
| **8.6A** | $SO_3 + H_2O$ | $\rightleftarrows$ | $H_2SO_4$ |

From a practical standpoint reaction 8.4A is irreversible but the other two reactions are not. Which reaction is rate limiting would be determined experimentally, but let's suppose it is reaction 8.5A and that the rate is found to be independent of the concentration of $O_2$. The rate equation for this **first order** reaction is then simply:

**Rate = k[$SO_2$]**

The real case here in a power plant is a good deal more complicated than the above analysis. All three states of matter will be present, the temperatures, gas pressures and reactant/product concentrations will vary greatly from the combustion chamber through to the emission stack.

### Multi-Step Reaction Energetics

Reaction sequences involving multiple steps may have interesting energetic profiles; that is, each step can be expected to have its own activation energy barrier. By way of example let's look at the following two reactions:

| | | | | |
|---|---|---|---|---|
| 8.7A | $CH_4 + H_2O$ | $\rightleftarrows$ | $CO + 3H_2$ | (steam reforming reaction) |
| 8.8A | $CO + H_2O$ | $\rightleftarrows$ | $CO_2 + H_2$ | (water gas shift reaction) |

Both of these reactions are very important industrial sources of hydrogen gas, a critical resource for many different industries. However, that is not the focus here. Rather the $CO_2$ synthesis is. In essence this is a two-step reaction wherein natural gas ($CH_4$) is converted into $CO_2$ via the CO intermediate. Both reactions are reversible, and endothermic, but 8.7A likely rate determining overall depending upon the experimental conditions chosen by the operator.

Figure 8.4 is a qualitative energy profile for this reaction. The first transition state requires the higher activation energy ($E_{a1}$) and is rate limiting for the overall reaction. This represents the steam reforming reaction 8.7A. The second transition state, a2, is of lower energy, but still must be overcome for the whole reaction to proceed to completion.

## 8.4A Catalysts

***Catalysts*** are compounds that are designed to lower the energy activation ($E_a$) barrier of a chemical reaction. Here are some other key properties that are associated with catalysts.

- Catalysts have no effect upon the energetics of a reaction. That is ΔG, ΔH, and ΔS are the same with or without a catalyst being present.
- Catalysts come in many forms, states, and chemistries.
- Catalysts are essential in biological system, for example enzymes are catalysts.

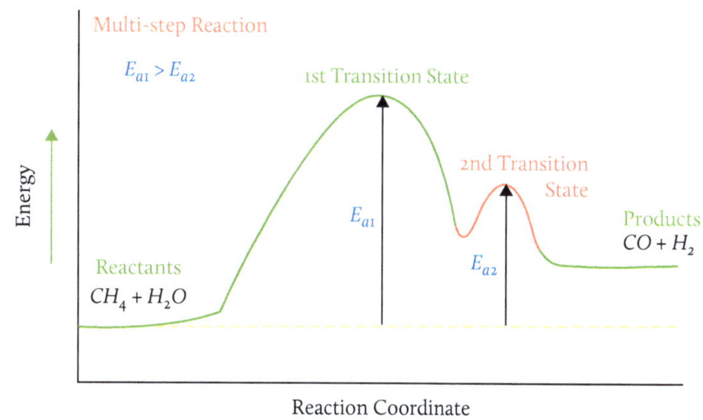


**FIGURE 8.4.** Multi-Step Reaction Profile.

Solid state inorganic catalysts are essential in a variety of chemical processes including catalytic converters.

Catalysts can be either heterogeneous or homogeneous.

Catalysts are *not* consumed as the reaction proceeds. However, over time catalysts lose their effectiveness, and must be replaced.

Catalyst chemistries are complex; and reaction mechanisms tend to be difficult to unravel.

The rate (kinetics) of a reaction, in both the forward and reverse directions, is increased by the presence of a catalyst.

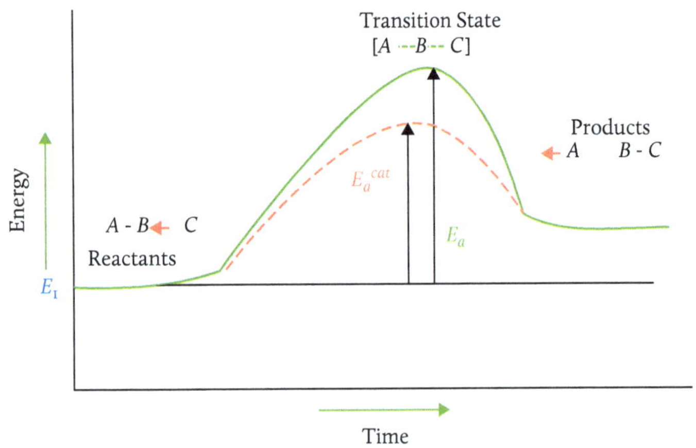


FIGURE 8.5. The Effect of a Catalyst on Reaction Energetics.

Figure 8.5 shows the effect of a catalyst upon the energetics of a one-step endothermic reaction. The same sort of change is seen with exothermic reactions. Here the red dashed line is the profile in the presence of a catalysts. The two key points concerning this plot are that the $E_a$ is lowered substantially, but that the energies of the reactants and products are unchanged. Associated to these points is that the degree of $E_a$ lowering is catalyst dependent

Two examples of catalysis will be described, a heterogeneous solid-gas phase reaction and an enzymatic process.

***Heterogeneous Catalysis:*** *Hetereogeneous catalysis involves the use of different states of matter; gas, liquid or solid.* Figure 8.6 shows a

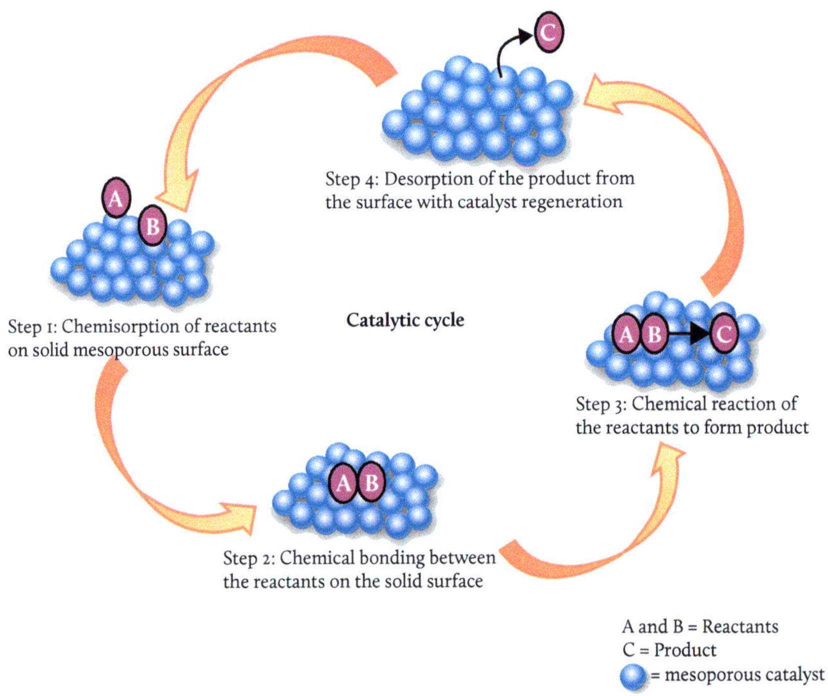


FIGURE 8.6. Heterogeneous Catalysis Mechanism.

generic catalytic cycle for a heterogeneous reaction. In this cycle reactants **A** and **B** **adsorb** onto the surface of a solid catalyst. In the remaining steps they react on the surface to form product C which then desorbs from the catalyst surface completing the reaction cycle, which can then repeat. The number of times this cycle can occur is known as the **turnover number** of the catalyst. The bigger this number is, the better as eventually the catalyst will itself react and die thus stopping the process.

***Enzymes:*** Enzymes are high molecular weight proteins that act as catalysts in biological systems of all kinds. Each enzyme has an **active site** where the catalysis occurs. The active site 3D structure is critically important to the enzyme's function, and only molecules called **substrates** will fit into the active site for the desired reaction to occur.

By way of example Figure 8.7 shows the cleavage hydrolysis of the carbohydrate sucrose into simpler carbohydrates, glucose and fructose. The sucrose is the substrate which fits selectively into the active site of the enzyme sucrose, where the reaction occurs and the products are released. The enzyme and its active site remain intact.

Active sites are crucial in a great many biological processes, good and bad. The last section dealing with kinetics describes a couple of these.

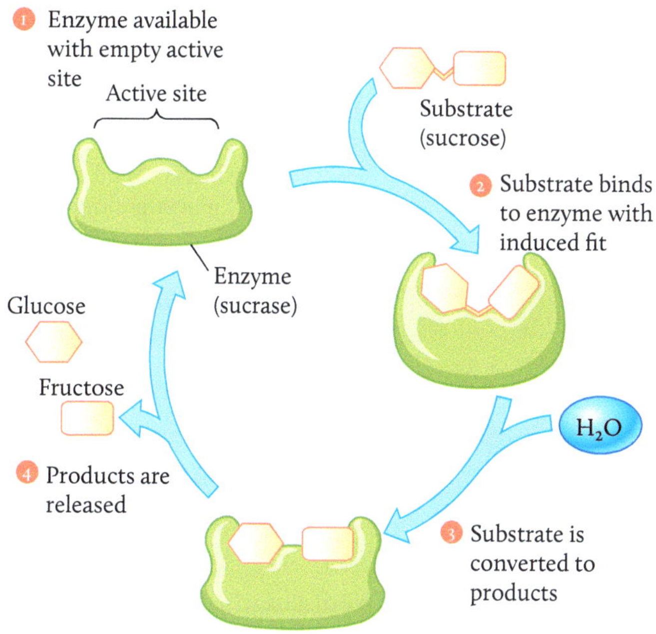


FIGURE 8.7. Enzymatic Conversion of Sucrose into Glucose and Fructose—Mediated by the Enzyme Sucrase.

## 8.5A Reaction Kinetics—Practical Examples

Basically, the first half of this chapter has outlined the nature of chemical kinetics, a complex and fascinating topic. In this last section we would like to describe a couple of examples of kinetic processes that have very different time frames, and cover a very wide range of chemistries and applications.

***Explosives:*** The practical requirements for an explosive are exacting. In addition to its effectiveness as an explosive a material must be safe to handle, store and transport, and must be cost-effective. Here we deal with just the necessary kinetics. Clearly, the chemical reaction must occur virtually instantaneously upon demand, but only then. Let's look at one classic example of an explosive, RDX:

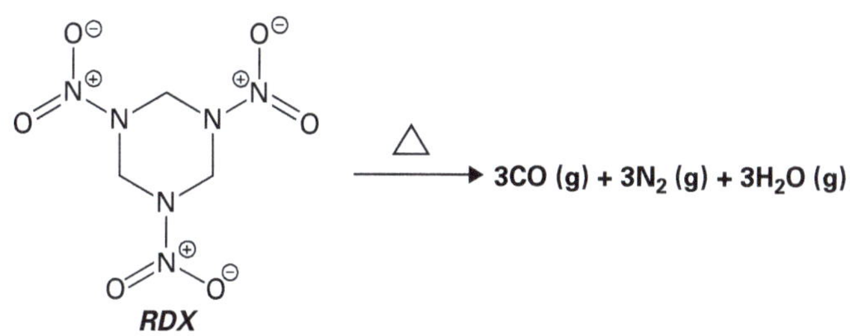


CHEMICAL STRUCTURE 8.1. RDX structure.

RDX is a solid, but note the products are all gaseous and there are a lot of them. The mechanism shown here is deceptively simple and leaves out a great deal in terms of what actually goes on mechanistically in getting from reactants to products. In fact, the mechanistic steps are not well understood as atmospheric oxygen is crucial to the chemistry too.

The reaction is very exothermic, but like most explosives, the key property for explosivity is the generation of lots of gas molecules very, very quickly. The pressure increase is phenomenal as is the resultant destructive power.

So, in terms of kinetics how fast is this reaction? It is temperature dependent, but as a frame of reference the molar rate constant is $\sim 1 \times 10^{20}$ $sec^{-1}$. In other words the reaction occurs with the awesome concomitant pressure generation in **0.000,000,000,000,000,000,01 sec.**

### Rock Weathering Chemistry

At the other extreme is the geologic time frame involved in the weathering of rocks. Here too the chemistry is complex; however, after an initial reaction of a granite surface to form what is called a weathering rind, the chemical reaction of the underlying rock surface is glacially slow in the absence of physical processes. For example a buried volcanic rock in the western United States will see a weathering rate measured in millions of years.

The interested reader is also referred to the chapters herein dealing with Energy System, Agricultural Chemistry, Biochemistry, and Toxicology and Pharmacology where reaction kinetics are featured.

# Part B: Equilibria

## 8.1B Dynamic Chemical Systems

We discuss chemical systems in static terms. For example we write water as $H_2O$ or at best a bunch of water molecules together, but we lose sight of the essential fact that molecules and atoms are anything but motionless. They twist, they turn, their bonds bend and stretch, and the electrons move at lightning-fast velocities. What goes on, the types of movements, the state of the matter, physical and chemical properties, and the types of reactions undertaken are all proportional to the system's total energy and energy distribution. The chemical system may or may not be at equilibrium. Let's now look at this topic more closely.

## 8.2B Chemical Systems at Equilibrium

Newton's 1st law of motion in physics deals with inertia; namely that a body at rest tends to stay at rest whereas one in motion tends to continue that way if unimpeded. Newton's 3rd Law of Motion states that when a body exerts a force on a second body, the second body exerts an equal force in direct opposition. Chemical (molecular) systems are similar in that when a steady state or *equilibrium* is reached they tend to stay that way if possible; resisting any changes imposed on the system. By definition, ***equilibrium*** is a state in which a reversible chemical reaction exists in a state where the relative concentrations of reactants and products remains unchanged. This also means that the rates of the forward and reverse reactions are equal and the ratio of the product to reactant concentration ratios remains constants.

Equation 8.1 shows an important example of a possible equilibrium reaction, namely the Haber reaction synthesis of ammonia:

$$\textbf{8.1B} \qquad 3H_2\,(g) \;+\; N_2\,(g) \rightleftarrows 2NH_3\,(g) \;+\; \textbf{heat}$$

Note here the two arrows pointing in opposite directions. This is chemical symbol used to denote a chemical reaction is reversible process, that is, one in which the reaction can proceed in both directions. . What are the relative concentrations at equilibrium? That depends upon the reaction being studied, and other experimental variables.

There are two more important points involving the reaction as shown above which relate to reaction thermodynamics (Chapter 6). One point is that it is an exothermic reaction where heat is released, that is, heat ($\Delta H$) is a product of the reaction. The second point is that the entropy ($\Delta S$) is negative as four moles of gas reactants yield just two moles of product gas. These factors work in opposite directions in terms of spontaneity.

What is an ***equilibrium*** in reaction chemistry? A chemical reaction system in equilibrium is one in which both reactants and products are present and their apparent concentrations are unchanging as well. Notice that the word apparent is underlined. A very important point here is that just because a system is at equilibrium does not mean *no* reactions are occurring. Quite the reverse is true. Reactants are converting to products and products are reversing course to form reactants. What equilibrium does mean is that the measured concentrations of reactants and products are static as a function of time.

Figure 8.8 shows two possible reaction profiles for the Haber process shown in equation 8.8:

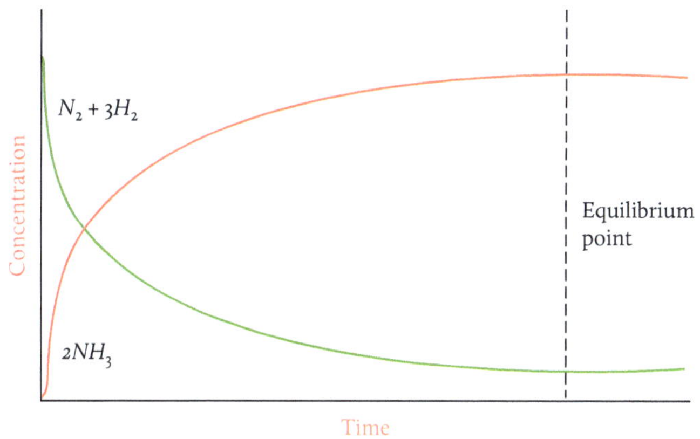


**FIGURE 8.8.** Qualitative Reaction Profiles for the Haber Process.

The green curve here represents a starting point wherein just $N_2$ and $H_2$ are present at the correct stoichiometry to begin forming $NH_3$. At some point in time (dotted line) equilibrium is reached wherein the molar ratio of $N_2$:$H_2$:$NH_3$ is 1:3:2 is established and maintained. The red line represents the reverse reaction starting with just $NH_3$ being present, but followed by a dissociation of $NH_3$ to form the equivalent equilibrium with $H_2$ and $N_2$.

## 8.3B Homogeneous and Heterogeneous Equilibria

Many reactions are run in one phase of matter, and these are called **homogeneous reactions.** The liquid phase is by far the most commonly used or observed, but all three phases are important. An example of a homogeneous reaction is the neutralization of an aqueous solution of HCl with an NaOH solution to form a salt and water as products:

**8.2B** $$\mathbf{HCl\ (aq) + NaOH\ (aq) \longrightarrow H_2O\ (aq) + NaCl\ (aq)}$$

Notice here the one directional arrow showing that this reaction is not readily reversible, but it is homogeneous as all reactants and products are in the aqueous phase.

Those equilibria involving more than one phase are called **heterogeneous reactions.** By way of example, the catalytic converter in a car reacts exhaust gases on a catalytic solid state surface (Pt or Pd supported on a ceramic) to form product decomposition gases (eg. $SO_2$, $CO_2$, $NO_2$), as well as liquids (especially $H_2O$), and solid particulates as products.

Both homogeneous and heterogeneous phase equilibria are well known, and are subject to the constraints of Le Chatelier's Principle.

## 8.4B Le Chatelier's Principle

Now that we have established what an equilibrium is, what happens if we impose a stress or change upon the system? By stress we mean that the equilibrium is shifted to a position that is not in equilibrium. For example, in equation 8.1B let's assume the reaction is being run in a closed vessel, equilibrium has been established, and further that the reactants (3 moles $H_2$, 1 mole $N_2$) and products (2 moles $NH_3$) are all in the gas phase. What happens if we introduce another 1 mole of $N_2$ gas? All of a sudden the mole ratio $N_2$:$H_2$:$NH_3$ is no longer 1:3:2. However, the system wants to remain at the 1:3:2 ratio. To do so it adjusts according to ***Le Chatelier's Principle*** which states:

> *When a chemical equilibrium is subjected to a change in the concentration of a reactant or product, or equivalently, a change in temperature, pressure or volume, it will adjust such that the imposed change is minimized, with the original equilibrium being reestablished, if possible.*

So, for our Haber reaction 8.1 example given above what happens if we add an extra mole of $N_2$ to the reaction mixture? The equilibria is now out of kilter, and the system will attempt to bring it back to the same equilibrium. To do so means using up part of the just added $N_2$ plus some of the originally present $H_2$ to form more $NH_3$. This shifts to the reaction to the right (product) side. Key here too is to recognize that the mole ratios will no longer be whole numbers in the balanced equation. Also note that the $H_2$ amount decreases. Clearly, if a reactant runs out that is the end of this process.

Here's another scenario. What if we remove $NH_3$ instead of adding $N_2$? Once again the equilibrium is disturbed, Le Chatelier applies. In this the reaction shifts the same way, towards product side replenishing the lost ($NH_3$) and the amounts of $N_2$ and $H_2$ decrease.

---

**PRACTICE** What happens if we add $NH_3$ instead of remove it? What happens if we remove $H_2$?

---

Now a third scenario. Recalling that heat is also a product of the reaction, what happens if we heat the system up? Just like adding $NH_3$ added heat will shift the reaction back to the reactants side. In this case, heat is a product so if heat is added, it needs to be "used up". This causes the shift to the reactants as heat will be consumed in the process of forming more reactants.

From a practical standpoint how is Le Chatelier's Principle an important concept? The synthesis of ammonia is one of the most chemical processes known. Without the Haber process we would simply not have enough nitrogen based fertilizers to feed the world's population. By removing the ammonia product as it forms in reaction 8.1 we can force the reaction to the right and make more and more $NH_3$ faster and faster. It is LeChatelier's principle that is employed to optimize ammonia synthesis in this process.

## Effects of Disturbing an Equilibrium—A Qualitative Analysis

As mentioned above a reaction equilibrium can be rendered unstable in three ways. Let's look at these three more closely from a qualitative perspective. Sections 8.6–8.8 extend this discussion to quantitative examples.

***Raising or lowering the temperature of the system:*** An **exothermic** reaction results in heat being given off, in other words heat is a product. Raising the temperature increases the heat thus forcing the action back to the the reactants side (for example reaction 8.1). An **endothermic** reaction requires the addition of heat. In this case, heat is a reactant. Adding heat into thissystem will force the reaction to the right forming more products in order to consume the added heat. For example reaction 8.3B, he splitting of water into $H_2$ and $O_2$:

$$\textbf{8.3B} \qquad 2H_2O\,(g) \;+\; \text{heat} \;\rightleftharpoons\; 2H_2\,(g) + O_2\,(g)$$

***Changing the pressure/volume*** (for systems involving gases): Solids and liquids are essentially incompressible, but gases certainly are so changes in volume (as thus pressure) can have an impact on reactions involving gases. Take equation 8.4B as an example. Let's assume here the gases are contained in a closed vessel at a volume of 1L and 1 atmosphere of pressure and an equilibrium is established. Now the pressure is increased to 2 atmospheres by decreasing the volume of the container thus disturbing the equilibrium. Le Chatelier says the pressure increase should be minimized. Recall Chapter 5 and PV = nRT. R and T are constant so $P \alpha n/V$. By decreasing the total number of moles of gas (n) P will also be minimized. The reaction shows 2 moles of gas on the reactant side and 3 on the product side, thus the reaction will shift back to the reactant side seeing as there are fewer moles of gas on that side of the equation.

---

**PRACTICE** If the vessel volume of the following reaction is made smaller which way will the reaction shift towards reactants or products?

---

$$\textbf{8.4B} \qquad 2SO_2\,(g) \;+\; O_2\,(g) \;\rightleftharpoons\; 2SO_3\,(g)$$

***Change in concentrations of reactants or products:*** This topic is a bit more complicated and involves the equilibrium constant, an important concept which follows through into the discussion of weak acids in the following chapter, but let's look at it in a broader sense now in the next section.

## 8.5B The Equilibrium Constant

Let us examine reaction 8.4B more closely in terms of expressions relating the relative concentrations of reactants and products.

$$8.5B \quad 2SO_2\,(g) + O_2 \underset{k_r}{\overset{k_f}{\rightleftharpoons}} 2SO_3\,(g)$$

In equation 8.5B we have added in forward and reverse rate constants, $k_f$, and $k_r$, respectively to equation 8.4B. The forward and reverse rate equations (see earlier discussion of kinetics in this chapter) are:

**Rate forward = $k_f[SO_2]^2[O_2]$ and Rate reverse = $k_r[SO_3]^2$**

At equilibrium these rates are equal to each other, so:

**$k_f[SO_2]^2[O_2] = k_r[SO_3]^2$**

Rearranging this equation:

$$\frac{k^r}{k^f} = [SO_3]^2 / ( [SO_2]^2[O_2]) = K_c$$

Here we introduce $K_c$, the ***equilibrium constant*** (often just given as K) which is equal to the ratio of the two rate constants.

***Equilibrium Constant (K):*** A number that reflects the ratio products to reactants for a chemical reaction at equilibrium at a given temperature.

Appendix 5 contains a list common equilibrium constant values. An important point here is that now we have related a constant K, specific to this reaction, to reactant and product concentrations which can be determined in the laboratory. The practical usefulness of this should be apparent.

## 8.6B Calculations and the Equilibrium Constant

We now have a quantitative relationship between the concentrations of the reactants and products and a reaction constant, $K_c$. Calculations using this relationship are pretty straightforward once the requisite sequence of steps are known. Like many areas of science the best way to learn this topic is to practice.

There is another key point that tends to get forgotten along the way. Solids and liquids concentrations do not change and so substances that are in these states do not show up in the calculations. Especially important to keep in mind is that the solvent, usually water, is present in massive quantities compared to the reactants and products thus we ignore it in the calculations. In the following examples solids will be identified with (s), liquids with (l). Do not include these in the equilibrium expressions.

So here we go with some examples beginning with the steps to be followed in each:

1. Write out a ***balanced*** chemical reaction.
2. Ignore solids and liquids in step.
3. Write down the associated equilibrium rate constant equation.

4. Plug in the known values given in the problem using dimensional analysis to make sure the calculations and units will be correct at the end.
5. Solve for the unknown, either a concentration of reactant or product or the $K_c$.

### Example 1 (weak acid)

Carbonic acid, $H_2CO_3$, is arguably the most important weak acid which dissociates as follows in water to form an equilibrium:

**8.6B** $$H_2CO_3\,(aq) + H_2O\,(l) \rightleftharpoons HCO_3^-\,(aq) + H_3O^+\,(aq):$$

What is the concentration of $H_3O^+$ at equilibrium if $K_c = 4.2\text{x}10^{-7}$ and $[H_2CO_3]$ = 0.10M and $[HCO_3^-] = [H_3O^+] = 0$ at the start?

After equilibrium is established the product ion concentrations must be equal to each other. Why?

Per steps 2 and 3 the equation is balanced and we ignore $H_2O$ (l). Per step 3:

$K_c = \dfrac{[HCO_3-][H_3O+]}{[H_2CO_3]}$ now to simplify set $[HCO_3^-] = [H_3O^+] = x$ at equilibrium and put in what is given in the problem to yield:

$4.2 \times 10^{-7} = \dfrac{x^2}{0.10}$ therefore $x^2 = 4.2 \times 10^{-8}$. Thus $x = [H_3O^+] =$ $\boxed{\mathbf{2.05 \times 10^{-4} M}}$

There is an assumption here which does not always hold true. Namely that the amount of carbonic acid that dissociates is vanishingly small compared to the significant figures involved with the calculation. This is a good assumption here where K is ~$10^{-7}$, a value so low that the numerator (product concentrations) is vanishingly small compared to the denominator (reactants). The results confirm the validity of the assumption here. The amount of $H_2CO_3$ dissociated is equal to $\mathbf{[H_3O^+] = 2.05 \times 10^{-4}M}$, inconsequential compared with the starting acid concentration of **0.10M**.

### Example 2 (gas phase aka homogeneous)

Ethanol ($CH_3OH$) is produced commercially from carbon monoxide and hydrogen as follows:

**8.7B** $$H_2\,(g) + CO\,(g \rightleftharpoons CH_3OH\,(g)$$

At equilibrium in a 6L vessel and 327°C the amounts of reactants and products present are 0.22 mol $CH_3OH$, 1.53 mol $H_2$, and 0.85 mol CO. What is $K_c$

1. Balance the equation:

$$2H_2\,(g) + CO\,(g) \rightleftharpoons CH_3OH\,(g)$$

2. All gases, no (l) or (s).
3. $K_c = \dfrac{[CH_3OH]}{[H_2]2[CO]}$ The concentrations are: $[CH_3OH] = \dfrac{0.22mol}{6L} = 0.037M$

   $[H_2] = \dfrac{1.53mol}{6L} = 0.255M$; $[CO] = \dfrac{0.85mol}{6L} = 0.142M$; therefore

4. $\mathbf{K_c} = \dfrac{[CH_3OH]}{[H_2]2[CO]} = \dfrac{0.037}{(0.255)2(0.142)} =$ $\boxed{\mathbf{4.01}}$

### Example 3 (mixed phases aka heterogeneous)

Carbon monoxide, CO, is an essential reactant in the manufacturing of many important pure metals, including the most important one, Fe. One way to make CO is the simple (on paper) reaction of coke (from coal) with $CO_2$ (from the air):

**8.8B** $$CO_2\,(g) + C\,(s) \rightleftharpoons CO\,(g)$$

At 1000K $K_c = 1.9$ while at 300K $K_c = 0.133$. You are the industrial engineer responsible for making CO. Is the reaction endothermic or exothermic? Which temperature to you run the reaction at? Why?

a. At 1000K 10kg of $CO_2$ are reacted with 100kg of C in a 3000L kettle (reaction vessel). How much CO is produced?
Step 1. A balanced equation is:

$$O_2\,(g) + C\,(s) \rightleftharpoons 2CO\,(g)$$

Step 2: C(s) is a solid so it is not part of the equilibrium constant expression.

Step 3: $K_c = \frac{[CO]2}{[CO2]}$ Now 10kg $CO_2$ = 10,000g; $\frac{10000g}{44g\,/\,mol}$ = 227 mol $CO_2$.

$[CO_2] = \frac{227\ mol}{3000L}$ = 0.076M

Step 4: $K_c = 1.9 = \frac{[CO]2}{0.076M}$ therefore $[CO]^2 = 0.143$

Taking the square root: **[CO] = 0.38M.**

b. What weight of C is consumed?
Per the balanced equation for every 1 mole of $CO_2$ consumed 1 mol of C is consumed as well, therefore:

(227 mol C)(12.0g/mol) = 2724g **~ 2.7 kg**

### The Solubility Product and Its Constant, $K_{sp}$

All compounds have inherent solubilities in a given solvent, generally water being the solvent of interest. Like the rest of chemistry the solubility properties of a given compound are determined by electron distributions, ionic and covalent bonds, and the types/strengths of bonding between atoms, molecules and ions. Closely related to the equilibrium constant is the solubility product constant, $K_{sp}$. The good news here that the logic is parallel, and the calculations are done exactly the same way. Appendix 5 lists $K_{sp}$ values for some common ionic salts. Let's just do a couple of examples.

All ionic salts have at least some solubility in water, but the solubilities vary over many orders of magnitudes. For example, the $K_{sp}$ for calcium phosphate, $Ca_2(PO_4)_3$, is $2.0 \times 10^{-29}$, a vanishing small number reflecting the following solubility equation:

**8.9B** $$Ca_3(PO_4)_2\,(s) \rightleftharpoons 2PO_4^{-3}\,(aq) + 3Ca^{+2}\,(aq)$$

The solubility product constant calculation is done exactly the same way as done for $K_c$. That is:

$$K_{sp} = [PO_4^{-3}]2[Ca^{+2}]3$$

Just like with equilibrium constant calculations solids and liquids are not part of the $K_{sp}$ expression. So, for this example say we know $[Ca^{+2}] = 1.0 \times 10^{-5}M$, can we calculate $[PO_4^{-3}]$? Certainly. Here you go:

$$K_{sp} = 2.0 \times 10^{-29} = [PO_4^{-3}]^2 (1 \times 10^{-5})^3$$

$$\frac{2.0 \times 10^{-29}}{1 \times 10^{-15}} = [PO_4^{-3}]^2$$

$2.0 \times 10^{-14} = [PO_4^{-3}]^2$ taking square root $\mathbf{[PO_4^{-3}] = 1.4 \times 10^{-7}M}$

A question that arises is why do we care about such calculations? A good question and this example is a very good one for an explanation. $(Ca_3(PO_4^{-3})_2$ is the primary component of our bones and teeth, clearly structural elements we want to have low solubilities in water. Also, phosphate rock $(Ca_3(PO_4^{-3})_2$ being a significant component of this rock) is virtually the only geological source for phosphorus in the world. Phosphorus is essential to all of life, as in all forms of life plant and animal, as we know it. The extraction of P depends crucially on the water solubility properties of the minerals that are mined.

**$K_{sp}$ Example Calculations** Magnesium hydroxide, $Mg(OH)_2$, has been used for a long time as an antacid for neutralizing stomach acids. One reason it is useful is its modest solubility in water. This property allows it to react (heterogeneously) slowly prolonging the relief of discomfort. The $K_{sp} = 1.8 \times 10^{-11}$ at 298K. What is the concentration of $Mg^{+2}$ if $[OH] = 9 \times 10^{-3}M$?

Following steps 1-4 given for the $K_c$ problems above:

8.10B $$Mg(OH)_2\,(s) \rightleftharpoons Mg^{+2}\,(aq) + 2OH^-\,(aq)$$

$$K_{sp} = 1 \times 10^{-11} = [Mg^{+2}][OH^-]^2 = [Mg^{+2}](9 \times 10^{-3})^2$$

$$[\boldsymbol{Mg}^{+2}] = 1 \times 10^{-6}\,\boldsymbol{M}$$

Here's another interesting example of the useful application of solubility products.

Silver sulfide, $Ag_2S$, is an important geological source of silver metal, one of the beautiful metals and useful in industry as well as in jewelry. The $K_{sp}$ here is $\mathbf{6 \times 10^{-51}}$ Wow, talk about an ionic compound whose ions just love one another

What is different here is the solubility equation in water:

8.11B $$Ag_2S(s) + H_2O\,(l) \rightleftharpoons 2Ag^{+1}\,(aq) + OH^-\,(aq) + HS^-\,(aq)$$

In this case: $K_{sp} = [Ag^+]^2[OH^-][SH^-] = 6 \times 10^{-51}$

Suppose we are given that $[Ag^+] = 2 \times 10^{-12}$ M, what is the concentration of the $SH^-$ ion?

***Solution:*** The key here is to recognize that $[OH^-] = [SH^-]$ since for each $Ag_2S$ dissolved you get one of each of the ions. Let's set this concentration equal to x. Then:

$$6 \times 10^{-51} = (2 \times 10^{-12})^2 x^2$$

$$x^2 = \frac{6 \times 10 - 51}{4 \times 10 - 24} = 1.5 \times 10^{-24} M$$

$$x = [SH^-] = 1.2 \times 10^{-12}\,\boldsymbol{M}$$

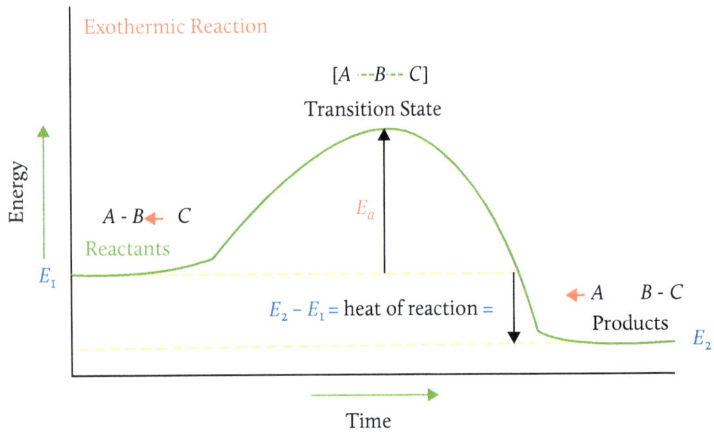


**FIGURE 8.9.** Reaction Progress Graph.

# Chapter 8 Summary

## Part A: Kinetics (Reaction Rates)

***Reaction Rates:*** How fast or slow a reaction proceeds; that is, how does the concentration of reactants and/or products change as a function of time. Figure 8.9 shows the sequence of a typical exothermic reaction.

The **transition state** is a transition from reactants AB + C to A + BC. This requires the addition of an energy, the **activation energy, $E_a$**, the magnitude of which is characteristic of the specific reaction being studied. $E_2$-$E_1$ is the energy change in going from reactants to products.

### Factors Affecting Reaction Rates

*Pressure (Gas Phase/Volume);* The higher a system's pressure, the more likely molecular collisions will occur.

*Temperature:* The higher the temperature the greater will be a system's kinetic energy.

*Concentration:* Higher concentrations will result in more molecular collisions.

*Molecular Orientation:* The molecules must be oriented in 3D space to react instead of just collide.

*Catalyst:* The presence of a catalyst can speed up a reaction by lowering the activation energy barrier. It does not affect energetics otherwise.

*Rate Law Expression;* **Rate = k[reactant]$^a$**

Where **k** is the rate constant for a given reaction. "**a**" is the order of the reaction. [ ] is the reactant concentration in molarity.

*Rate Limiting Step:* The step in a multistep reaction sequence that determines the overall reaction rate (kinetics).

*Catalysts:* Lower the activation energy (Ea) barrier, but not the overall rewhereaction energetics. Ideally the catalysts is not consumed in a reaction.

## Part B: Equilibrium

*Chemical Reaction Equilibrium:* Where in a chemical reaction the reactants and products are bothe present. The reaction can be made to shift either way, towards the eractants or towards the products, by control of the experimental variables.

*Homogeneous Equilibria:* The system is all of a single phase.

*Heterogeneous Equilibria:* The system contains multiple phases, gas, liquid and/or solid.

*Le Chatelier's Principle:* When a chemical equilibrium is subjected to a change in the concentration of a reactant or product, or equivalently, a change in temperature, pressure or volume, it will adjust such that the imposed change is minimized, with the original equilibrium being reestablished, if possible.

*Equilibrium Constant (K):* A number that reflects the ratio products to reactants for a chemical reaction at equilibrium.

*Solubility Product Constant ($K_{sp}$):* A number that reflects the relative solubilities of products and reactants in a solution at equilibrium.

# CHAPTER 8 EXERCISES

## Kinetics

1. What factors affect the rate of a chemical reaction?
2. What is a transition state?
3. The Haber process is being run where the concentration of $N_2$ decreasing at the rate of 2.2 M/min. What are the rates at which $H_2$ is reacting and $NH_3$ is being formed?

   $$N_2 + 3H_2 \longrightarrow 2NH_3$$

4. How do the coefficients in a balanced equation relate to the overall reaction order?
5. Given the following balanced equation and tabled information:

   $$CH_3Br + OH^- \longrightarrow CH_3OH + Br^-$$

| Experiment | $[CH_3Br]$ (M) | $[CH_3Br]$ (M) | Initial Rate (M/s) |
|---|---|---|---|
| #1 | $1.5 \times 10^{-3}$ | $1.5 \times 10^{-3}$ | $2.2 \times 10^{-4}$ |
| #2 | $3.0 \times 10^{-3}$ | $1.5 \times 10^{-3}$ | $8.8 \times 10^{-4}$ |
| #3 | $1.5 \times 10^{-3}$ | $3.0 \times 10^{-3}$ | $4.4 \times 10^{-4}$ |

   a. What is the order of the reaction in $CH_3Br$?

   b. What is the order of the reaction in $OH^-$?

   c. What is the order of the whole reaction?

6. Given the data shown in experiment #1 of problem #5, what is the rate constant, k, associated with this reaction?
7. Label the numbered parameters shown in the following energy-time graph (figure 8.10).
8. An aqueous sample of $H_2O_2$ is heated to decomposition yielding $H_2O$ and $O_2$ products. The decrease in hydrogen peroxide peroxide concentration is followed yielding the following data. Fill in the rates of reactions in the table.

   $$H_2O_2 \longrightarrow H_2O + O_2$$

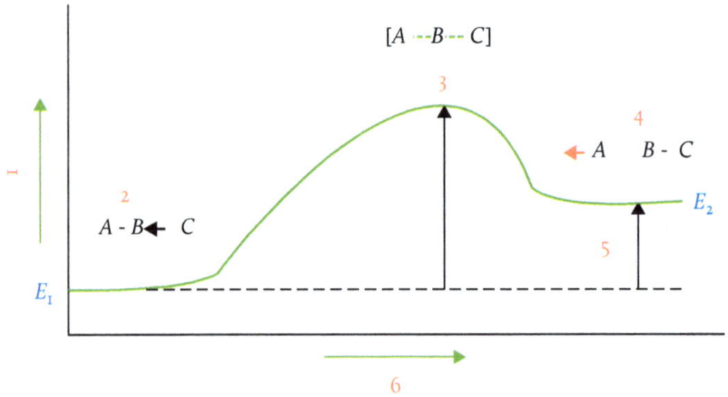


**FIGURE 8.10.** Energy-Time Graph.

TABLE 8.1

| Time (s) | $[H_2O_2]$ (M) | Rate (M/s) |
|---|---|---|
| 0 | 0.16 | |
| 30 | 0.08 | |
| 60 | 0.06 | |
| 90 | 0.04 | |
| 120 | 0.04 | |

9. The following reaction is found to be first order in NO and zero order in $O_2$.

$$2N_2O \quad + \quad O_2 \longrightarrow 4NO$$

a. What is the rate expression for this reaction if $k = 1.8 \times 10^{-4}$?

b. Based upon your answer to a, fill in the initial rates of reactions in the table below

TABLE 8.2

| Experiment | $[N_2O]$ (M) | $[O_2]$ (M) | Rate (M/s) |
|---|---|---|---|
| #1 | 2.4M | 2.4M | |
| #2 | 2.4M | 4.8M | |
| #3 | 4.8M | 2.4M | |
| #4 | 0 | 4.8M | |
| #5 | 4.8M | 0 | |

10. A car's catalytic converter consists of a solid state Pt or Pd catalyst supported on an inert solid support, usually silica-based. As the exhaust gases come through the tailpipe residual hydrocarbons are decomposed which decreases the resulting air pollution. The finer the particle sizes of the solid catalyst is the more efficient is this process. Why?

## Equilibria

1. The equilibrium constant for the first dissociation of carbonic acid in water at 25oC is $K = 4.3 \times 10^{-7}$ ? What is favored in this equilibrium reactants or products?

   $$H_2CO_3\,(aq) + H_2O\,(l) \longrightarrow HCO_3^-\,(aq) + H_3O^+\,(aq)$$

2. What is the equilibrium expression for the reaction given in problem #1?
3. If the initial concentration of carbonic acid in problem #1 is 0.1M, what is the concentration of the bicarbonate anion ($HCO_3^-$) after equilibrium is reached?
4. Write the equilibrium expressions for the following reactions:

   a. $Sn\,(s) + 2Br_2\,(g) \longrightarrow Sn(Br)_4\,(s)$

   b. $2SO_2\,(g) + O_2\,(g) \longrightarrow 2SO_3\,(g)$

   c. $CO_2\,(g) + 2S\,(l) \longrightarrow CS_2\,(g) + O_2\,(g)$

   d. $C_3H_6\,(g) + H_2\,(g) \longrightarrow C_3H_8\,(g)$

5. Write the equilibrium or solubility product expressions for the following reactions:

   a. $2NO\,(g) + O_2\,(g) \longrightarrow 2NO_2\,(g)$

   b. $H_2\,(g) + Se\,(s) \longrightarrow H_2Se\,(l)$

   c. $Si\,(g) + 2Cl_2\,(g) \longrightarrow SiCl_4\,(g)$

   d. $H_2C{=}CH_2\,(g) + I_2\,(l) \longrightarrow C_2H_2I_2\,(s)$

6. Write solubility product expressions for the following reactions:

   a. $Ca^{+2}\,(aq) + 2SO_4^{-2}\,(aq) \longrightarrow CaSO_4\,(s)$

   b. $NO_3^-(aq) + H^+(aq) \longrightarrow HNO_3\,(aq)$

   c. $AgCl\,(s) \longrightarrow Cl^-(aq) + Ag^+(aq)$

   d. $SrSO_3(aq) \longrightarrow SO_3^-(aq) + Sr^{+2}(aq)$

7. Hydrogen peroxide ($H_2O_2$) is a very weak acid at 25°C with a $K = 2.4 \times 10^{-12}$. Here is the equilibrium:

   $$H_2O_2\,(aq) + H_2O\,(l) \rightleftharpoons H_3O^+\,(aq) + HOO^-$$

   If the initial concentration of hydrogen peroxide is 0.22M, what is the equilibrium concentration of $HOO^-$?
8. Hydrocyanic acid (HCN) is a very weak acid at 25°C with a $K = 2.4 \times 10^{-10}$. Here is the equilibrium:

   $$HCN\,(aq) + H_2O\,(l) \rightleftharpoons H_3O^+\,(aq) + CN^-$$

   If the initial concentration of HCN s 1.66M, what is the equilibrium concentration of $CN^-$?

9. What is the pH of the solution in problem #7 at equilibrium and 25°C?

10. What is the pH of the solution in problem #8 at equilibrium and 25°C?

11. Hydrogen sulfide ($H_2S$) is a weak acid at 25°C, but one that has two ionizations rather than just one as follows:

$$\text{Ionization 1: } H_2S\,(aq) + H_2O\,(l) \rightleftharpoons H_3O^+\,(aq) + HS^- \quad K_1 = 9.5 \times 10^{-8}$$

$$\text{Ionization 2: } HS^-\,(aq) + H_2O\,(l) \rightleftharpoons H_3O^+\,(aq) + S^{-2} \quad K_2 = 1 \times 10^{-19}$$

Note here the two ionization constants, $K_1$ and $K_2$ with the latter being 11 orders of magnitude smaller!

a. Why is $K_2$ so much smaller than $K_1$?

b. If the initial concentration of hydrogen sulfide is 4.0M, what is the equilibrium concentration of $HS^-$?

c. Given the result of part b, what is the concentration of $S^{-2}$?

d. What is the pH of the solution?

12. What is the pH of a 0.66M acetic acid solution at equilibrium and 25°C? (see Appendix 5 for K).

13. What is the $[OH^-]$ of problem #12?

14. What is the pH of a 2.50M benzoic acid solution at equilibrium and 25°C? (see Appendix 5 for K)

15. What is the $[OH^-]$ of problem #14?

CHAPTER 9

# Acid-Base Chemistry

Acids and bases are likely the most common chemicals you might come across in daily life. Acids are present in citrus fruits, soda and vinegar. Bases can be found in household cleaning products, soap and some baking products. Lemon juice tastes sour due to the acids presents while the bases in soaps cause them to taste bitter.

The first definitions of an acid and a base were due to the work of Svante Arrenhius. This scientist made observations of acids reacting with water and bases reacting with water. He discovered that acids increased the concentration of hydrogen ions present in water. Bases, he found, increased the concentration of hydroxide ions in water. While this model is true for acids and bases reacting with water, it falls short when discussing acids and bases reacting with compounds other than water. In this chapter we will discuss the **Bronsted Lowry** Model for acids and bases as it works when discussing reactions in water as well as reactions with other compounds.

## 9.1 BrØnsted-Lowry Acids and Bases

The main issue with the Arrenhius model for acids and bases is that it does not readily explain acid base reactions in non-aqueous solutions. Reactions between acids and bases can occur in the gas phase and in solvents other than water. Johannes Brønsted and Thomas Lowry independently noted a common behavior that occurred during acid-base reactions occurring in other mediums. During these reactions, the scientists noted, a proton ($H^+$) is transferred. More specifically, an acid donates or gives a proton to a base. For this reason, a ***Bronsted-Lowry acid*** *is defined as a proton donor. A* ***Bronsted-Lowry base*** *is defined as a proton acceptor.* Moving forward in this text, we will be using the Bronsted-Lowry definition for acids and bases.

Let's first consider a reaction between hydrochloric acid (HCl) and ammonia, a base ($NH_3$). The reaction that occurs is explained by the following chemical equation:

$$HCl(g) + NH_3(g) \longrightarrow NH_4Cl(s)$$

As the equation depicts, HCl is the acid and donates a proton to $NH_3$, the base. In this case the final product is solid ammonium chloride, an ionic salt. The white powder that is released is the ammonium chloride product of the reaction.

## Conjugate Pairs

As we have discussed, when an acid reacts with a base and proton is transferred from the acid to the base. In the reaction between water and HCl below, water is the base and HCl is the acid.

$$\mathbf{HCl(aq) + H_2O(l) \longrightarrow Cl^{-}(aq) + H_3O + (aq)}$$

When water accepts the proton from HCl, a ***hydronium ion*** is formed (**$H_3O^+$**) as well as a chloride ion. The products of the reaction, $Cl^-$ and $H_3O^+$, are called ***conjugates.*** In this case, $Cl^-$ is the conjugate base and $H_3O^+$ is called the conjugate acid. For HCl $Cl^-$ is always the conjugate base so $HCl/Cl^-$ are acid-base conjugate pairs and $H_3O^+/H_2O$ are acid-base conjugate pairs. Conjugate pairs are related by the gain and loss of a proton.

When water reacts with a base like ammonia, it will act as the acid instead of the base.

$$\mathbf{NH_3(aq) + H_2O(l) \rightleftharpoons NH_4^{+}(aq) + OH^{-}(aq)}$$

In this case, water donated the proton to ammonia. This means water is the acid in this reactions while ammonia is the base. In this case, $NH_4^+/NH_3$ are an acid-base conjugate pair and $H_2O/OH^-$ are an acid base conjugate pair. This demonstrates water's ability to act as an acid or a base. Any substance that can act as an acid or a base is called ***amphoteric.*** Some examples of amphoteric substances are $HPO_3^-$, $HSO_4^-$ and $HCO_3^-$.

# 9.2 pH and the pH Scale

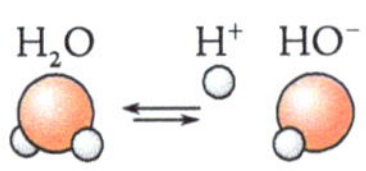


FIGURE 9.1. Autoionization of Water.

As just discussed, water can act as an acid or a base. By extension, two water molecules can transfer protons between themselves! One water molecule can donate a proton to a different water molecule. This process is known as the ***autoionization of water*** and is depicted by the equation below (Fig. 9.1).

This reaction exists at an equilibrium where the forward rate of the reaction is equal to the reverse rate for the reaction. This example represents a very important application of the concepts just presented in Chapter 8, Equilibrium. Per this chapter the reaction can be represented by the expression:

$$\mathbf{K_w = [H_3O^+][OH^-]}$$

The extent to which pure water auto-ionizes varies as a function of temperature. For our purposes we will consider water at room temperature (25°C) where the $K_w$ is constant and equal to $1 \times 10^{-14}$. As long as the temperature is at 25°C, the $K_w$ will remain at this value. Please note these values are for pure water meaning water free from any dissolved substances.

## Calculating $H^+$ and $OH^-$ Concentrations

We can use the expression for the autoionization of water to calculate the concentrations of $H^+$ ($H_3O^+$) and $OH^-$ in water solutions. In pure water, the concentration of $H_3O^+$ must be the same as $OH^-$. Why? Because every time one molecule of water donated a proton (forming $OH^-$) a second molecule accepted one (forming $H_3O^+$). This means the ratio of $OH^-$ to $H_3O^+$ is 1:1 in pure water.

You will recall that:

$$K_w = [H_3O^+][OH^-]$$

And in pure water, $[H_3O^+] = [OH^-]$ so:

$$K_w = [H_3O^+][H_3O^+] = [H_3O^+]_2$$

At 25°C (room temperature) the Kw is $1 \times 10^{-14}$ so:

$$[H_3O^+]^2 = 1 \times 10^{-14}$$

Taking the square root:

$$[H_3O^+] = = 1 \times 10^{-7}\ M$$

And since $[H_3O^+] = [OH^-]$:

$$[OH^-] = = 1 \times 10^{-7}\ M$$

It is important to take note of this calculation. There are two important points to take home from this. First, the concentrations of hydronium ions and hydroxide ions in water is equal when water is pure. Second, the concentrations of hydronium ions at room temperature is $1 \times 10^{-7}$ M as is the concentration of hydroxide ions. These concentration are only true for *pure* water. Water with any added acid or base will not have this same concentration of ions. The $K_w$ for any sample of water at room temperature will always be $1 \times 10^{-14}$. However, when acid or base is added to that same the concentrations of hydroxide and hydronium ions will change while keeping the $K_w$ the same. So, if the concentration of hydronium ions increases, it follows that the concentration of hydroxide ions will decrease to maintain the $K_w$ and vice versa.

***Neutral*** solutions are those solutions where the hydronium ion concentration and hydroxide ion concentration are equal. In solutions where the concentrations of hydronium ions and hydroxide ions are not equal can be either acidic or basic. An ***acidic*** solution is a solution with a concentration of hydronium ions greater than the concentration of hydroxide ions. In a ***basic solution,*** the concentration of hydronium ions less than the hydroxide ion concentration (Figure 9.2).

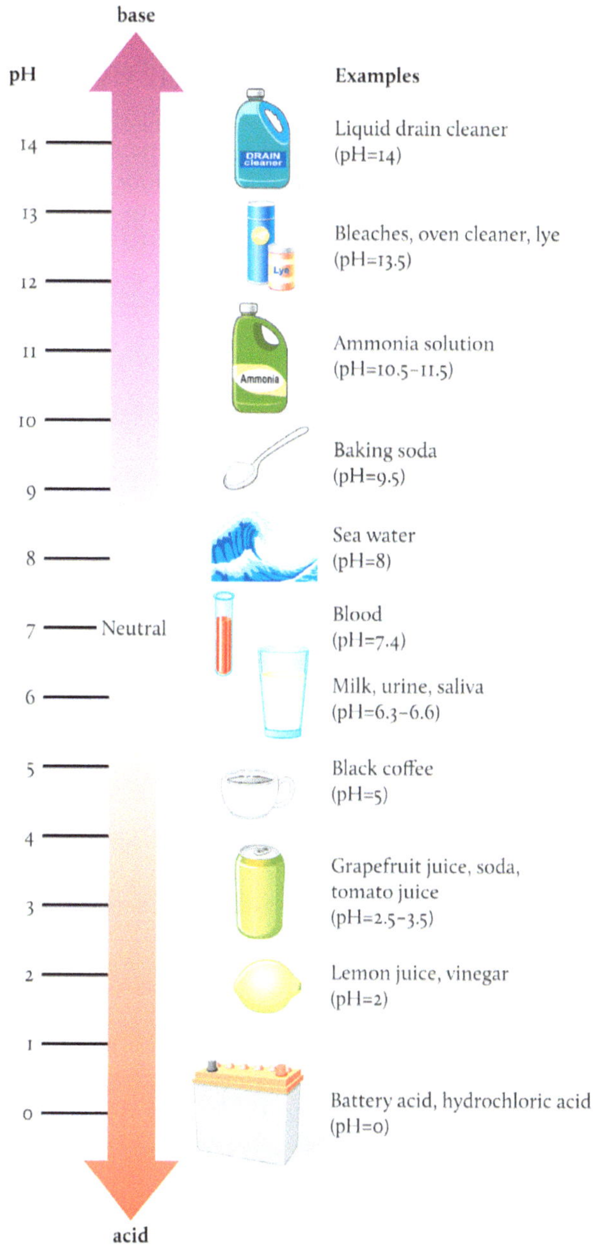


**FIGURE 9.2.** The pH of Common Substances.

## Determining pH

The calculations for hydroxide ions and hydronium ions yields concentration values that involve numbers that are very small making them difficult to work with. Dealing with numbers in scientific notation are difficult, so a method to make the numbers not as awkward was developed. The ***pH scale*** converts the concentrations of hydronium ions from very small molar concentrations to numbers that are larger, do not involve scientific notation, making them easier to handle. The pH scale uses common logarithms (log base 10) to convert values for concentrations of hydronium ions to values that are easier to work with. In this text "log" will stand for logarithm base 10 (in algebraic terms log = $\log_{10}$).

**pH = –log $[H_0O^+]$**

Recall that in pure water at 25°C the concentration of hydronium ions is $1\times10^{-7}$ M. Let's convert this to pH:

$$pH = 7 - \log[1\times10^{-7}] = 7$$

Because this pH value represents the concentration of ions in pure water, a pH of 7 is considered a neutral solution. *As the concentration of hydronium ions increases from* $1\times10^{-7}$*, the value for pH will decrease* and as the concentration of hydronium ions decrease from $1\times10^{-7}$ M, the pH will *increase.* This being the case for acidic solutions, where the $[H_3O^+] > [OH^-]$, the pH will be less than 7. While basic solutions, where $[H_3O^+] < [OH^-]$, the pH will be greater than 7. Figure 9.3 depicts the pH scale with corresponding concentrations of ions and also correlates the relative pH's for common substances.

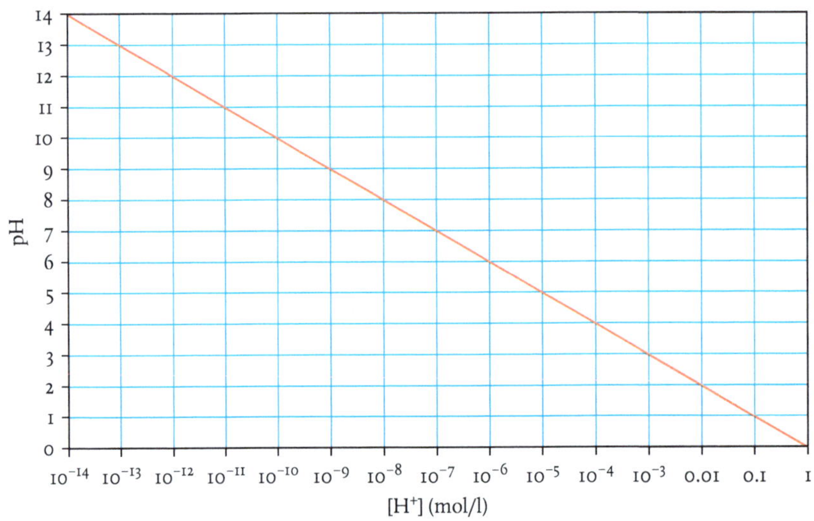


**FIGURE 9.3.** Change in pH Relative to Concentration of $H^+$.

The pH of most substances will fall between 0 and 14. However, solutions with a hydronium ion concentration greater than 1 M will have a negative value for pH. Solutions with a hydroxide ion concentration greater than 1M will have a pH greater than 14.

As an alternative the pH scale, chemists will sometimes use pOH scale when reporting acid/base properties of a solution. The only difference for this scale is that it is reported with respect to the hydroxide ion concentration.

**pOH = –log $[OH^-]$**

A neutral solution will have a pOH of 7 because the concentration of $[H_3O^+] = [OH^-]$ so the values for pOH and pH will be the same. However, in the case of the pOH scale, basic solutions will have a pOH < 7 will acidic solutions will have a pOH > 7.

By taking the antilog of the pH and pOH, one can determine the concentrations for hydronium ions and hydroxide ions, respectively. The antilog of the pH yields the hydronium ion concentration and the antilog of the pOH yields the hydroxide ion concentration.

$[H_3O^+] = 10^{-pH}$
$[OH^-] = 10^{-pOH}$

There is a very useful relationship between pH and pOH that can be used for simplifying calculations for ion concentrations. Recalling that $K_w = [H_3O^+][OH^-]$, we can create a new expression:

$$-\log([OH^-][H_3O^+]) = -\log Kw$$

Or:

$$-\log([H_3O+]) + (-\log [OH^-]) = -\log Kw$$

We know that at 25°C the $K_w = 1 \times 10^{-14}$, so:

$$-\log([H_3O^+]) + (-\log [OH^-]) = -\log 1 \times 10^{-14}$$

Finally, we can substitute pH and pOH where appropriate:

$$pH + pOH = \log 1 \times 10^{-14}$$

Solving the logarithm we get:

$$pH + pOH = 14$$

This expression tells us that (at 25°C) the sum of the pH and the pOH *must* be equal to 14. We can use this expression to determine the pH from the pOH and vice versa.

# 9.3 Strong Acids and Strong Bases

We have just discussed pH and learned that a solution with a pH less than 7 is acidic while a solution with a pH greater than 7 is basic. Can pH tell us whether an acid or a base is strong? To answer this question, it is important to know what the definition is for a strong acid and a strong base.

## Strong Acids

The definition for a strong acid is actually quite simple. It is an acid that dissociates completely once dissolved in water.

$$HCl(g) + H_2O(l) \longrightarrow H_3O^+(aq) + Cl^-(aq)$$

When writing an equation for the reaction of a strong acid with water, a one directional arrow (→) is used to indicate the absence of a reverse reaction. So, for a strong acid, all of the protons from the acid are donated to the base, in this case water. This also means that if we know the concentration of acid that we start out with, we also know the concentration of hydronium ions formed one the reaction is complete. Again, this is because *all* of the protons from the strong acid will be donated to water molecules.

There are seven common strong acids including **HCl, HBr, HI, HNO3, $HClO_3$, $HClO_4$ and H2SO$_4$**. Sulfuric acid, $H_2SO_4$, is a diprotic acid. It can donate up to two protons to water. In this case, it is the first deprotonation event that occurs to completion, thus giving it a classification as a strong acid. The second deprotonation event is not a full deprotonation event, and is characteristic of a weak acid, a topic we will discuss later.

## Strong Bases

Similarly, for a strong base, when it is dissolved in water, it too will dissociate completely.

$$NaOH(s) \longrightarrow Na^{+}(aq) + OH^{-}(aq)$$

Again, a one directional arrow (→) is used in the chemical equation to denote that there is no reverse reaction and the strong bases will dissociate completely in water. Strong bases include the metal hydroxides from column IA of the periodic table as well as the hydroxides formed with Ca, Sr and Ba. It is important to note that the last three hydroxides listed are not very soluble.

As with strong acids, if we know the concentration of the strong base, we can determine the concentration of hydroxide ions that base produces. For the alkaline metal hydroxides, like NaOH, the concentration of the base is equal to the concentration of hydroxide ions produced. However, with the alkaline earth hydroxides, including $Ca(OH)_2$, the concentration of the base is not equal to the concentration of hydroxide ions produced. To explain this, it is best to first considered what happens when these bases dissolve in water. Take a look at the chemical equation for the dissociation of $Ca(OH)_2$ in water:

$$Ca(OH)2(s) \longrightarrow Ca^{2+}(aq) + 2OH^{-}(aq)$$

Notice the stoichiometry for this equation. For every 1 unit of $Ca(OH)_2$ that dissolves in water, 2 hydroxide ions are produced. So in this case, the concentration of hydroxide ions produced is *twice* the concentration of the original base. It is always important to write out chemical equations for calculations involving chemical reactions, including reactions for acids and bases. The stoichiometry will give the mole ratio for reactants and products involved in the reaction.

TABLE 9.1 **The Common Strong Acids and Strong Bases**

| Acids | Bases |
|---|---|
| HCl | LiOH |
| HBr | NaOH |
| HI | KOH |
| $HClO_4$ | RbOH |
| $HNO_3$ | CsOH |
| $H_2SO_4$[a] | $Mg(OH)_2$<br>$Ca(OH)_2$<br>$Sr(OH)_2$<br>$Ba(OH)_2$ |

[a] $H_2SO_4$ ionizes in two distinct steps. It is a strong acid only in its first ionization.

## Calculations with Strong Acids and Strong Bases

As already mentioned, by knowing the concentration of a strong acid or base, we know the concentration of hydronium ions or hydroxide ions because these reactions and dissociation events go to completion. We can use this information to easily determine the pH of strong acid or base solution.

### Example 9.1 Strong Acids

What is the pH of a 0.020 M solution of HCl at 25°C?

First recall the reaction:

$$HCl(g) + H_2O(l) \longrightarrow H_3O^{+}(aq) + Cl^{-}(aq)$$

If the concentration of HCl = 0.020 M and all of the HCl reacts to form $H_3O^+$, then $H_3O^+$ = 0.020M.

$pH = -\log [H_3O^+]$

$pH = -\log[0.20M]$

$pH = 1.699 = 1.70$

(note the pH values are generally reported to two decimal places)

### #2: Strong Acids

What is the concentration of $HNO_3$ when the pH of the solution is 2.56 at 25°C?

First, write the chemical equation for the reaction:

$HNO_3(aq) + H_2O(l) \longrightarrow H_3O^+(aq) + NO_3^-(aq)$

In this case, we do start with a pH but want to know a concentration so the appropriate equation is:

$[H_3O^+] = 10^{-pH}$
$[H_3O^+] = 10^{-2.56}$
$[H_3O^+] = 0.003$ M

We know that $[HNO_3] = [H_3O^+]$, so:

$[HNO_3] = 0.003$ M

Before moving onto calculations involving strong bases, it is important to emphasize how sulfuric acid is unique. It is a diprotic acid, so it can donate up to two protons, not just one. This means that the concentration of $H_2SO_4$ is not the same as $H_3O^+$ for solutions of this acid. We will discuss this acid in more detail when we get to weak acids and bases.

### Example 9.2 Strong Bases

What is the pH of a 0.250 M solution of NaOH at 25°C?

First, let's consider the chemical equation:

$NaOH(s) \longrightarrow Na^+(aq) + OH^-(aq)$

This being a strong base, all of the sodium hydroxide will dissociate so [NaOH] = $[OH^-]$ = 0.250 M

$pOH = -\log [OH^-]$
$pOH = -\log [0.250 \text{ M}]$
$pOH = -\log [OH^-]$
$pOH = 0.60$

Know we know the pOH but need to determine pH. Remember at 25°C:

$pH + pOH = 14$
$pH = 14 - pOH$
$pH = 14 - 0.60$
*pH = 13.40*

What is the pH of a 0.125 M solution of $Ca(OH)_2$ at 25°C?

### #2: Strong Bases

First, let's consider the chemical equation:

$$Ca(OH)_2(s) \longrightarrow Ca^{2+}(aq) + 2OH^-(aq)$$

This is a strong base, however, for every 1 unit of $Ca(OH)_2$ that dissociates, 2 ions of OH- are produced, so:

$[OH^-] = 2 \times [Ca(OH)_2]$
$[OH^-] = 2 \times [0.125\ M]$
$[OH^-] = 0.250\ M$

We can also approach this using stoichiometry:

$$\frac{0.125\ \cancel{mol\ calcium\ hydroxide}}{1\ L\ of\ solution} \times \frac{2\ mol\ hydroxide}{1\ \cancel{mol\ calcium\ hydroxide}} = 0.250 \frac{mol\ hydroxide}{1\ L\ solution}$$

$= 0.250\ M\ OH^-$

With the concentration of hydroxide we can determine pOH:

$pOH = -\log [OH^-]$

$pOH = -\log[0.250\ M]$

$pOH = -\log[OH^-]$

$pOH = 0.60$

Know we know the pOH but need to determine pH. Remember at 25°C:

$pH + pOH = -14$

$pH = 14 - pOH$

$pH = 14 - 60$

$pH = 13.40$

Take note, this pH is the same as that for the NaOH solution that had twice the initial concentration. Does this make sense? It does because, each $Ca(OH)_2$ produces 2 $OH^-$ ions, doubling concentration of $OH^-$ produced and yielding the same pH.

### #3: Strong Bases

If the pH of a solution of KOH at 25°C is 12.50, what is the concentration of KOH?

First, let's consider the chemical equation:

$$KOH(s) \longrightarrow K^+(aq) + OH^-(aq)$$

In this case we have pH and want a concentration, we can get OH- concentration from pOH:

$[OH^-] = 10^{-pOH}$

Let's first convert pH to pOH:

$pH + pOH = 14$

$pOH = 14 - pH_p$

$pOH = 14 - 12.50$

$pOH = 150$

Now we can solve for hydroxide concentration:

$$[OH^-] = 10^{-pOH}$$

$$[OH^-] = 10^{-1.50}$$

$$[OH^-] = 0.32M$$

Based on the chemical equation the ratio of $[OH^-]$ to [KOH] is 1 to 1 so:

$$[OH^-] = [KOH] = 0.32$$

# 9.4 Weak Acids and Weak Bases

Any acid or base that is not strong, is considered *weak.* This being the case, a majority of the acids and bases we encounter are actually weak. So what does it mean to be a weak acid or base? Many wrongly assume the definition for a weak acid or base is linked to the pH. In truth, it is possible for a solution of a weak acid to be lower than a solution of a strong acid as pH depends on concentration. If a concentration of a weak acid is significantly higher than the concentration of a solution of a strong acid, the strong acid solution could have a higher pH. So what makes an acid or base weak? Let's find out.

## Weak Acids

As we learned, strong acids will completely deprotonate when added to water. So what happens to weak acids? ***Weak acids** partially dissociate* in water. The extent to which a weak acid deprotonates is a property of a given acid. Each weak acids, at 25°C, reaches an equilibrium. Every weak acid has a characteristic ***acid ionization constant*** ($K_a$) at a given temperature. Here is a chemical equation for hydrofluoric acid (HF), a weak acid:

$$\mathbf{HF(aq) + H_2O(l) \rightleftharpoons H_3O^+(aq) + F^-(aq)}$$

The only difference in this chemical equation from that of a strong acid is the use of a bidirectional arrow (⇄). This arrow indicates that the reaction exists at an equilibrium and does not proceed to "completion" with all of the reactants completely converting to products. This has a few consequences including the fact that the concentration of $H_3O^+$ in solution will not be equal to the concentration of HF. The calculations for concentration of $H_3O^+$ and the solution pH is much more complex for weak acids than strong ones.

We have already discussed the strong, diprotic acid, $H_2SO_4$. At this time, we can better understand what happens when it dissolves in water. The first deprotonation event will go to completion with all of the protons being donated to water. The second deprotonation event; however, will not go to completion and only a portion of the protons will be donated to water. The chemical reactions for this process are shown below:

$$\mathbf{H_2SO_4(aq) + H_2O(l) \rightarrow H_3O + (aq) + HSO_4^-(aq)}$$

$$\mathbf{HSO_4^-(aq) + H_2O(l) \rightleftharpoons H_3O + (aq) + SO_4^{2-}(aq)}$$

Note that the first reaction is represented with a one directional arrow while the second is represented by a two directional arrow. The amount of protons sulfuric acid releases upon dissolution in water is somewhere between 1 and 2 per molecule of the acid. We know that one will be completely transferred to water while the second is involved in an equilibrium reaction.

## Weak Bases

Much like weak acids, weak base only partially ionize in water and their reactions also exist at an equilibrium. During these reactions a base will accept a proton but will also donate it back to the conjugate base in a reversible process. Ammonia, carbonate and methylamine are all examples of weak bases.

$$NH_3(aq) + H_2O(l) \rightleftharpoons OH^-(aq) + NH_4^+(aq)$$
$$CO_3^{2-}(aq) + H_2O(l) \rightleftharpoons OH^-(aq) + HCO_3^-(aq)$$
$$CH_3NH_2(aq) + H_2O(l) \rightleftharpoons OH^-(aq) + CH_3NH_3^+(aq)$$

Again, note the use of the bidirectional arrow for the chemical equations, indicating these reactions exist at equilibrium. The ***base ionization constant,*** $K_b$, is associated with the equilibrium for a particular weak base when it reacts with water at a given temperature.

## Calculations with Weak Acids and Weak Bases

Calculations involving weak acids and bases are more complex than those involving the strong acids and bases. This is simply because of the reversibility of the reactions. Given that the conversion from the weak acid or weak base is not complete, we cannot say that the concentration of the acid is equal to the hydronium ion concentration or the concentration of the base is equal to the hydroxide ion concentration. To determine the pH of a weak acid or base solution, the $K_a$ or $K_b$ must be used. Table 9.2 provides the $K_a$ and $K_b$ values for several acids and bases.

To start with these calculations, on must first consider the reaction involved.

$$HF(aq) + H_2O(l) \rightleftharpoons H_3O^+(aq) + F^-(aq)$$

The set up for the equilibrium constant here should be:

$$K_a = \frac{[F-][H3O+]}{[HF]}$$

*Products always go in the numerator and reactants in the denominator.* Water is a liquid present in massive quantities; whose concentration in essence doesn't change, and therefor it is left out of the equation. The above equations leads us to the generic equation for $K_a$. The following set up for $K_a$ can be used when approaching calculations involving any weak acid.

$$K_a = \frac{[conjugate\ base][hydronium\ ion]}{[acid]}$$

So what does the equilibrium constant equation have to do with pH? If you look at it closely, you will see that a component of the equation is the concentration of hydronium ion which we can use to determine pH. So, if we manipulate the equation we can solve for the concentration of hydronium ion and then get pH:

$$[H_3O^+] = \frac{Ka[acid]}{[conjugate\ base]}$$

TABLE 9.2 **Weak Acid and Weak Base Equilibrium Constants ($K_a$ and $K_b$)**

| Name of Acid | Acid | $K_a$ | Name of Base | Base | $K_b$ |
|---|---|---|---|---|---|
| Sulfuric acid | $H_2SO_4$ | large | hydrogen sulfate ion | $HSO_4^-$ | very small |
| Hydrochloric acid | $HCl$ | large | chloride ion | $Cl^-$ | very small |
| Nitric acid | $HNO_3$ | large | nitrate ion | $NO_3^-$ | very small |
| Hydronium ion | $H_3O^+$ | 1.0 | water | $H_2O$ | $1.0 \times 10^{-14}$ |
| Hydrogen sulfate ion | $HSO_4^-$ | $1.2 \times 10^{-2}$ | sulfate ion | $SO_4^{2-}$ | $8.3 \times 10^{-13}$ |
| Phosphoric acid | $H_3PO_4$ | $7.5 \times 10^{-3}$ | dihydrogen phosphate ion | $H_2PO^{4-}$ | $1.3 \times 10^{-12}$ |
| Hexaaquairon(III) ion | $Fe(H_2O)_6^{3+}$ | $6.3 \times 10^{-3}$ | pentaaquahydroxoiron(III) ion | $Fe(H_2O)_5OH^{2+}$ | $1.6 \times 10^{-12}$ |
| Hydrofluoric acid | $HF$ | $7.4 \times 10^{-4}$ | fluoride ion | $F^-$ | $1.4 \times 10^{-11}$ |
| Formic acid | $HCO_2H$ | $1.8 \times 10^{-4}$ | formate ion | $HCO^{2-}$ | $5.6 \times 10^{-11}$ |
| Benzoic acid | $C_6H_5CO_2H$ | $6.3 \times 10^{-5}$ | benzoate ion | $C_6H_5CO_2^-$ | $1.6 \times 10^{-10}$ |
| Acetic acid | $CH_3CO_2H$ | $1.8 \times 10^{-5}$ | acetate ion | $CH_3CO_2^-$ | $5.6 \times 10^{-10}$ |
| Hexaaqua aluminum ion | $Al(H_2O)_6^{3+}$ | $7.9 \times 10^{-6}$ | pentaaquahydroxoaluminum ion | $A1(H_2O)_5O^{H2}+$ | $1.3 \times 10^{-9}$ |
| Carbonic acid | $H_2CO_3$ | $4.2 \times 10^{-7}$ | hydrogen carbonate ion | $HCO3^-$ | $2.4 \times 10^{-8}$ |
| Hydrogen sulfide | $H_2S$ | $1 \times 10^{-7}$ | hydrogen sulfide ion | $HS^-$ | $1 \times 10^{-7}$ |
| Dihydrogen phosphate ion | $H_2PO_4^-$ | $6.2 \times 10^{-8}$ | hydrogen phosphate ion | $HPO_4^{2-}$ | $1.6 \times 10^{-7}$ |
| Hypochlorous acid | $HClO$ | $3.5 \times 10^{-8}$ | hypochlorite ion | $ClO^-$ | $2.9 \times 10^{-7}$ |
| Ammonium ion | $NH_4^+$ | $5.6 \times 10^{-10}$ | ammonia | NH3 | $1.8 \times 10^{-5}$ |
| Hydrocyanic acid | $HCN$ | $4.0 \times 10^{-10}$ | cyanide ion | $CN^-$ | $2.5 \times 10^{-5}$ |
| Hexaaquairon(II) ion | $Fe(H_2O)_6^{2+}$ | $3.2 \times 10^{-10}$ | pentaaquahydroxoiron(II) ion | $Fe(H_2O)_5OH^+$ | $3.1 \times 10^{-5}$ |
| Hydrogen carbonate ion | $HCO_3^-$ | $4.8 \times 10^{-11}$ | carbonate ion | $CO_3^{2-}$ | $2.1 \times 10^{-4}$ |
| Hydrogen phosphate ion | $HPO_4^{2-}$ | $3.6 \times 10^{-13}$ | phosphate ion | $PO_4^{3-}$ | $2.8 \times 10^{-2}$ |
| Water | $H_2O$ | $1.0 \times 10^{-14}$ | hydroxide ion | $OH^-$ | 1.0 |
| Hydrogen sulfide ion | $HS^-$ | $1 \times 10^{-19}$ | sulfide ion | $S^{2-}$ | $1 \times 10^{5}$ |

**Strong Acids:** $HCl$, $HBr$, $HI$, $HNO_3$, $H_2SO_4$, $HClO_4$
**Strong Bases:** $NaOH$, $KOH$, $Ca(OH)_2$, $Ba(OH)_2$

So now, all we need to know if the $K_a$, which are known constants (Table 9.2), the concentration of the acid and the concentration of the conjugate base at equilibrium. The last part is the tricky part. To do this the following steps should be followed:

a. Determine the initial concentrations of each of the reactants and products. You can ignore water, because it is not involved in the equilibrium constant. Usually, you will start with some reactant but no products, but this is not always the case.
b. When equilibrium concentrations are known, determine how much the change in the concentration that occurs from the initial concentration to reach that equilibrium amount.

c. Use stoichiometry to adjust changes to the other reactions components. Remember, if reactants are used, they are used to form products. If a reactant decreases by x amount, products will increase by stoichiometric equivalents.
d. Combine the initial concentrations with the change to determine the equilibrium concentrations for each component. For this you will, most likely, subtract from the initial concentration of the reactants and add to the initial concentrations of the products (you will use reactants to form products).
e. Plug values into the equilibrium constant and solve.

Let's perform some practice problems to illustrate this process.

## 9.5 Weak Acid Calculations

a. **What is the Ka of a weak acid, HA, with an initial HA concentration of 0.175 M and a pH of 2.70?**

The easiest way to approach this problem is to fill information we know into the table below.

1. First we enter the initial concentrations. For HA its 0.175 and for the products it is zero, because products have yet to form.
2. Next we can determine the concentration of $H_3O^+$ at equilibrium because we know the pH. So we can determine the concentration and add it to the table as well.

$$[H_3O^+] = 10^{-pH}$$
$$[H_3O^+] = 10^{-2.70}$$
$$[H_3O^+] = 0.002\ M$$

3. Next, we can determine the change in the concentration of $[H_3O^+|$ because we know both the initial and final amounts. The concentration increase is by 0.002 M.
4. Using stoichiometry, all ratios are 1:1 so the concentration of the A- also increased by 0.002 and the concentration of HA decreased by 0.002 (because reactants were used to form products).

TABLE 9.3 Weak Acid Equilibrium Constant Calculation

| | HA(aq) + | $H_2O(l) \rightleftarrows$ | $H_3O^+(aq)$ + | $A^-(aq)$ |
|---|---|---|---|---|
| Initial Concentration (M) | 0.175 | | 0 | 0 |
| Change in concentration (M) | −0.002 | | +0.002 | +0.002 |
| ***Equilibrium concentration (M)*** | ***0.175–0.002 = 0.173*** | | ***0.002*** | ***0.002*** |

Note here the $H_2O$ concentration does change; however, since it is present in vast quantities the change of 0.002 M becomes unimportant once significant figures are considered.

Now that the table is complete, we can use the equilibrium concentrations to calculate the $K_a$

In this case:

$$K_a = \frac{[A^-][H_3O^+]}{[HA]}$$

$$K_a = \frac{[0.002][0.002]}{[0.173]}$$

$$K_a = 2.31 \times 10^{-5}$$

There are no units used for the $K_a$ or $K_b$.

b. **What would the pH be for a 0.750 M solution of hydrocyanic acid (HCN) which has a Ka of $4 \times 10^{-10}$ at 25°C?**

1. Set up a table.
2. First enter initial concentrations, 0.750 M for HCN and 0 M for the products.
3. Seeing as we have no equilibrium values, we do not know the change, but we can represent it with a variable, "$x$". Again, products will increase by $x$ amount, while reactants will decrease by $x$ amount.
4. Ratios are 1:1 so stoichiometry is easy here, everything either increases or decreases by the same amount.
5. Now solve for the equilibrium amounts.

TABLE 9.4 **Weak Acid Equilibrium Constant Calculation**

| | $HCN(aq)$ + | $H_2O(l)$ ⇄ | $H_3O^+(aq)$ + | $CN^-(aq)$ |
|---|---|---|---|---|
| Initial Concentration (M) | 0.750 | | 0 | 0 |
| Change in concentration (M) | $-x$ | | $+x$ | $+x$ |
| ***Equilibrium concentration (M)*** | ***0.750 – x*** | | ***x*** | ***x*** |

1. Let's plug the values into the equilibrium expression.

$$K_a = \frac{[A^-][H_3O^+]}{[HA]}$$

$$K_a = \frac{[x][x]}{[0.750 - x]}$$

We know that $K_a = 4.0 \times 10^{-10}$ so:

$$K_a = \frac{[x][x]}{[0.750 - x]} = 4.0 \times 10^{-10}$$

$$[x][x] = 4.0 \times 10^{-10} \times [0.750 - x]$$

$$[x]^2 = 4.0 \times 10^{-10} \times [0.750 - x]$$

Now here is a little trick we can use. When the $K_a$ is $10^{-5}$ or smaller, we can assume that the value for x is small compared to the initial concentration of the acid and thus $0.750 - x \approx 0.750$. For example, 0.750 – 0.000005 is still about 0.750 M. Follow? So then we can say:

$$[x]^2 = 4.0 \times 10^{-10} \times [0.750]$$

To determine pH, we need the concentration of hydronium ions which is equal to x, so let's solve for x:

$$[x]^2 = 4.0 \times 10^{-10} \times [0.750]$$

$$[x] = \sqrt{(4.0 \times 10^{-10} \times 0.750)}$$

$$[x] = 1.73 \times 10^{-5}\,M$$

Now for the pH:

$$pH = -\log [H_3O^+]$$
$$pH = -\log [1.73 \times 10^{-5}\,M]$$
$$pH = 4.76$$

# 9.6 Acids, Bases, Salts, and Buffers

You now know that when an acid dissolves in water, it will lower the pH of the solution. When a base dissolves in water it will increase the pH of a solution. In addition to these effects on pH, certain salts can lower the pH of a solution while others can increase the pH of a solution. This indicates that certain salts, or ionic compounds, have acid and/or base properties.

## Anions and pH

As you already know, a salt is comprised of a cation and an anion. Certain anions, if they are the conjugate base of a weak acid, have the ability to act as a base once dissolved in solution. Take for example the ionic compound sodium acetate:

$$NaCH_3COO(s) \longrightarrow Na^+(aq) + CH_3COO^-(aq)$$

Once this compound dissolves in water, the ions will separate forming one $Na^+$ ion and $CH_3COO^-$ ion. For now, let's focus on the anion, $CH_3COO^-$, it is the conjugate base of acetic acid, a weak acid.

$$CH_3COOH(aq) + H_2O(l) \rightleftharpoons H_3O^+(aq) + CH_3COO^-(aq)$$

Because acetate ($CH_3COO^-$) is the conjugate base of a weak acid it can accept a proton to form acetic acid. As depicted in the equation above, it can take the proton from a hydronium ion when an acetic acid solution is prepared. When a sodium acetate solution is prepared something a little different will happen. First, recall, the sodium acetate will dissociate:

$$NaCH_3COO(s) \longrightarrow Na^+(aq) + CH_3COO^-(aq)$$

Next, the acetate can accept a proton from a water forming hydroxide and acetic acid.

$$CH_3COO^-(aq) + H_2O(l) \rightleftharpoons OH^-(aq) + CH_3COOH(aq)$$

This will increase the concentration of $OH^-$ in water. As we already learned, an increase this concentration will result in a increase in the pH of the solution.

The behavior applies to any salt whose anion is a conjugate base of a weak acid. It will not apply to salts containing a conjugate base of a strong acid, like HCl. The HCl reaction with water is not reversible. This means that the reaction to form HCl from water and chloride ions is NOT favorable and will not occur. Thus, there will not be an increase in $H_3O^+$ concentration and the resultant drop in pH.

## Cations and pH

As with anions, there are some cations from salts that can alter the pH when dissolved in water. The easiest example to use is ammonium chloride ($NH_4Cl$). We have already reviewed the reaction of the cation, ammonium, earlier in this chapter. Ammonium is the conjugate acid to the weak base ammonia. The reaction for the dissociation of ammonium chloride and subsequent reaction with water are shown below.

$$NH_4Cl \longrightarrow NH_4^+(aq) + Cl^-(aq)$$
$$NH_4^+(aq) + H_2O(l) \rightleftharpoons H_3O^+(aq) + NH_3(aq)$$

As the second reaction depicts, ammonium will donate a proton to water. This results in the formation of $H_3O^+$. This will increase the concentration of $H_3O^+$ in water. As we already learned, increase this concentration will decrease the pH of a solution. The behavior applies to any salt whose cation is a conjugate acid of a weak base, here $Cl^-$. It will not apply to salts containing a conjugate acid of a strong base, like NaOH.

Another consideration for influence of salts on pH is the effect of certain metal cations. Many metals can cause a decrease in the pH of a solution, most notably small, highly charged metals, such as $Fe^{3+}$ and $Al^{3+}$. Table 9.2 (above) gives the $K_a$ values for several metals. The reason these metals are capable of reducing pH is due to the manner in which they interact with water molecules. As the metals dissolve in water, water molecules surround the metal cation with the electron rich oxygens oriented towards the metal. The bonds of between the hydrogen and oxygen of these water molecules are weakened. This make the hydrogens in the weakened bonds "acidic" or ready to be donated to another, free water molecule. The new water molecule forms $H_3O^+$ and this leads to a drop in pH.

## Buffer Solutions

Buffer solutions are very common and incredibly important to life. What is a buffer solution? *A **buffer** is a solution that is able to withstand large changes in pH when small amounts of strong acid or base are added.* The purpose of a buffer is to keep the pH of a solution within a specific range. Take our blood for example which has a buffer. Human beings cannot withstand drastic changes in pH. Think about what might happen when we drink a large glass of orange juice but had no buffering system? It is likely that our pH could drop dramatically causing severe damage.

How is a buffer made? It is comprised of a weak acid and a salt containing its conjugate base or a weak base and a salt containing its conjugate acid. Acetic acid and sodium acetate are used to make a buffer solution. $NH^{4+}$ and $NH_3$ could be used to make a buffer solution. There are many other examples of buffer solutions. Each one will maintain a pH in a specific range but will ultimately do the same thing: resist large changes in pH.

So what happens when acid or base are added to a buffer solution? Let's use the acetate buffer to explain. So in acetate buffer, there will be both acetate ($CH_3COO^-$) and acetic acid ($CH_3COOH$). These two components play an important role in the functionality of the buffer solution.

First consider the addition of a strong acid to a solution, normally, in unbuffered water the following would happen:

$$HA(aq) + H_2O(l) \longrightarrow A^-(aq) + H_3O^+(aq)$$

The increase in hydronium leads to a decrease in the pH. However, if a buffer, like acetate is present, look at what happens:

$$HA(aq) + CH_3COO^-(aq) \longrightarrow CH_3COOH(aq) + A^-(aq)$$

The acetate ion accepts the proton instead of water. This leads to the formation of more acetic acid, but the concentration of hydronium ion does not change. If the hydronium ion concentration remains the same then the pH will remain the same. Tricky right?

Now let's consider the effect of added base. When base is added to water, the hydroxide ion concentration increases. This leads to an increase in pH.

$$A^-(aq) + H_2O(l) \longrightarrow OH^-(aq) + HA$$

In the presence of a buffer; however, the base can react with acetic acid.

$$A^-(aq) + CH_3COOH(aq) \longrightarrow CH_3COO^-(aq) + HA(aq)$$

Once again, the reaction with the buffer prevents a change in pH. In this case by reacting with the weak acid component of the buffer instead of water and hydroxide ions are not produced.

All buffers work in a similar manner to acetate buffer. They all react with acids to prevent formation of $H_3O^+$ and with bases to prevent the formation of $OH^-$. Each buffer will keep the pH within a certain range. There is a limit to the amount of acid or base that can be added to solution a buffer solution before it stops working. At some point, once a significant amount of acid or base is added, the buffer will run out of the necessary components to resist pH change and thus become nonfunctional.

## Neutralization and Titration Reactions

When acids and bases are mixed with one another the reaction that occurs is called a neutralization reaction. The newly formed mixture will have a pH somewhere between the pH's of the original acid base solutions (Fig. 9.4). Consider the reaction that occurs between HCl and NaOH:

$$HCl(aq) + NaOH(aq) \longrightarrow NaCl(aq) + H_2O(l)$$

The generic reaction that occurs between an acid and a metal hydroxide base is:

**Acid + Base ⟶ Salt + Water**

In general, all acid base reactions will produce a salt. Those involving the metal hydroxides will produce water. Here is another example.

$$CH_3NH_2(aq) + HCl(aq) \longrightarrow Cl^-(aq) + CH_3NH_3^+(aq)$$

Notice that, in both of the reactions listed, neither $OH^-$ nor $H_3O^+$ is produced. This indicates that the resulting solution. If you were to look of the definition for neutralize in Webster's Dictionary, you would see that it means to "stop something from being harmful or effective". It is fitting that the name given to reactions between acids and bases is "neutralization reactions". During the reactions both the acid and the base are rendered ineffective in regards to their change of pH. They acid and the base no longer react with water, but react with each other instead.

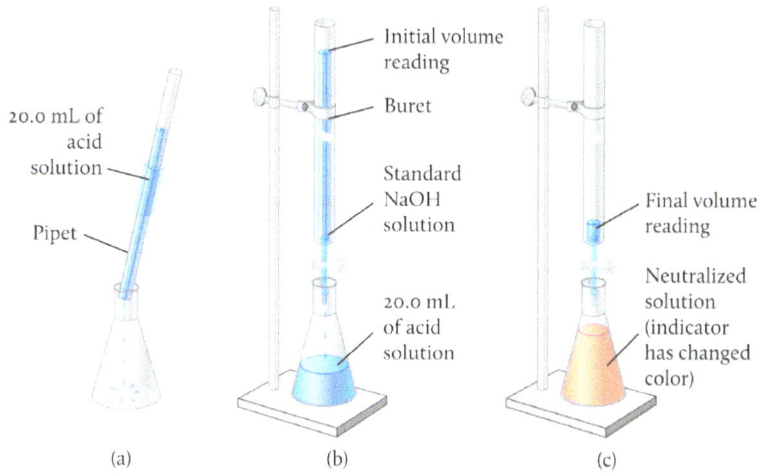


**FIGURE 9.4.** Titration Apparatus Setup.

## Titration Reactions

A titration procedure involves the slow, controlled addition of one solution to another. Through this process, the concentration of an unknown solution can be determined via solution stoichiometry. Titration reactions can be performed on acid-base solutions, and other solutions as well so long as a method exists to determine the end point of the reaction. In terms of an acid base titration, if a base solution with a known concentration is added to an acid solution with an unknown concentration, the end point would be when all of the base in the unknown is consumed.

The setup for a titration, involves the use of a burette (Figure 9.5). This instrument is a long glass cylinder that is open on one end and has a stopcock at the other end. The length of a burette is graduated so the volume of solution can be measured. The burette is usually suspended above a flask or beaker containing the solution with an unknown concentration. In the burette is the solution of known concentration that can be slowly added to the solution below by opening the stopcock. Once the endpoint is reached, the stopcock is closed and the volume of solution added is measured through the graduations on the burette.

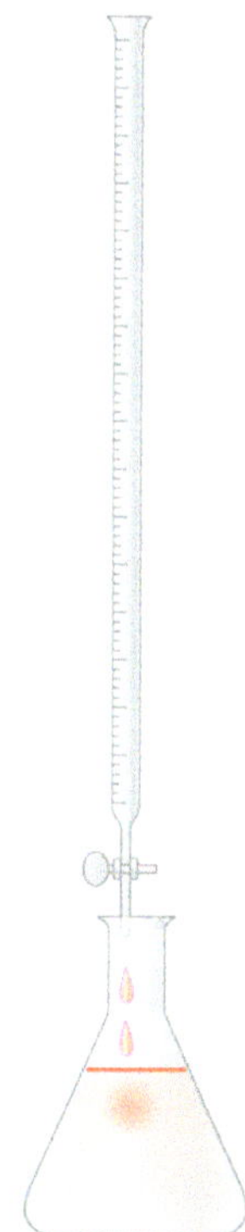

**FIGURE 9.5.** Titration.

The other necessary component to this setup is a method for detecting the endpoint. There are many possible methods for endpoint detection. In the case of acids and bases there are two common methods. The first, arguably more accurate method, is to monitor the pH change over the course of the titration to find the equivalence point. The second, easier, method is to use something called an *indicator.*

The equivalence point of a reaction is the when the moles of the acid are equal to the moles of base. For any given acid-base titration, the equivalence point will be reached at a certain pH. The point can be determined by continually measuring pH and volume of base as it is added to an acid solution. The, using the data collected, creating a plot pH versus volume of added base. This type of plot is called a titration curve. The volume of added base needed to reach the equivalence point and the pH at which the equivalence point is reached can be determined from the titration curve as shown (Figure 9.6).

Once the pH of the equivalence point is known an indicator can be selected as a method for determining the endpoint of a titration. These indicators change color at a certain pH. A very common indicator is phenolphthalein. This indicator will change

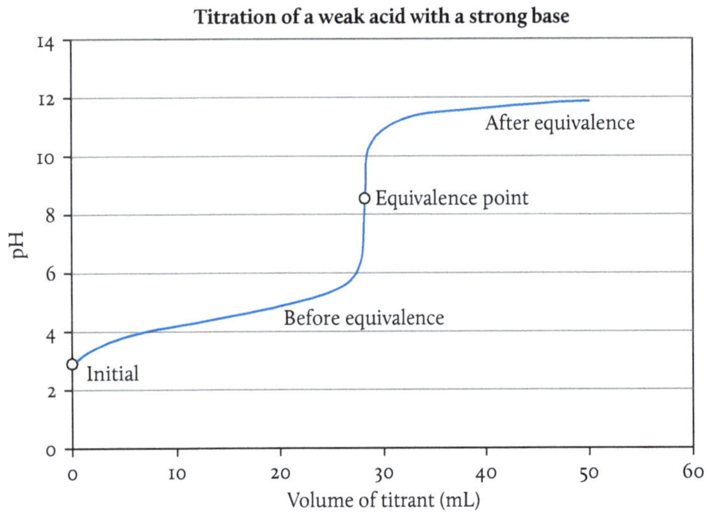


FIGURE 9.6. Titration Curve.

from colorless to pink over a specific pH range of 8.3-10.0. Another indicator, methyl red, changes colors in the pH range of 4.2 to 6.0. Ideally, when performing a titration, you want to choose an indicator that changes precisely at the equivalence point. However, it is not necessary for the indicator to be this precise. This being the case, any indicator that changes color within this range of pH values is a sufficient choice for a titration. The key is for the color change to happen in the rapid change of pH. As you can see in for the titration depicted, methyl red is outside of the rapid pH change associated with the equivalence point. This makes it a poor choice for an indicator. Phenolphthalein sits right in the rapid pH change making is a good choice for an indicator in this case.

# Chapter 9 Summary

A ***bronsted-Lowry acid*** is a proton ($H^+$) donor.

A ***bronsted-Lowry base*** is as a proton ($H^+$)acceptor.

When acids and bases react, conjugate pairs are formed. These consist of a conjugate base and a conjugate acid.

The *conjugate acid* is formed when a proton is added to a base. A ***conjugate base*** is formed when an acid gives up a proton. One example is the pair HCl/$Cl^-$. The acid is HCl and $Cl^-$C is its conjugate base.

*Amphoteric* substances can act as an acid or a base.

Water can ***autoionize.*** In this case, one molecule of water acts as a acid and donates a proton to a second molecule of water acting as the base.

$$H_2O(l) + H_2O(l) \rightleftharpoons H_3O^+(aq) + OH^-(aq)$$

The equilibrium constant for the autoionization of water is the $K_w$. Its value at 25C is $1 \times 10^{-14}$.

$$K_w = [H_3O^+][OH^-]$$

*pH* is a measure of the acidity or basicity of a substance. At 25$^0$C a pH greater than 7 is consistent with a basic solution. A pH less than 7 is an acidic substance. Substances with a pH of 7 are neutral (neither acidic or basic).

*Strong* acids and bases fully dissociate when dissolved in water to form their conjugate bases or conjugate acids.

*Weak* acids and bases only partially dissociate when dissolved in water to form their conjugate bases or conjugate acids. Weak acid and base reactions exist at an equilibrium. A majority of acids and bases are weak.

There are 7 strong acids: HCl, HBr, HI, $H_2SO_4$, $HNO_3$, $HClO_3$ and $HClO_4$.
The 8 strong bases are the alkali metal hydroxides (NaOH, KOH, LiOH, CsOH and RbOH) and the of the alkaline earth metal hydroxides ($Ca(OH)_2$, $Ba(OH)_2$ and $Sr(OH)_2$).

***Buffer*** solutions are able to resist large changes in pH upon the addition of an acid or a base to the solution.

Buffers are prepared by forming a solution consisting of a weak acid and its conjugate base or a weak base and its conjugate acid. One example is an acetate buffer made by combining acetic acid (a weak acid) with its conjugate base acetate.

***Neutralization reactions*** are reactions that occur between an acid and a base. The low pH of the acid solution and the high pH of the base solution can "neutralize" each other and bring the pH to 7.

***Titration*** is a process by which a solution with an unknown concentration is determined by a controlled reaction between two solutions. In the case of an acid and a base, the solutions are combined until an ***equivalence point*** is reached where the solutions are neutralized.

An ***indicator*** is a weak acid or base that changes color when a specific pH is reached. These are used to determine the equivalence point during titration reactions.

## CHAPTER 9 EXERCISES

1. What is an acid?
2. What is a base?
3. What is the difference between strong acids and weak acids?
4. What is the difference between strong bases and weak bases?
5. Is it true that a weak acid will ALWAYS have a higher pH than a strong acid? Explain your answer?
6. Name the 7 strong acids.
7. Name the 8 strong bases.
8. Circle the correct answer to complete each sentence below:

   a. When $[H_3O^+] > [OH^-]$ the solution is acidic/basic/neutral.

   b. When the pOH is greater than 7, the solution is acidic/basic/neutral.

   c. An acid donates/accepts a proton.

   d. Adding a base to water will increase/decrease the pH.

   e. If the $[OH^-]$ is equal to $1 \times 10^{-7}$ at 25°C the solution will be acidic/basic/neutral.

9. Answer the following regarding water:

a. Give the chemical equation for the autoionization of water. Include symbols for the states of matter (s,l,g,aq).

b. What are the molar concentrations of $H_3O^+$ and $OH^-$ when a solution is neutral at 25°C.

c. What is the equation for the equilibrium constant ($K_w$) for the reaction?

10. Complete the table below:

| Acid | Conjugate Base | Base | Conjugate Acid |
|---|---|---|---|
| | $H_2O$ | | $NH_4^+$ |
| $HCO_3^-$ | | $H_2PO_4^-$ | |
| $CH_3NH_3^+$ | | | $HCO_3^-$ |
| | $HSO_4^-$ | $CH_3COO$- | |
| HCl | | | $HNO_3$ |

11. ***Complete*** the reactions below and ***label*** acid and the base.

a. $H_2O(l) + H_2O(l) \leftrightarrows$

b. $NH_3 + HNO_3 \rightarrow$

c. $H_2SO_4 + OH^- \rightarrow$

d. $H_2O + HCH_3NH_2^+ \leftrightarrows$

e. $HCH_3COO + CH_3NH_2 \leftrightarrows$

12. A solution of $Ca(OH)_2$ has a concentration of 0.250 M.

a. What is the chemical equation for this reaction?

b. What is the pH of this solution?

c. What would the mass of $Ca^{2+}$ ions be in a 250 mL solution?

d. What is the percent dissociation for this solution?

13. Pyridine ($C_5H_5N$) is a weak base.

a. Write a chemical equation for its reaction in water.

b. Complete the table below to determine the pH of an aqueous solution of 0.25 M pyridine ($C_5H_5N$)? $K_b = 1.7 \times 10^{-9}$

| | | | | | |
|---|---|---|---|---|---|
| Initial Concentration | | | | | |
| Equilibrium Concentration | | | | | |

c. Write the Kb equation for the reaction.

d. What is the $[OH^-]$?

14. You are presented with a 0.035 M aqueous solution of barium hydroxide.

a. What is the reaction for the dissociation of barium hydroxide in water?

b. What is the concentration of hydroxide ions?

c. What is the concentration of hydronium ions?

d. What is the pH of the solution?

15. A 0.128 M solution of uric acid ($HC_5H_3N_4O_3$) has a pH of 2.39.

a. What is the $K_a$ of uric acid?

b. What is the percent dissociation?

16. Consider solutions of HClO and HCl.

a. How are strong acids and weak acids different?

b. True or False, a weak acid ALWAYS has a lower pH than a strong acid.

c. For the solution of HClO, it has a concentration of 0.5 M. If the $K_a$ of HClO is $3.0 \times 10^{-8}$, what is the pH of this solution?

d. The solution of HCl has a concentration of $5 \times 10^{-5}\ M$. What is the pH of this solution?

e. What is the percent dissociations for both HClO?

f. What is the percent dissociation for HCl?

17. Indicate whether the following solutions will be acid, basic or neutral.

| | |
|---|---|
| a. NaCl | ACIDIC / BASIC / NEUTRAL |
| b. $NH_4Cl$ | ACIDIC / BASIC / NEUTRAL |
| c. NaF | ACIDIC / BASIC / NEUTRAL |
| d. $Cu(NO_3)_2$ | ACIDIC / BASIC / NEUTRAL |
| e. $NaHCO_3$ | ACIDIC / BASIC / NEUTRAL |

18. What is the pH of a 0.01 M benzoic acid solution ($C_6H_5COOH$)? $K_a = 6.5 \times 10^{-5}$

19. Answer the following regarding buffer solutions:

a. From the list of chemicals below, select a ***suitable pair*** to produce a buffer solution.

HCl, $CH_3COOH$, $NH_4Cl$, $Ba(NO_3)_2$, KBr, $HNO_3$, $CaCl_2$, $Na(CH_3COO)$, $NH_3$

b. What is the purpose/function of a buffer?

20. You are presented with two separate solutions. One solution contains 0.030 M HCl and the other contains 0.030 M acetic acid ($CH_3COOH$).

a. Write the chemical equation for HCl in water.

b. What is the pH of the HCl solution?

c. Write the chemical equation for acetic acid in water. ($K_a = 1.8 \times 10^{-5}$)

d. What is the pH of the acetic acid solution? (try using a table)

# Credits

Fig. 9.1: Source: https://commons.wikimedia.org/wiki/File:Autoprotolyse_eau_simplifiee.svg.

Fig. 9.2: Copyright © 2013 by OpenStax College, (CC BY-SA 3.0) at https://commons.wikimedia.org/wiki/File:216_pH_Scale-01.jpg.

Fig. 9.3: Copyright © 2004 by Peo, (CC BY 3.0) at https://commons.wikimedia.org/wiki/File:PH-skala.jpg.

Table 9.1: Source: http://docplayer.net/docs-images/40/172716/images/20-0.jpg.

Table 9.2: Source: http://images.slideplayer.com/17/5318186/slides/slide_14.jpg.

Fig. 9.4: Source: http://wps.prenhall.com/wps/media/objects/3311/3390507/blb0406.html.

Fig. 9.5: Copyright © 2005 by BrokenSegue, (CC BY-SA 3.0) at https://commons.wikimedia.org/wiki/File:Titration.png.

Fig. 9.6: Copyright © 2011 by Quantumkinetics, (CC BY 3.0) at https://commons.wikimedia.org/wiki/File:Titration_of_weak_acid_with_strong_base.PNG.

# Oxidation-Reduction: Electrochemistry

## How chemistry really works—it's all about electrons, almost.

Electricity is inherently mystifying. It is so useful that we take it for granted, and in a more fundamental sense it is crucial to our survival as a species. Even if we choose to ignore biological systems, computer systems will simply not work in the absence of electron transfer processes.

The majority of important chemical reactions involve a transfer of electrons wherein one chemical compound or element loses electrons while another gains them. This process is called **oxidation-reduction** or **redox**.

## 10.1 Electrochemistry

*Electrochemistry is the study of chemical processes that involve electron transfers*. Reactions can be either spontaneous, that is run downhill energetically (release energy); or nonspontaneous wherein energy is supplied to make the reactions go forward. An important point involving spontaneous reactions is that just because they will occur without assistance, this does not mean they will occur quickly. Many such reactions occur over years or even eons in time elapsed.

There are two fundamental types of electrochemical processes which are in a sense mirror images of each other:

***Voltaic Cells:*** These are electrochemical systems wherein energy is stored in reactants, which upon contact will undergo spontaneous reactions that produce chemical products and a flow of electrons that can be harnessed to do work. A familiar example of this class is batteries (Figure 10.1).

FIGURE 10.1. Tesla Electric Drive Car.

***Electrolytic Cells:*** Here the reverse process occurs wherein electrons are supplied to make a reaction occur in a nonspontaneous manner. Energetically the overall process runs uphill, that is, consumes energy. An important example here is the manufacture of aluminum metal.

We will look at a number of very important oxidation-reduction processes later in this chapter as well as later chapters. Before doing that it is essential to first understand the chemical underpinnings of redox processes in general.

# 10.2 Oxidation States or Formal Charges

It is crucial one understands the essence of oxidation-reduction when compared to other chemical processes. For example, the reaction of hydrogen and oxygen to make water is a redox reaction, that is, it involves the transfer of electrons from hydrogen to oxygen:

$$H_2 \quad + \quad 1/2O_2 \longrightarrow H_2O \qquad \textbf{(10.1)}$$

On the other hand the reaction of hydrochloric acid with a water molecule is *not* a redox process:

$$HCl \quad + \quad H_2O \longrightarrow Cl^- \quad + \quad H_3O^+ \qquad \textbf{(10.2)}$$

In the latter reaction a proton, $H^+$, is transferred from HCl to $H_2O$, but *no electrons are being transferred.*

Oxidation-reduction is a very simple process on paper as shown in equation 9.1 above. But what is really going on here in a chemical sense? It involves electrons being transferred from the hydrogen atoms to the one oxygen atom. This can be represented by the following where the curved arrows represent electron (shown as dots) transfer from H to O:

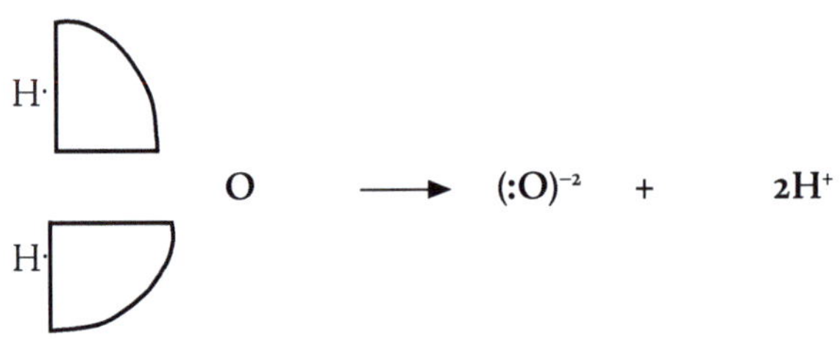


## Formal Charges (AKA Formal Oxidation Numbers or Oxidation States)

Since oxidation-reduction involves electron transfer it is convenient to look at how the electrons are distributed among the elements in a formal sense. That is, we ignore (for now) that the electrons in $H_2O$ are shared covalently, and assign oxidation numbers, also known as formal charges or oxidation states, to the elements involved. This is where the concept of **electronegativity** comes into play in a big way. Recall that electronegativity reflects the relative strength of the attraction of a given nucleus for electrons. Understanding this fundamental concept (described fully in Chapter 2) is crucial to understanding oxidation-reduction.

## Pure, Unreacted Elements

By definition any element that is in its pure, unreacted state has a formal charge of 0. Note that this rule includes the seven diatomic species $H_2$, $O_2$, $N_2$, $F_2$, $Cl_2$, $Br_2$, and $I_2$.

## Oxygen and Hydrogen

Because these two elements are so common in chemistry it is worthwhile examining them individually.

***Oxygen:*** In the vast majority of compounds of all types oxygen has a formal oxidation state of −2. A notable exception to this rule is the important class of compounds known as peroxides which contain a weak O-O single bond. In this case each oxygen has a formal charge of −1. The familiar example of this is the parent compound hydrogen peroxide: $H_2O_2$, more accurately written as HOOH.

***Hydrogen:*** Hydrogen almost always has a formal oxidation state of +1 in compounds. The exception here is metal hydrides, such as NaH or $CaH_2$ wherein H has a formal charge of −1.

## Ionic Compounds

By definition the bonding in these compounds involves a complete electron transfer, so assigning formal charges is easy. Take for example potassium chloride, KCl:

*KCl:*

Cation $K^{+1}$ so the formal charge is simply the valence of +1
Anion $Cl^{-1}$ so the formal charge is simply the valence of −1

*MgO:*

Cation $Mg^{+2}$ so the formal charge is simply the valence of +2
Anion $O^{-2}$ so the formal charge is simply the valence of −2

*$AlBr_3$:*

Cation $Al^{+3}$ so the formal charge is simply the valence of +3
Anion $Br^{-1}$ and there are three Br ions so the formal charge of each one is the individual valence of −1

## Covalent Compounds

The situation is more interesting and complex here, and this is where a basic understanding of electronegativity really comes into play. The two most electronegative elements are F and O, in that order. Covalently bound fluorine is very rare in nature, but oxygen is certainly not. Thus when oxygen occurs in a compound formula it almost invariably has a formal oxidation number of −2. This fact can be used to deduce the oxidation numbers of other elements in a given compound. The same is true for hydrogen which almost always has a charge of +1. Keeping these points in mind we can introduce a simple set of ***rules for assigning formal oxidation numbers*** in covalent compounds as follows:

1. Fluorine always has a −1 charge.
2. The sum of the charges of the elements in a neutral (uncharged) compound, including polyatomic ions, equals 0.

3. The sum of the charges of the elements in a charged compound, including polyatomic ions, is equal to the charge on the ion whether it be plus or minus.
4. In general (but not always!) the order of the elements indicates relative electronegativity of the respective elements. For example, the formula for sulfuric acid is $H_2SO_4$. Here oxygen is the most electronegative element followed by sulfur and finally hydrogen. A key exception here is $OH^-$ where O is still −2.
5. The range of possible formal oxidation numbers for any element is narrow, basically from −7 to +7.

Some examples:

*$CO_2$:*

$O^{-2}$ ***each*** O has a formal charge of −2
$C^{+4}$ since there are two Os here, each −2 and $CO_2$ overall is neutral (no charge) so C has to be +4

*CO:*

$O^{-2}$ the O has a formal charge of −2
$C^{+2}$ since there is just one O at −2 and CO is neutral so C has to be +2

*$NO_3^{-1}$:*

$O^{-2}$ ***each*** O has a formal charge of −2
$N^{+5}$ since there are three Os here, each −2 so together the Os are −6. The nitrate ion $NO_3^-$ has a −1 charge overall leaving a −5 to be neutralized so N has to be +5.

*$SO_4^{-2}$:*

$O^{-2}$ ***each*** O has a formal charge of −2
$S^{+6}$ since there are four Os here, each −2 so together they are −8. The sulfate ion $SO_4^{-2}$ has a −2 charge leaving a −6 to be neutralized so the S has to be +6.

*$HClO_4$:*

$O^{-2}$ ***each*** O has a formal charge of −2
H is +1.
$Cl^{+7}$ since there are four Os here, each −2 so together they are −8. H is +1. $HClO_4$ has no charge so the Cl has to be +7.

**PRACTICE** Assign formal oxidation numbers to the elements is the following compounds or ions:

# 10.3 Balancing Redox Reactions Using the Half Reaction Method

Almost surely the reader questions why we have gone through the exercise of assigning formal oxidation numbers to the elements in formulas. To this point in the book we have balanced equations that could be done simply by inspection, but they are not always this easy. The method of balancing equations using half reactions is a powerful tool for accomplishing this for more complicated equations. By way of example how about balancing this equation? *We assume here that the solution is aqueous and acidic.*

## Balancing in Acidic Solutions

$$MnO_4^{-1} + C2O_4^{-2} \longrightarrow Mn^{+2} + CO_2 \qquad (10.3)$$

Pretty simple looking reaction, but not so simple to balance. *Two rules must be abided by for balancing any equation:*

***The elemental masses on each side of the equation must be equal.*** For this example the number of moles of Mn to the left of the arrow must equal the number to the right of the arrow. Similarly the numbers of Cs and Os must be equal on each side etc.

***The total charge on one side of the equation must equal that on the other side.*** That is, the way the equation above is written the charge sum on the left hand side is −3 $(-1-2=-3)$, the sum on the right is +2. Unequal so this needs to be corrected.

Hereto we can formulate a straightforward set of rules for balancing such equations as follows:

1. Split the overall reaction into two half reactions, one being the oxidation (loss of electrons, $e^-$); the other the reduction (gain of electrons, $e^-$). What element is oxidized; which element is reduced?
2. Treating each half reaction separately, first balance the elements besides O and H.
3. Balance the oxygen O by adding water molecules to side with fewer Os.
4. Balance hydrogen H by adding $H^+$ as necessary.
5. Balance charge by adding $e^-$ (−1 charges) as necessary. Note here that the net charge on each side of the equation does not have to be zero, rather just has to be the same.
6. Multiply through the half reaction equations as required using a common denominator so that after completing step 5 the electrons will cancel out of the equation.
7. Add the half reactions together and simplify the result.

**CAUTION** It is crucial to remember that in balancing equations you cannot change a subscript in a formula without changing the molecule itself. For example, adding an O to CO makes $CO_2$, a very different compound.

Let's now apply these steps to balancing equation 10.3

***Step 1:*** Assign formal oxidation numbers to the elements, and determine what is oxidized and reduced.

$\mathbf{MnO_4^{-1}}$ Each **O** is **−2**, but there are 4 therefore: $(-2)\times(4)=-8$. The whole ion has a −1 charge therefore the **Mn** formal charge is **+7**

$\mathbf{Mn^{+2}}$ **Mn** formal charge here is simply **+2**

This is a reduction, that is a gain of 5 electrons with each $e^-$ having a −1 charge. Mn changes from a +7 to a +2 charge.

$C_2O_4^{-2}$ Each **O** is **−2**, but there are 4 therefore: $(-2) \times (4) = -8$. The whole molecule (ion) has a **-2** charge therefore the **C** formal charge is **+3**

$CO_2$ Each **O** is **−2**, and there are 2 therefore: $(-2) \times (2) = -4$. The whole molecule has no charge therefore the **C** formal charge is **+4**

This is an oxidation, that is, a loss of 1 electron having a −1 charge. C goes from +3 to +4

Now write out the half reactions and balance using steps 2-7 above as necessary.

*For the reduction half reaction these are the steps:*

$$MnO_4^{-1} \longrightarrow Mn^{+2}$$

***Step 2:*** Mn is balanced already.

***Step 3:*** Add 4 water molecules to the right hand side of the equation to balance the number of oxygen atoms:

$$MnO_4^{-1} \longrightarrow Mn^{+2} + 4H_2O$$

***Step 4:*** Add 8H+ to the left hand side of the equation to balance H (4x2 = 8):

$$8H^+ + MnO_4^{-1} \longrightarrow Mn^{+2} + 4H_2O$$

***Step 5:*** Note that the total ionic charge on the left side is +7 (8-1) while on the right side it is +2. To make the charge the same on both sides add 5 $e^-$ (each a -1) to the left side:

$$5e^- + 8H^+ + MnO_4^{-1} \longrightarrow Mn^{+2} + 4H_2O \quad \textbf{(10.4)}$$

This half reaction is done.

*For the oxidation half reaction these are the steps:*

$$C_2O_4^{-2} \longrightarrow CO_2$$

***Step 2:*** To balance C place a coefficient of 2 in front of $CO_2$. You cannot change the $CO_2$ to $C_2O_2$ as they are different compounds!

$$C_2O_4^{-2} \longrightarrow 2CO_2$$

***Step 3:*** The Os are balanced.

***Step 4:*** The Hs are balanced (none present!).

***Step 5:*** Add 2e− to the right hand side to give a balanced −2 charge on both sides of the equation:

$$C_2O_4^{-2} \longrightarrow 2CO_2 + 2e^- \quad \textbf{(10.5)}$$

This reaction too is done, but the e− need to cancel out at the end therefore we need a common denominator for **2e−** and **5e−** the simplest of which is 10. Multiply the whole equation 10.4 through by 2, and 10.5 by 5

$$2(5e^- + 8H^+ + MnO_4^{-1} \longrightarrow Mn^{+2} + 4H_2O) \quad \textbf{(10.4)}$$

$$5(C_2O_4^{-2} \longrightarrow 2CO_2 + 2e^-) \quad \textbf{(10.5)}$$

Then add them together to yield the final balanced equation.

$$10e^- + 16H^+ + 2MnO_4^{-1} + 5C_2O_4^{-2} \rightarrow 2Mn^{+2} + 8H_2O + 10CO_2 + 10e^-$$

Check the result as proof:

2 Mn on both sides of the equation.
28 O on both sides of the equation.
10 C on both sides of the equation.
16 H on both sides of the equation.
Net charge is +4 on both sides of the equation.

## Balancing in Basic Solutions

The example above is a reaction that is run in acidic solution. It is rare that a reaction is run under conditions where the pH is neutral, but what about solutions that are basic? This is easily done as follows:

1. First balance the equation under acidic conditions as shown in the example above.
2. Add OH– to both sides of the equation until the H+ are gone.

For the example above this would be:

$$16OH^- + 16H^+ + 2MnO_4^{-1} + 5C_2O_4^{-2} \rightarrow 2Mn^{+2} + 8H_2O + 10CO_2 + 16OH^-$$

Which after simplification is equivalent to:

$$8H_2O + 2MnO_4^{-1} + 5C_2O_4^{-2} \rightarrow 2Mn^{+2} + 10CO_2 + 16OH^-$$

**PRACTICE** Balance the following reaction equation using the method of half reactions:

$$CR_2O_7^{-2} + H^+ + I^- \longrightarrow Cr^{+3} + I_2 + H_2O$$

# 10.4 The Activity Series

Ohm's Law is fundamental and simple. It relates the three key variables of electricity, voltage, current, and resistance:

$$V = IR$$

*Where V = voltage (volts), I = current (amps), R = resistance (ohms)*

Useful electricity is driven by the movement of electrons from point A to point B as current. It is a given that every element in the periodic table can gain and lose electrons (to yield a current), but what makes this fact especially interesting is that there are specific voltages for each element at which this favorably occurs. These voltages are directly related to the activity, or more accurately put, the reactivity of the elements. Let's examine some of the most important metals; Li (lithium), Ag (silver), Zn (zinc), Al (aluminum), Mg (magnesium), Cu (copper), and Fe (iron). All of these metals have valence electrons, but their willingness to give them up varies over a wide range. This willingness can be quantified in the ***Activity Series*** as shown in Table 10.1:

TABLE 10.1 **Reduction Potentials of Common Metals**

| Reactivity | Metal | Ion formed | Voltage1 |
|---|---|---|---|
| Most active | Li | $Li^{+}$ | −3.04V |
| | Mg | $Mg^{+2}$ | −2.36V |
| | Al | $Al^{+3}$ | −1.66V |
| | Zn | $Zn^{+2}$ | −0.76V |
| | Fe | $Fe^{+2}$ | −0.45V |
| **Reference1** | **H** | $\mathbf{H^{+}}$ | **0** |
| | Cu | $Cu^{+2}$ | +0.34V |
| Least active | Ag | $Ag^{+1}$ | +0.80V |

*Referenced to the hydrogen ion reaction:* $H_2(g) \longleftrightarrow 2H^{+}(aq) + 2e^{-}$

These numbers are also referred to as ***reduction potentials***. Appendix 3 contains a more extensive list of reduction potentials. Note that the hydrogen ion reduction to form elemental hydrogen gas is used as a reference reaction that all of the other elemental activities are based upon. For example, two Li metal atoms will rapidly give up their lone valence electrons to $2H^{+}$ to yield two $Li^{+}$ ions and a molecule of elemental $H_2$.

The rest of this chapter deals with important applications and use of the Activity Series in batteries, fuel cells, capacitors, and corrosion.

## 10.5 Voltaic Cells (AKA Galvanic Cells)

In a voltaic cell a chemical reaction runs spontaneously via a transfer of electrons between reactants. Recall a spontaneous reaction is one that given time, which may be very long or very short, the reaction will run. That is, it is favored to run energetically downhill. When this occurs we can use the electrons being transferred to do useful work. There are many ways to tap into this electrical energy source, and we will briefly look at just a few of the most important ones now.

### Batteries

A battery stores potential energy in the form of chemical reactants. In principle this energy is stored indefinitely until a load is connected, whereupon the chemical energy is converted to electrical energy which performs useful work.

There are a large number of battery types and designs; however, they all share some essential components. The components used are determined by the application of interest and the constraints and performance requirements therein.

## Electrodes

***Anode:*** This is where oxidation (loss of $e^-$) occurs, and electrons flow from here through wires to the load where they do work, and then continue on to the:

***Cathode:*** This is where reduction (gain of $e^-$) occurs, and electrons flow from the load through wires to form a new product or products.

## Electrolyte

As the electrons move negative charge from anode to cathode, this loss of minus charge must be compensated for by supplying anions to the anode or the battery will shut down instantly. The same is true at the cathode but in reverse; that is the negative charges being received via electrons must be neutralized with cations. The anions and cations are pre-dissolved and available in the electrolyte by design.

A specific example of a battery, there are many, is the Daniell cell, a simple and practical battery that was first constructed in 1836.

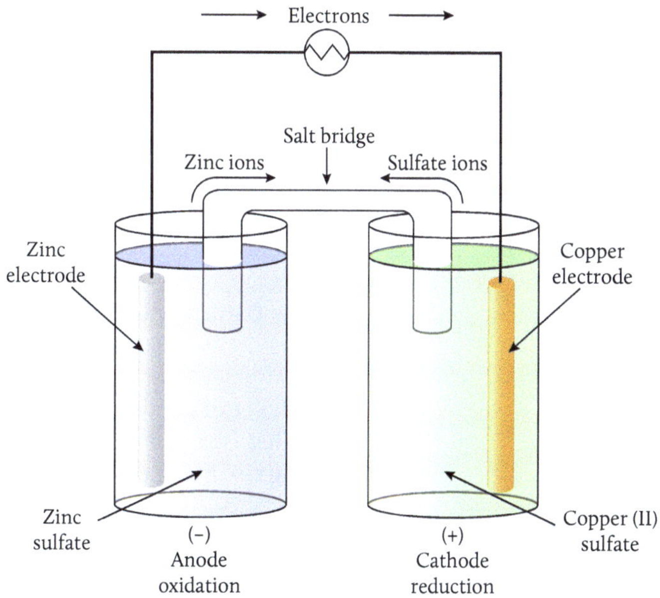


**FIGURE 10.2.** Daniell Cell.

### Schematic of Daniell cell

In the Daniell cell (Figure 10.2) zinc (Zn) is the more reactive metal and acts as the solid state anode by giving up electrons (oxidized) per this reaction:

$$\mathbf{Zn\ (s) \longrightarrow Zn^{+2}\ (aq) + 2e^-}$$

The cathode is a solid piece of copper metal. As electrons are transferred from the anode to the cathode $Cu^{+2}$ ions, already dissolved in the electrolyte, are reduced on the surface of the cathode to form copper metal:

$$\mathbf{Cu^{+2}\ (aq) + 2e^- \longrightarrow Cu\ (s)}$$

As the reaction proceeds $Zn^{+2}$ ions are continuously lost via dissolution in the electrolyte. In addition, the battery performance is limited by the rate at which the $Cu^{+2}$ ions are transported through the electrolyte solution and common in contact with the cathode surface.

Although appearing simple on paper this reaction is remarkable complex. The actual reactions are taking place on the electrode surfaces (Figure 10.3).

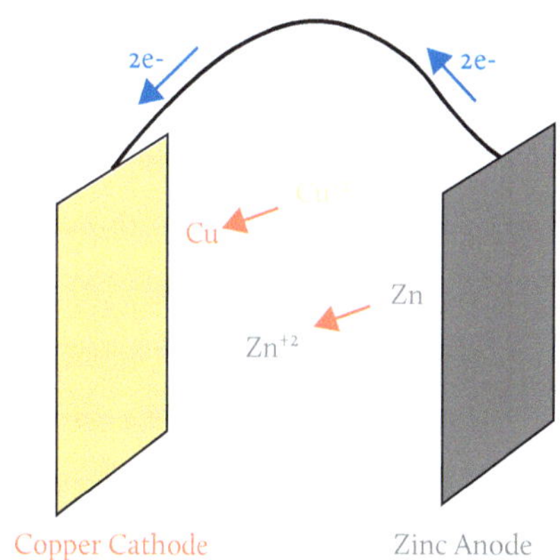


**FIGURE 10.3.** Copper Cathode and Zinc Anode.

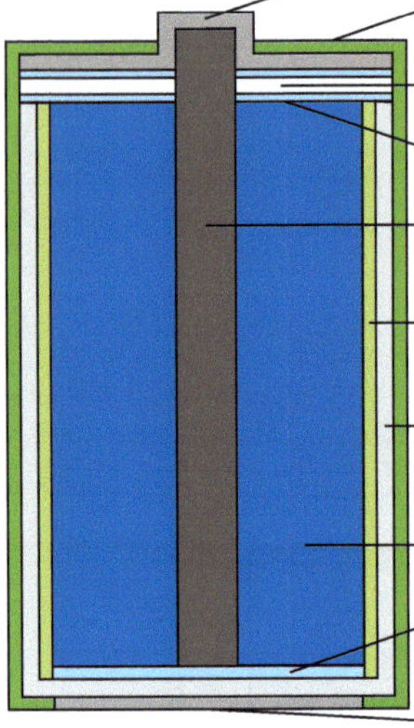


**FIGURE 10.4.** Dry (D) Cell Schematic.

### Schematic of an Zinc/$MnO_2$ Battery

These primary (non-rechargeable) cells (Figure 10.4) are one of the most important batteries ever developed. The familiar dry cells, AA, AAA, C, and D are based on this chemistry which is given below. Here Zn does double duty as anode and outer casing. Inside is a carbon rod which serves as a current collector. In between these components are the electrolyte and cathode.

## Ohm's Law Revisited

An understanding of the relationship between voltage, current and resistance ($V = IR$) is fundamental to the design of useful electrochemical systems and applications.

***Voltage (V)***: The cell voltage (E°cell) is set by the choice of anode and cathode materials, and can be calculated using the half-cell voltages (reduction potentials) such as those given in Table 9.1 above using the following equation:

$$E^\circ cell = E^\circ red(cathode) - E^\circ ox(anode)$$

For the Daniell cell $E^\circ_{cell}$ is then:

$$E^\circ cell = +0.34V - (-0.76V) = +1.10V$$

The choice of anode and cathode materials pretty much fixes what the cell voltage will be. That is, a zinc anode-copper cathode cell will always give a voltage of ~1.10V everything else being equal.

**PRACTICE** One of the most promising high energy batteries is based upon a lithium anode and a sulfur cathode. Given the data contained in Appendix 3, what is the theoretical voltage of this battery?

***Current (I):*** The cell current is much more variable than the voltage. Recalling that for our cell above the reactions are all taking place on the surface of the electrodes this suggests the chemical and physical properties of those surfaces are critical in terms of the current achieved. For example, it should be intuitive that the larger the surface area of the electrodes the higher the current will be. This is a good assumption, but there are many other factors that come into play. The good news is you have a lot of variables present to optimize the cell performance. The bad news is the same.

***Resistance (R):*** Resistance is merely the sum of the impediments to electronic and ionic flow. Like current there are many variables involved. For both electrons and ions one can think of the charges going from one point to another via a torturous path. Electron movements tend to be very fast indeed, but not so for ions moving through solution. Relative to electrons the ions are huge and must deal with interactions with solvent molecules and other ions as they move through the electrolyte. Other factors which affect resistance are electrical connections and temperature.

## Primary & Secondary Batteries

***Primary batteries*** are ones that cannot be recharged. Basically the energy is stored in the form of chemical reactants that are physically separated until a load is applied and the electrodes connected to complete the circuit. A common example here is flashlight batteries wherein the reactants remain inert until the switch is flipped whereupon the anode and cathode are electrically connected through the light bulb. Electrons flow, the reaction proceeds and light is emitted from the bulb. Once either one of the reactants is consumed the reaction stops, and the light goes out.

***Secondary batteries*** can be charged and discharged many times. Upon discharge these batteries act exactly like primary batteries. Upon being connected to a power source the reactions can be reversed. In essence the recharge process involves a cell inversion; that is, the anode becomes the cathode and the cathode becomes the anode. Note that the recharge process requires energy input. Secondary batteries tend to be significantly more complicated that primary ones.

## Important Battery Types

There are many, many different types of batteries that have been evaluated. We will look at s of some of the most important types. As mentioned above the heterogeneous nature of the reactions, transient nature of the intermediates, and the transport properties involved make the exact processes difficult to assess.

Batteries do not generally perform well at high or low temperatures. From a practical standpoint the temperature range limits are between –20 and +40°C. The closer the temperature is to 25°C the better will be the performance.

## Primary Batteries

### *Zn-MnO$_2$ (Alkaline) Battery*

This most important primary battery is in widespread use worldwide and has been known since 1866, but the common alkaline battery has only been in use since 1955. Before that zinc-carbon batteries were the standard, but they have much shorter useful lifetimes. The basic electrochemistry is the same, but the electrolyte is acidic rather than basic, as well as the cell design itself.

Uses for these types of batteries include a very wide range of electronic devices especially in the design/construction forms of AAA, AA, C, D and dry cells.

The half reactions are:

*Anode reaction*

$$Zn(s) + 2OH^-(aq) \longrightarrow ZnO(s) + H_2O(l) + 2e^-$$

*Cathode reaction*

$$2MnO_2(s) + H_2O(l) + 2e^- \rightarrow Mn_2O_3(s) + 2OH^-(aq)$$

Overall reaction:

$$Zn(s) + 2MnO_2(s) \rightleftharpoons ZnO(s) + Mn_2O_3(s)$$

***Electrolyte:*** Aqueous potassium hydroxide (KOH).

The key performance characteristics here are: **Voltage = +1.45V; Specific power: 50 watts/kg.**

**Zn-$Ag_2O$ Battery** Developed in 1960 these batteries are expensive because of the silver, but possess relatively long lifetimes, a higher voltage, and a steady, non-fluctuating voltage. They are thus widely employed in important specialty markets.

Important uses for these types of batteries in the form of coin (button) cells include hearing aids and small portable devices such as watches.

The half reactions are:

*Anode reaction*

$$Zn(s) + 2OH^-(aq) \longrightarrow ZnO(s) + H_2O(l) + 2e^-$$

*Cathode reaction*

$$Ag_2O(s) + H_2O(l) + 2e^- \longrightarrow 2Ag(s) + 2OH^-(aq)$$

Overall reaction:

$$Zn(s) + Ag_2O\ (aq) \rightleftharpoons 2Ag + ZnO(s)$$

***Electrolyte:*** Aqueous potassium hydroxide (KOH).

The key performance characteristics here are: **Voltage = +1.55V; Specific power: 200 watts/kg.**

**Li/$I_2$ Primary Batteries** These non-rechargeable batteries are interesting because they represent the polar opposite of the secondary lithium batteries to be described, and reflect the versatility of lithium battery technologies.

The heart pacemaker application has drastically different performance requirements compared with that of the electric car. Clearly, in the former case very low voltages and currents are a prerequisite, but so is a very long service life. Li/$I_2$ batteries have lifespans of about 10 years. The chemistry is straightforward:

***Anode (Li/C):*** $Li \longrightarrow Li^+ + e^-$

***Cathode ($CoO_2$):*** $I_2 + 2e^- \longrightarrow 2I^-$

***The net reaction is:*** $2Li + I_2 \longrightarrow 2Li^+ + 2I^-$

***Electrolyte:*** In this case a solid state electrolyte is used, polyvinylpyridine:

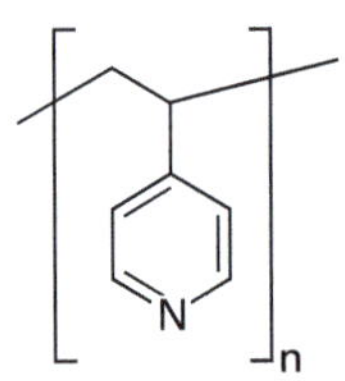


**CHEMICAL STRUCTURE 10.1.** Electrolyte Structure.

*Polyvinylpyridine (PVP); Here n represents a polymer of n number of units.*

The key performance characteristics here are: Voltage = 2.8V; Specific power: 240 watt-hr/kg

## Secondary Batteries

***Lead Acid Battery:*** Familiar to all of us as the key battery in vehicles this battery chemistry was invented in 1859 (Figure 10.5).

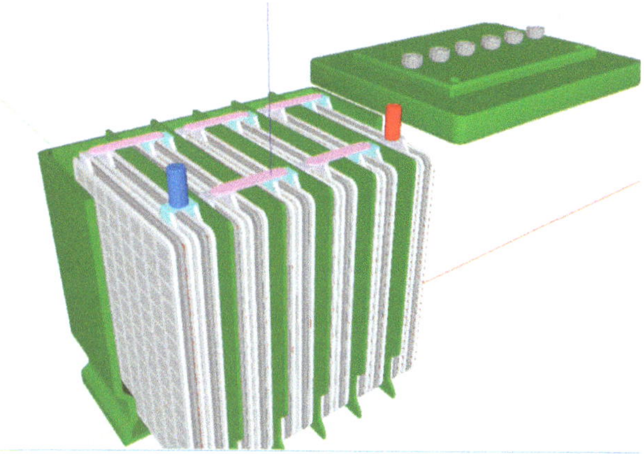

FIGURE 10.5. Lead Acid Battery Configuration.

The chemistry is interesting because it involves basically the same materials as anode and cathode. The anode and cathode are Pb and PbO, respectively, and the electrolyte is sulfuric acid, $H_2SO_4$. The discharge chemistry is:

*Anode reaction*

$$Pb(s) + HSO_4^-(aq) \longrightarrow PbSO_4(s) + H^+(aq) + 2e^-$$

*Cathode reaction*

$$PbO_2(s) + HSO_4^-(aq) + 3H^+(aq) + 2e^- \longrightarrow PbSO_4(s) + 2H_2O(l)$$

The total reaction can be written as

$$Pb(s) + PbO_2(s) + 2H_2SO_4(aq) \longrightarrow 2PbSO_4(s) + 2H_2O(l)$$

The charging process is basically the reverse of these chemistries.

***Electrolyte:*** Concentrated (~98%) sulfuric acid ($H_2SO_4$).

The key performance characteristics here are: **Voltage = 2.0V; Cycle durability: 1000 charge discharge cycles. Specific power: 180 watts/kg.**

Lead acid batteries are easy to assemble and of low cost. They account for a large percentage of total battery usage worldwide. Since lead is so dense the specific power is low. Although lead is toxic and an environmental concern most of the lead metal is recycled. Two main uses for these batteries are in transportation and stationary power storage applications.

**Lithium Batteries** Lithium battery chemistries are diverse and complex. There are a range of lithium batteries available for many different applications. Originally developed as primary battery systems, since the mid-1970s the secondary Li battery applications and markets have blossomed. Figure 10.6 is a representation of metallic lithium intercalated with a carbon anode.

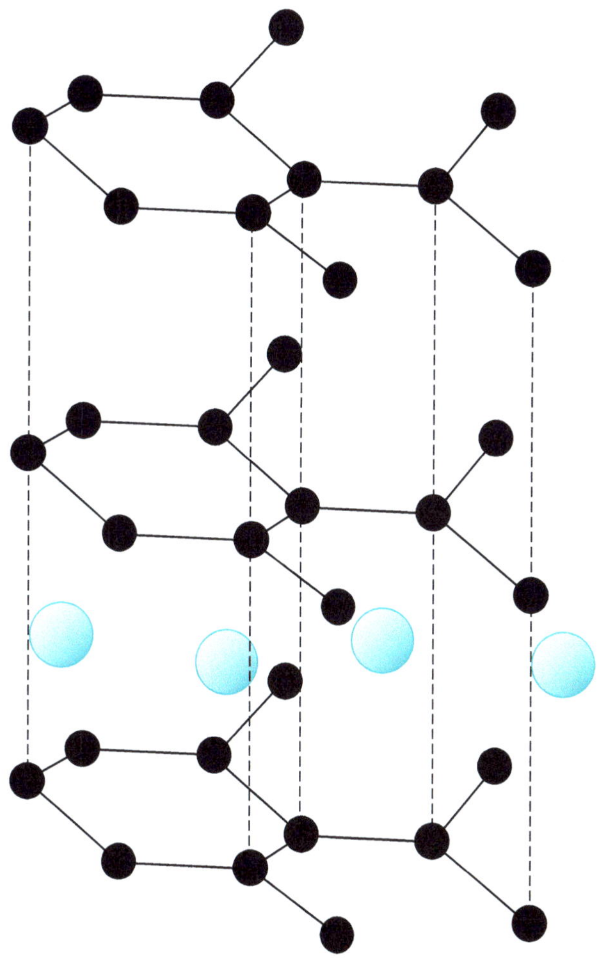

FIGURE 10.6. Li in Carbon Anode INTERCALATION.

The many applications for this versatile chemistry range from those that require very modest voltage and current requirements (pace maker battery) to those demanding tremendous electrochemical performance (electric car).

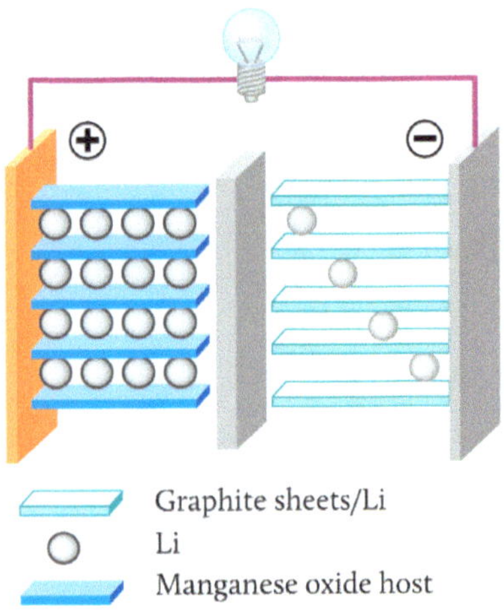


**FIGURE 10.7.** Li Battery Schematic.

**Lithium-Transition Metal Oxide Systems** The chemistry and engineering of lithium batteries is inherently complex for a number of reasons. As one of the most active metals lithium will readily, and even explosively, react with water. Thus aqueous electrolytes are not an option here. Sealed batteries are constructed in dry rooms in the absence of water and oxygen. To date the electrolyte of choice has been dimethylcarbonate, which is flammable, with lithium salts dissolved therein to conduct ions. In addition, lithium metal is not suitable as an anode for practical reasons, basically the formation of metallic dendrites that short out the batteries upon repeated charge-discharge cycling.

Figure 10.7 shows the basic anode-cathode constructions.

Here the anode is carbon into which lithium metal is intercalated (dispersed within the carbon 3D structure). Upon discharge the Li give up their electrons which travel through the load to the cathode where cobalt is reduced in formal charge. Simultaneously, $Li^+$ ions intercalate into the cobalt oxide cathode. The discharge reactions are:

| | | | |
|---|---|---|---|
| *Anode (Li/C):* | $xLiC_6$ | $\longrightarrow$ | $xLi^+ + xe^- + xC_6$ |
| *Cathode ($CoO_2$):* | $Li_{1-x}CoO_2 + xLi^+ + xe^-$ | $\longrightarrow$ | $LiCoO_2$ |
| The net reaction is: | $xLiC_6 + Li_{1-x}CoO_2$ | $\longrightarrow$ | $LiCoO_2 + xC6$ |

There are many other cathode chemistries used besides the one shown here based on cobalt.

***Electrolyte:*** Since lithium reacts vigorously with water an aqueous electrolyte is not possible. The solvent of choice today is a mixture of organic compounds such as diethylcarbonate, $CH_3CH_2C(O)CH_2CH_3$, in which is dissolved a lithium salt. These organic solvents have the disadvantage of being quite flammable, a significant safety concern.

The key performance characteristics here are: **Voltage = 3.6V; Cycle durability: 1000 charge discharge cycles. Specific power: 1000 watts/kg.**

## Nickel Batteries

*Ni-Cd:*

Invented in 1899 cadmium is the anode here, and nickel oxide is the cathode with the following discharge chemistries:

*Anode (Cd):*

$$Cd + 2OH^- \longrightarrow Cd(OH)_2 + 2e^-$$

*Cathode (NiO):*

$$NiO(OH) + 2H_2O + 2e^- \longrightarrow 2Ni(OH)_2 + 2OH^-$$

*The net reaction is:*

$$2NiO(OH) + Cd + 2H_2O \longrightarrow 2Ni(OH)_2 + Cd(OH)_2$$

Cadmium, and to a lesser extent nickel, are both of toxicological and environmental concern, hence the markets for these batteries have been in sharp decline. However, compared to other battery systems they have the distinction of being useful at temperatures down to −45°C.

***Electrolyte***: Aqueous potassium hydroxide is commonly used.

**The key performance characteristics here are: Voltage = 1.2V; Cycle durability: 500 charge/discharge cycles. Specific power: 200 watts/kg.**

*Ni-Metal Hydride:*

Because of significant environmental and toxicity issues with the cadmium in Ni-Cd batteries, their usage is in decline. Ni-metal hydride (MH) batteries perform in a similar manner, but are much safer. They have found widespread usage in electric car as well as aerospace applications. They are particularly distinguished by being able to be cycled thousands to hundreds of thousands of times and still perform well.

Invented in 1967 this battery has an intermetallic compound (M below), for example $LaNi_5$, as the anode with nickel oxide as the cathode. The anode chemistry is complex and interesting (to an extent meaning not well understood). It involves the intercalation of ***elemental hydrogen*** within the multi-metal structure as follows (M is the intermetallic species):

*Anode (Cd):* $MH + OH^- \longrightarrow M + H_2O + e^-$

The cathode chemistry is similar to that of the Ni-Cd system:

*Cathode (NiO):* $NiO(OH) + H_2O + e^- \longrightarrow Ni(OH)_2 + OH^-$

**The net discharge reaction is:** $NiO(OH) + MH \longrightarrow Ni(OH)_2 + M$

***Electrolyte:*** Aqueous potassium hydroxide is commonly used here as well

**The key performance characteristics here are: Voltage = 1.2V; Cycle durability: 20,000 charge/discharge cycles. Specific power: 1000 watts/kg.**

## Fuel Cells

Fundamentally, fuel cells can be considered to be batteries wherein the chemical reactants are constantly being fed, rather than having been stored, to an anode-cathode assembly where the oxidation-reduction reaction takes place. In this variation the electrodes are usually inert catalytic materials, such as platinum, which are not consumed during the reaction. In principle a fuel cell will run as long as the reactants are being supplied to the cell.

### Components of a Typical Fuel Cell (Hydrogen Fuel)

*Anode*

- Negative post of the fuel cell.
- Conducts the electrons that are freed from the hydrogen molecules so that they can be used in an external circuit.
- Etched channels disperse hydrogen gas over the surface of catalyst.

*Cathode*

- Positive post of the fuel cell
- Etched channels distribute oxygen to the surface of the catalyst.
- Conducts electrons back from the external circuit to the catalyst.
- Recombine with the hydrogen ions and oxygen to form water.

*Electrolyte*

- Proton exchange membrane (PEM).
- Specially treated material, only conducts positively charged ions.
- Membrane blocks electrons.

*Catalyst*

- Special material that facilitates reaction of oxygen and hydrogen.
- Usually platinum powder very thinly coated onto carbon paper or cloth.
- Rough & porous maximizes surface area exposed to hydrogen or oxygen.
- The platinum-coated side of the catalyst faces the PEM.

Many different reactants have been explored, but usually gaseous oxygen is being reduced at the cathode since it is ubiquitous and basically free from the atmosphere.

Here are some other important fuel cell configurations and chemistries:

*Alkaline (KOH) fuel cell (AFC)*

- One of the oldest designs. It has been used in the U.S. space program since the 1960s. The AFC is very susceptible to contamination, so it requires pure hydrogen and oxygen. Commercial prospects are improving due to materials advances.

*Phosphoric-acid fuel cell (PAFC)*

- The phosphoric-acid fuel cell has potential for use in small stationary power-generation systems. It operates at a higher temperature than PEM fuel cells, so it has a longer warm-up time. This makes it unsuitable for use in cars.

*Solid oxide fuel cell (SOFC)*

- These fuel cells are best suited for large-scale stationary power generators that could provide electricity for factories or towns. This type of fuel cell operates at very high temperatures (around 1,832 F or 1,000 C). This high temperature makes reliability a problem, but it also has an advantage: The steam produced by the fuel cell can be channeled into turbines to generate more electricity. This improves the overall efficiency of the system.

*Molten carbonate fuel cell (MCFC)*

- These fuel cells are also best suited for large stationary power generators. They operate at 1,112 F (600 C), so they also generate steam that can be used to generate more power. They have a lower operating temperature than the SOFC, which means they don't need such exotic materials. This makes the design a little less expensive.

*Fossil Fuel Based*

- The most important and familiar type of fuel cell. The three most commonly used fuels have been hydrogen ($H_2$), natural gas ($CH_4$), and methanol ($CH_3OH$). On paper hydrogen is the ideal choice because its combination with $O_2$ produces just water as the product (see figure 10.8).

$$2H_2 + O_2 \longrightarrow 2H_2O$$

*Figure 10.8 diagrams the workings of a hydrogen fuel cell.*

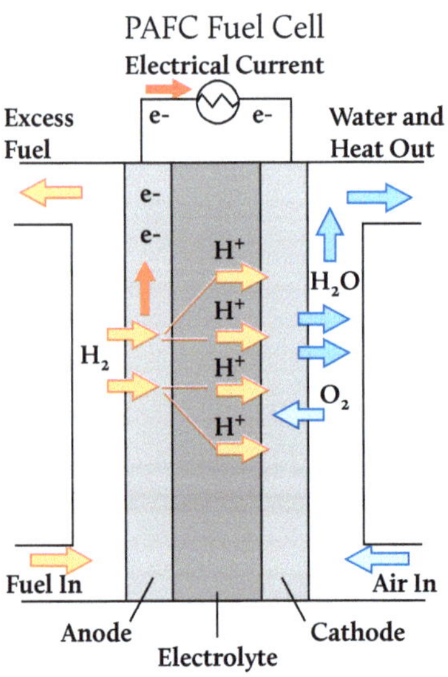


FIGURE 10.8. Basic Hydrogen Fuel Cell.

Hydrogen and oxygen gases are fed in the reaction chamber wherein inert anode and cathode are contained. Like batteries the two half reactions, oxidation of hydrogen and reduction of oxygen, occur on the catalyzed surfaces of the anode and cathode, respectively. When compared to combustion processes a fuel cell is not constrained by Carnot Cycle theoretical limitations, so much higher efficiencies are possible of the order of 40–70%.

It is not as simple as it appears. Although hydrogen is generated in substantial quantities via naturally occurring biochemical processes, it does not hang around. Being so small, volatile, and reactive that it is either consumed in other naturally occurring reactions or it just dissipates through the earth's atmosphere into space.

So to run a fuel cell on hydrogen you need to synthesize the $H_2$. There are many ways to do this, but as always practicality (economics) is of paramount importance. Four processes are commonly used to make hydrogen in large quantities as follows:

***Steam Reforming****—Endothermic reaction run at high temperatures (700-1100°C) to make* ***syngas:***

$$CH_4 + H_2O \longrightarrow 3H_2 + CO$$

***Coal Gasification***

$$C + H_2O + O_2 \longrightarrow H_2 + CO$$

***Water-gas Shift Reaction***

$$CO + H_2O \longrightarrow H_2 + CO_2$$

***Electrolysis***

$$H_2O \longrightarrow H_2 + CO_2$$

All of these processes require energy, but are carried out in large volumes annually.

## Capacitors

A capacitor is a device designed to separate plus (+) and minus (–) electrical charges in an efficient manner. They are also closely related to batteries and fuel cells, and are often paired with the same in circuits. Figure 10.9 shows the design of a basic capacitor:

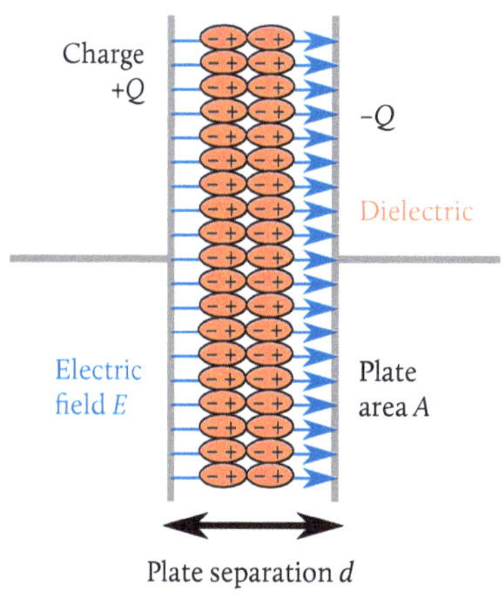


**FIGURE 10.9.** Basic Capacitor Design.

There are many different types and designs of capacitors, but they all share these fundamental components. **Anodes** and **cathodes**, and an electrical circuit wired to a load of some kind. Between the electrodes; however, there is a **dielectric**, a substance that is an electrical and ionic insulator. What the dielectric does is to allow the separation of electronic charge between two plates. The more insulating the dielectric, the larger the charge that can be separated. This is where chemistry comes to the fore in the design of the material to be used as the dielectric. The materials can be gas, liquid or solid. For example, air is an excellent dielectric, a silicate rock is OK, while seawater is a very poor candidate.

Typically capacitors are used where a pulse of power is required. It is the rare electrical device that does not have a capacitor included. High voltages are common, as are long term storage capacity of the same, while the energy densities possible are very low. In today's world the differences between capacitors and batteries has become blurred as a range of energy densities (a battery's strong suit) coupled with short term high power (a capacitor's strong suit) that are available.

# Chapter 10 Summary

*Electrochemistry:* the study of chemical processes that involve electron transfers.

*Voltaic Cells:* These are electrochemical systems wherein energy is stored in reactants, which upon contact, undergo spontaneous reactions that produce chemical products and a flow of electrons that can be harnessed to do work.

*Electrolytic Cells:* Here electrons are supplied to make a desired reaction occur in a nonspontaneous manner. Energetically the overall process runs uphill, that is, consumes energy.

*Oxidation Reduction Reactions:* Reactions that occur wherein electrons are transferred from one chemical species to another resulting in elements changes oxidation states (formal charges).

*Oxidation States (Formal Charges):* The assignment of formal charges or valences to elements within a chemical formula.

*Balancing Equations Using Half Reactions:* A method of balancing complex oxidation reduction reactions wherein the oxidation and reduction steps are separated and balanced, and then recombined into a full reaction equation.

*Activity Series:* The tabulation or ordering of pure of metals from most electrochemically reactive to least.

*Voltaic Cells:* Here an electrochemical reaction runs spontaneously via a transfer of electrons between reactants. The two primary types of voltaic cells are batteries and fuel cell:

*Batteries:* A battery stores potential energy in the form of chemical reactants. In principle this energy is stored indefinitely until a load is connected, whereupon the chemical energy is converted to electrical energy which performs useful work.

*Primary Batteries* are ones that are single usage and cannot be recharged.

*Secondary Batteries* can be recharged and used again.

*Fuel Cells:* These are basically batteries wherein the chemical reactants are constantly being fed to an anode-cathode assembly where the oxidation-reduction reaction takes place.

*Capacitor:* A device designed to separate plus (+) and minus (–) electrical charges in an efficient manner. Upon demand a pulse of electrical charge can be released through a load to do work. In terms of construction capacitors have similar components to batteries and fuel cells.

# CHAPTER 10 EXERCISES

1. Identify the formal oxidation number of each of the elements in the following elements and molecules:

   a. Xe

   b. CsBr

   c. $H_2SO_4$

   d. $NO_3^-$

   e. $PBrCl_2$

2. Identify the formal oxidation number of ***each*** of the elements in the following elements and molecules:

   a. $KClO_3$

   b. $Fe_2O_3$

   c. $I_2$

   d. $CS_2$

   e. $N_2O_4$

3. Balance the following reaction using the method of half reactions:

   $$I_2O_5(s) + CO(g) \longrightarrow I_2(s) + CO_2(g)$$

4. Balance the following reaction using the method of half reactions:

   $$Hg^{2+}(aq) + N_2H_4(aq) \longrightarrow Hg(l) + N_2(g)$$

5. Balance the following reaction using the method of half reactions:

   $$H_2S(aq) + NO_3^-(aq) \longrightarrow S(s) + NO(g)$$

6. Balance the following reaction using the method of half reactions:

   $$MnO_4^-(aq) + S_2^-(aq) \longrightarrow S(s) + MnO_2(s) + OH^-(aq)$$

7. Balance the following reaction using the method of half reactions:

   $$H_2O_2(aq) + Cl_2O_7(g) \longrightarrow ClO_2^-(aq) + O_2(g)$$

8. Balance the following reaction using the method of half reactions:

   $$Ba_2^+(aq) + H_2O_2(aq) + ClO_2(aq) \longrightarrow Ba(ClO_2)_2(s) + O_2(g)$$

9. What physical and chemical properties make lithium an excellent candidate for the anode of batteries?

10. What are the basic components of any battery system? Draw a voltaic cell with all of these components.

11. The electric drive has come to the forefront of future transportation technologies, especially for cars and trucks. You, a chemist and an engineer rolled into one, are the designer of new battery systems. What are the safety and other logistical constraints that you must consider?

12. You are designing a anode-cathode couple for a new battery system. You are not allowed to use the components of current large scale battery systems, including lithium, lead acid, nickel cadmium, nickel metal hydride, silver, and zinc-manganese dioxide. You want get get the largest open circuit voltage you can. What do you choose as anode and cathode materials?

13. You are a battery builder that is stuck in the desert in Nevada, a state rich in minerals. At hand readily available are metallic silver and iron as well as silver and iron salts of various kinds. Diagram a battery you could build. Show which way the electrons will flow. What voltage can you expect?

14. In problem #13 the voltage is set in stone by your choice of anode and cathode materials; however, the current is not. What might you do to maximize current flow?

15. What is the difference between a battery and a fuel cell?

16. Can a ***totally*** solid state fuel cell be constructed?

17. Which of the following compounds might be used as a fuel in a fuel cell

    a. Diethyl ether

    b. Silicon dioxide

    c. Phosphorous

    d. Methanol

    e. Carbon tetrachloride

18. What do a battery and a capacitor have in common?

19. Rank the following compounds in terms of highest to lowest dielectric strength (higher means the resistance to breakdown at a greater imposed voltage).

    a. Carbon

    b. Rubber

    c. Air

    d. Vacuum

    e. Titanium

    f. Silicon dioxide

    g. Silicon

20. Draw a double layer capacitor.

## Credits

Fig. 10.1: Source: https://commons.wikimedia.org/wiki/File:Roadster_2.5_windmills_trimmed.jpg.

Fig. 10.2: Copyright © 2012 by Hazmat2, (CC BY 3.0) at https://commons.wikimedia.org/wiki/File:Galvanic_cell_labeled.svg.

Fig. 10.4: Source: https://commons.wikimedia.org/wiki/File:Zinc-carbon-english.svg.

Fig. 10.5: Copyright © 2012 by KVDP, (CC BY-SA 3.0) at https://commons.wikimedia.org/wiki/File:SLA_battery.png.

Fig. 10.6: Source: https://commons.wikimedia.org/wiki/File:Intercalation(GraphitePotassium).svg.

Fig. 10.7: Source: https://commons.wikimedia.org/wiki/File:Basic_battery_charging.jpg.

Fig. 10.8: Source: https://commons.wikimedia.org/wiki/File:Fcell_diagram_pafc.gif.

Fig. 10.9: Copyright © 2008 by Papa November, (CC BY-SA 3.0) at https://commons.wikimedia.org/wiki/File:Capacitor_schematic_with_dielectric.svg.

CHAPTER 11

# Metallurgy, Alloys, and Corrosion

## Just suppose steel was suddenly not available. How would you go about making it? From scratch!

Metallurgy is the study of metals; how they are prepared and purified as well as what their physical and chemical properties are.

We take metals for granted. We are all familiar with the value and uses of important metals such as iron, copper and aluminum, but where do they come from? How are they synthesized? Not as familiar are the biologically important metals in ionic form, such as $Na^+$, $K^+$, $Mg^{+2}$, and $Ca^{+2}$, without which life as we know it would simply not be possible. Unknown to most of us are other biologically crucial trace metals, for example Co, Zn, Mn, and B that are necessary for processes involving enzymes and vitamins. There are even ultra-trace elements, including arsenic, that appear to be biologically important as well. Then there are the semimetals or metalloids whose properties bridge the gap between metals and non-metals, and that are essential components of all modern electronic devices. The metals world is one that is interesting and chemically complex.

## 11.1 Pure Metals

The previous chapter covered the topic of metal activities. Recall the reactivity order of some important metals is:

REACTIVE **Li** > **Mg** > Al > Fe > **Zn** > Cu > **Au** > **Pt** INERT

Highlighted in red are the three metals that are essential to society as we know it today. Although copper occurs on earth in pure form this circumstance is quite rare, and the other two (Al and Fe) never occur naturally in a pure metallic form. From a practical standpoint all three metals must be derived from their respective native **ores,** usually from the chemical form of an oxide or sulfide (Figure 11.1).

**Ores** are generally complex mixtures of chemical compounds, predominantly derivatives containing large amounts of $SiO_2$ and $Al_2O_3$. **Minerals** are a subset of the family of ores, traditionally defined as the components that are economically valuable. Much of the content of these mixtures is not commercially useful (at least today) and this material is referred to as **gangue** (rhymes with bang).

Note in Figure 11.1 Table 11.1 that the big three metals, Fe, Al, and Cu, occur mostly as oxides and sulfides. The separation of a pure and useful metal from a naturally occurring mixture is known as **metallurgy.** Metallurgical processes tend to be complex and involve a number of steps as follows:

1. Find an ore that is rich in the metal of interest.
2. Mine the ore.
3. Pretreat the ore to get rid of the gangue.
4. Reduce the mineral to form the metal of choice in a crude form.
5. Refine or purify the metal.
6. Alloying (maybe).

| Anion | Metal |
|---|---|
| None (Native Metals) | Au, Ag, Pt, Cu (rarely) |
| Oxide | Fe, Al, Mg, Sn, Si |
| Sulfide | Fe, Cu, Zn, Pb |
| Chloride | Li, K, Na |
| Carbonate | Li, Ca, Mg |
| Sulfate | Ca, Mg |
| Silicate | Al, Li |

FIGURE 11.1. Ore Sources—Common Metals.

The overall process involves a very large number of disciplines and skills, both technical and non-technical. Let's assume we have accomplished steps 1 and 2. Now what?

# 11.2 Metallurgy

Each metal is different in terms of processing, but some of the basic steps and chemistries have features in common.

## Ore Pretreatment

The mined ores must first be treated to remove the gangue. Depending on the class of compound of interest the processor can take advantage of the properties of the components of the ore to achieve separation. For example, sulfides (e.g. $Cu_2S$) tend to have higher densities than the gangue compounds so one technique is to first pulverize the ore and then use gravity to separate the lighter unwanted materials from the heavier minerals. Other techniques include flotation and agglomeration as well as the chemical conversion of sulfides, hydroxides and carbonates into the corresponding oxides. The oxide forms are generally more easily reduced than the sulfides in the refining process.

## Refinement to Free Metals: Oxidation-Reduction

*Metal refining is one of the most important applications of oxidation-reduction chemistry, and one of the most significant technological advances by humankind in history.*

Some native (occurring in elemental form) metals, notably gold, silver, tin, and copper, have been known and used by man since antiquity. The availability of these relatively pure metals to early mankind was a fortunate consequence of the activity series of metals (see Chapter 10). Figure 11.2 shows the basic techniques used to refine metals from their naturally occurring compounds.

| Metal | Metal Reduction Process |
|---|---|
| Al, Mg, Li | Electrolysis of the metal ions in ionic salt mixtures at high temperatures |
| Fe, Zn, Cu | Roasting with either C (coke) or carbon monoxide |
| Ag, Au, Pb, Pt | found in native metal forms or roasting of the sulfides at under mild conditions |

**FIGURE 11.2.** Basic Reduction Processes—Common Metals.

Although the detailed procedures used for each metal vary, there are just three fundamental refining techniques. In addition, there is a lot of overlap between the various processes.

| Reduction Potential[1] | Metal Reduction Half Reaction |
|---|---|
| −3.04V | $Li^+ + 1e^- \longrightarrow Li$ |
| −2.37V | $Mg^{+2} + 2e^- \longrightarrow Mg$ |
| −0.44V | $Fe^{+2} + 2e^- \longrightarrow Fe$ |
| −1.66V | $Al^{+3} + 3e^- \longrightarrow Al$ |
| −0.76V | $Zn^{+2} + 2e^- \longrightarrow Zn$ |

[1] The $H^+$ reduction to $H_2$ is the reference at 0V.

**FIGURE 11.3.** Reduction Potentials—Common Metals.

In each case a metallic cation is being reduced to form the free metal, an energetically demanding process. The other half reaction is an oxidation that supplies the necessary electrons. Although the electron sources vary depending upon the metal being reduced, carbon is generally the ultimate reductant. Even in those cases where the final step does not involve carbon, the source of energy can be traced back to solar power, in other words our sun.

We will now look at the basic processes used in refining some important metals. Figure 11.3 contains a list of the reduction potentials (activities) of the metals being discussed.

## Oxide Reduction Refinement Processes

### Iron

Iron is the most important and widely used structural metal, especially as the primary component of steels, which are alloys. Alloys are discussed in the next section of this chapter.

Iron use and production goes back to antiquity. The chemistry involved is simultaneously relatively straightforward (on paper) and complex (in practice). The overall process is fraught with safety issues, especially the requirement for lava-like temperatures.

The ores of choice are hematite, $Fe_2O_3$, and magnetite, $Fe_3O_4$; however, iron ores are finite resources, and many of the richest deposits have already been exploited. Today taconite, which is magnetite with a lot of silica ($SiO_2$) included, is a primary Fe source. There are a number of chemical steps involved in the manufacturing of iron as follows.

$$Fe_3O_4\,(s) + 4CO\,(g) \longrightarrow Fe\,(l) + 4CO_2\,(g) \quad (11.1)$$

The reductant CO required for this step is prepared from coke, which in turn is derived from coal via destructive distillation. Basically coke is a somewhat purified form of coal:

$$O_2\,(g) + 2C\,(s) \xrightarrow{\Delta} 2CO\,(g) \quad (11.2)$$

Figure 11.4 diagrams a typical blast furnace used to prepare liquid iron.

This process requires a great deal of energy, and releases large quantities of $CO_2$ into the atmosphere. The reason for having limestone ($CaCO_3$) present is to react with the **gangue** (basically silica; $SiO_2$) to form **slag.**

$$\mathbf{CaCO_3\,(s) \longrightarrow CaO\,(s) + CO_2\,(g)}$$
$$\mathbf{CaO\,(s) + SiO_2\,(s) \longrightarrow CaSiO_3\,(l)}$$

The slag is less dense than iron and floats on top of the liquid iron metal. The liquid slag protects the iron from oxidation with oxygen in the air which would basically be reversing the desired chemical process.

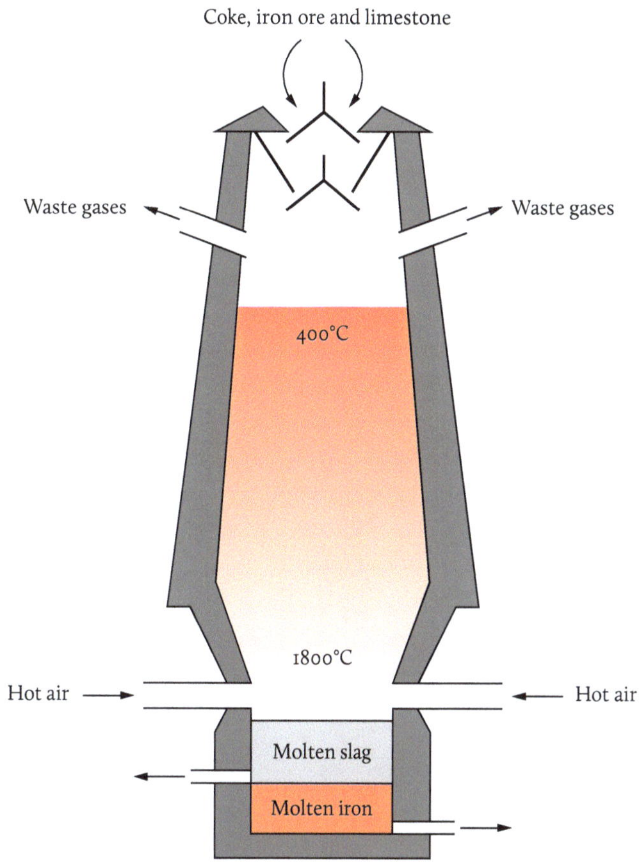


**FIGURE 11.4.** Blast Furnace Basics.

## Titanium

Titanium is becoming more important due to its exceptional corrosion resistance as well as having the highest strength to weight ratio of any metal. It does not occur naturally in pure form, and is prepared from the oxide by the *Kroll* process:

$$\mathbf{TiO_2(s) + 2C(s) + 2Cl_2(g) \longrightarrow TiCl_4\,(g) + 2CO\,(g)} \quad \mathbf{(11.3)}$$
$$\mathbf{TiCl_4\,(g) + 2Mg\,(s) \longrightarrow 2MgCl_2 + Ti\,(l)}$$

In the final step a more active metal, Mg in this case, is used to provide the electrons for the reduction of $Ti^{+4}$ to Ti, an energetically costly four electron process.

## Copper

As mentioned above copper occurs naturally as the pure metal, but this is rare, and those sources are gone. Commercially viable lodes of copper ores are relatively rare too. Geological survey projections suggest the metal will become scarce relatively soon.

There are two copper ores of particular significance, mixed metal sulfides such as chalcopyrite CuSFeS ($CuFeS_2$), and carbonates such as azurite $Cu_3(CO_3)_2(OH)_2$ Cu is prepared from the sulfide via the following steps:

$$\mathbf{CuFeS_2(s) + 3O_2(g) \longrightarrow FeO(s) + 2CuS(g) + 2SO_2\,(g)} \quad \mathbf{(10.4)}$$
$$\mathbf{2CuS\,(s) + O_2(g) \longrightarrow Cu_2S\,(s) + SO_2\,(g)}$$
$$\mathbf{Cu_2S\,(s) + O_2(g) \longrightarrow 2Cu\,(s) + SO_2\,(g)}$$

The copper metal produced is purified using an electrochemical process known as electrowinning. In essence this is a Daniell cell wherein pure copper is produced at the cathode.

The overall process is not nearly as energy intensive as the other metals described herein since Cu is a much less active metal, one that is therefore easily reduced (Figure 11.5).

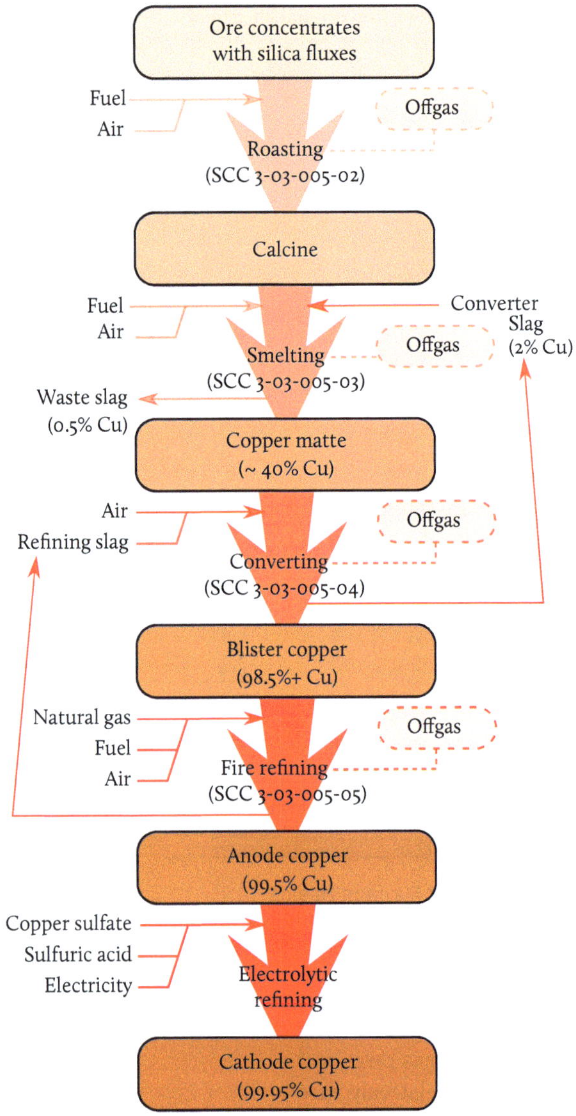


**FIGURE 11.5.** Copper Ore Processing.

### Zinc

Zinc is important especially as a metal for protecting iron structures from rusting. Galvanized steel is prepared by electroplating zinc (the more active metal) onto iron surfaces. Alternatively pure zinc is used as a sacrificial anode for protecting steel hulls, especially on boats and ships. Zn is also used in pharmaceutical applications, and is known to have biocidal properties.

The commercially most important zinc ore is zinc sulfide which is easily converted to the oxide by heating in air:

$$2ZnS\,(s) + O_2(g) \longrightarrow 2ZnO\,(s) + SO_2\,(g)$$

The $SO_2$ so produced can be immediately converted to sulfuric acid which is then used to make zinc sulfate ($ZnSO_4$):

$$ZnO\,(s) + H_2SO_4(aq) \longrightarrow ZnSO_4\,(aq) + H_2O\,(l)$$

Zinc metal can then be prepare from the aqueous $ZnSO_4$ via electrolysis, a process described in the next section.

## Electrolytic Reduction Processes

### Aluminum

Aluminum is arguably the most important commercial metal after iron. With a density of 2.7g/cm³ it is less than a half as dense as iron (7.9 g/cm³), and possesses a high strength to weight ratio. Aluminum's uses are widespread; especially notable applications being aircraft, cans, and aluminum foil. As one of the most active metals it is prepared by an electrolysis process. In essence electrolysis is the same as the charging process undertaken in batteries. That is, electrochemical energy is supplied to run a reaction in a non-spontaneous direction.

Aluminum is the third most common element on earth after oxygen and silicon. The most important aluminum ore is the alumina known as bauxite, which is a mixture of aluminum oxide (alumina, $Al_2O_3$) and aluminum hydroxide [$Al(OH)_3$] in varying ratios and containing a number of impurities.

Pure aluminum metal is produced from alumina using the Hall-Heroult process, the first step of which is shown in Figure 11.6, dissolution of aluminum oxide. This process was initially developed in the late 1800s.

There are a number of procedures for converting crude bauxite to alumina. One of the most important is the Bayer process which involves four steps: digestion, clarification, precipitation and calcination. The first three steps result in the aluminum being combined in the form of the hydroxide, $Al(OH)_3$, which upon calcination at temperatures of ~1000°C is converted to alumina in the form of a fine powder:

$$2Al(OH)_3\,(aq) \longrightarrow Al_2O_3\,(s) + 3H_2O\,(l)$$

The alumina is then converted to aluminum metal at 1000°C with carbon again being the sacrificial reductant:

$$2Al_2O_3\,(l) + 3C\,(s) \longrightarrow 4Al\,(l) + 3CO_2\,(g)$$

The electrolyte used is a molten salt, cryolite ($Na_3AlF_6$). A voltage of 4-6V is required, and current loads are very high at 30,000 to 300,000 amperes. Once again this process is very energy intensive.

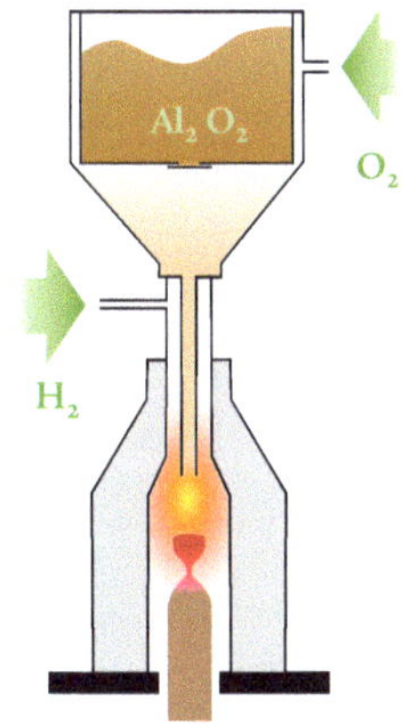


FIGURE 11.6. Hall Heroult Process—1st Step.

## Lithium

Until very recently lithium would have been at best a footnote in this chapter; however, over the last several decades the development of lithium batteries has dramatically changed this situation. As the most active of metals it must be prepared by electrolysis after being converted from the carbonate form to lithium chloride, LiCl:

$$Li_2CO_3\,(aq) + 2HCl\,(aq) \longrightarrow 2LiCl\,(aq) + H_2CO_3\,(aq)$$

The $Li_2CO_3$ can be prepared from oxide ores, for example Spodumene $LiAlF_2O_6$, by a number of different processes depending on the type of ore.

The electrolysis is run in a LiCl-KCl liquid electrolyte at ~400-500°C, and is an energy intensive process:

$$2LiCl\,(aq) \longrightarrow 2Li\,(l) + Cl_2\,(g)$$

The anode is graphite, where $Cl_2$ is produced and collected, while the cathode is mild steel where liquid lithium metal collects. The voltage required is 3.6V.

**PRACTICE** Suppose you want to make aluminum foil. What steps could you take to make it? Gulp, but access to internet information is allowed here.

1. Find a good geologist. He/she finds the lode of ore rich in aluminum. What is the preferred ore?
2. What now?
3. ?????????

## Silicon

Silicon is a metalloid or semiconductor with an electrical conductivity between that of a true metal and non-metal. It is a key chemical component of virtually every important electronic device in use today. As the second most common element on earth, after oxygen, it is widely available in the form of rocks. The conversion of silica ($SiO_2$) into pure elemental silicon is straightforward as follows:

$$SiO_2\,(s) + 2C\,(s) \xrightarrow{2350^\circ C} Si\,(l) + 2CO\,(g)$$

The carbon is supplied by using low ash coal. The tremendous temperatures need to accomplish this conversion is achieved using an electric arc.

# 11.3 Alloys

Before we can discuss corrosion we need to briefly look at the topic of alloys. An ***alloy*** is a mixture of two or more relatively pure elements, one being a metal, that which after mixing retains the properties of a metal. The distinction between an alloy and an impure metal is hazy, but generally an alloy is designed to possess improved chemical and/or physical properties in one way or another.

Alloys also have been known since ancient times. Examples include bronze which is a mixture of copper and tin. Bronze is much harder than either copper or tin alone, and was a tremendous technological advance of the Bronze Age.

Alloys is a branch of science of its own covering a great many metal-element combinations and methods of preparation. We will focus here on just a few iron and aluminum alloys that are of substantial commercial and practical importance. Table 11.1 contains a brief listing of iron and aluminum alloys.

TABLE 11.1 **Some Important Alloys of Iron and Aluminum**

| *Iron Type* | *Additives (wt%)* | *Properties* | *Uses* |
|---|---|---|---|
| *Carbon Steels* | *Carbon (~1–2%)* | *Very hard, low cost, poor corrosion* | *Cars, trucks, sheet steel, tools* |
| *Stainless Steels* | *16%Cr, 8%Ni, 0.2% C* | *Superior corrosion resistance, high cost* | *Cutlery, tools* |
| *Silicon Steels* | *3% Si, 0.5% C* | *High strength, magnetic* | *Magnets* |
| *Manganese Steels* | *14% Mn, 0.5% C* | *Wear resistance, high strength* | *Railroad track, armor plating* |
| ***Aluminum Alloy #*** | ***Additives (wt%)*** | ***Properties*** | ***Uses*** |
| *2024* | *4% Cu* | *High strength, tough, poor corrosion resistance* | *Aircraft* |
| *3004* | *1.5% Mn* | *Food grade* | *Cans, utensils, heat exchangers* |
| *6061* | *0.6% Si, 1% Mg* | *Excellent corrosion resistance* | *Marine, Trucks* |
| *7075* | *5.5% Zn, 2% Mg* | *Very high strengths, poor corrosion* | *Aircraft* |

As can be seen in the table, the addition of relatively small amounts of additives can result in very substantial improvements or degradations in properties. When pure, both iron and aluminum are soft, ductile metals that possess at best modest physical properties. In the case of iron just 1–2% of carbon homogeneously distributed throughout the iron matrix yields an alloy with the very high strength, but the poor corrosion resistance characteristic of carbon steels (think of cars rusting away). On the other hand, a relatively large amount of chromium is required to impart the corrosion resistance familiar to us in stainless steels. However, Cr is relatively rare and expensive. Tradeoffs in properties by alloy type is a given.

Aluminum alloys also span a range of properties. Aluminum is not generally thought of as being susceptible to corrosion, but it is, especially in the marine environment.

Even in aircraft, one of aluminum's most important applications, corrosion is an issue, especially under exposure to salt-laden air. We will now discuss the important subject of corrosion.

## 11.4 Corrosion

We are all familiar with the deleterious effects of corrosion in our lives. In the usual sense corrosion refers to the chemical reaction and deterioration of a metal (usually Fe) with oxygen and water present in the environment to form oxides ($Fe_2O_3$, FeO; see Figure 11.7).

The costs to society of corrosive processes are enormous. In the United States alone the costs are many billions of dollars a year. The three metals of primary interest to society, in order, are iron (Figures 11.8 and 11.9), aluminum and copper. The first two are active metals that readily react with oxygen to form the oxides as the final product:

$$4Fe\,(s) + 3O_2\,(g) \longrightarrow 2Fe_2O_3\,(s)$$
$$4Al\,(s) + 3O_2\,(g) \longrightarrow 2Al_2O_3\,(s)$$

FIGURE 11.7. Bridge Corrosion.

Simple oxidation-reduction chemistry on paper, but in practice the reaction details are fairly complex, and very dependent upon the environmental conditions, such as acidity or basicity.

By way of example, iron is oxidized giving up $e^-$ to oxygen:

$$2Fe \longrightarrow 2Fe_2^{+} + 4e^-$$
$$O_2 + 4e^- + 2H_2O \longrightarrow 4\underline{OH}^-$$

Note that hydroxide ion, $OH^-$, is formed here. Under acidic conditions the hydroxide is taken up by the acidic $H^+$ to form water. In turn this forces the reactions to the right (LeChatalier's Principle) which accelerates the rusting process. Most metals corrode faster under acidic conditions.

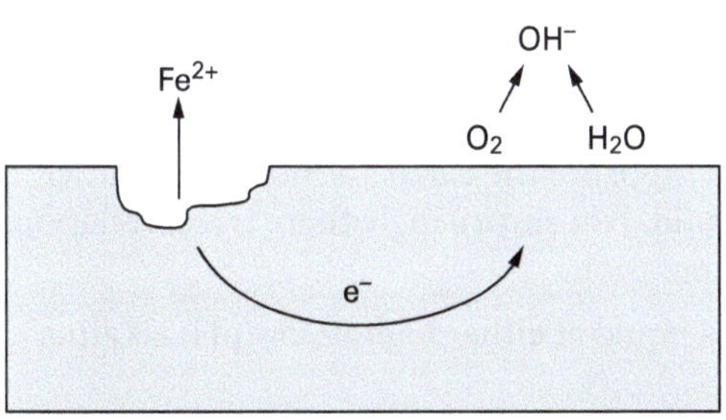


FIGURE 11.8. Iron Corrosion Electrochemistry.

FIGURE 11.9. Steel Corrosion.

There are other reactions occurring as well. In the presence of water, ubiquitous in our world:

$$4Fe_2^{+} + O_2 \longrightarrow 4Fe_3^{+} + 2\,O_2^{-}$$

Also these:

$$Fe_2^{+} + 2H_2O \rightleftharpoons Fe(OH)_2 + 2H^{+}$$
$$Fe_3^{+} + 3H_2O \rightleftharpoons Fe(OH)_3 + 3H^{+}$$

And these dehydration reactions:

$$Fe(OH)_2 \rightleftharpoons FeO + H_2O$$
$$Fe(OH)_3 \rightleftharpoons FeO(OH) + H_2O$$
$$2FeO(OH) \rightleftharpoons Fe_2O_3 + H_2O$$

To make matters worse, the presence of salts containing such ions as $Ca^{+2}$ and $Na^{+}$ also accelerate the rusting rates since they serve as excellent electrolytes.

As mentioned before aluminum is a much more reactive metal than is iron; however, it appears to react, or rust away, far more slowly than iron does. This fact is due to the nature of the oxidation products. Aluminum oxide forms a tight and dense thin film on the surface of the aluminum metal which prevents water and oxygen from reaching the unreacted metal underneath. As we all know the iron oxide rust flakes off easily exposing more iron metal to reaction.

Aluminum does rust, however. In fact aluminum metal has been used as a rocket fuel because of vigorous and energetic redox reactions such as that with water:

$$2Al\ (s) + 6H_2O\ (l) \longrightarrow 2Al(OH)_3\ (s) + 3H_2\ (g)$$
$$2Al(OH)_3\ (s) \longrightarrow Al_2O_3\ (s) + 3H_2O\ (l)$$

Remember here too that for each Al oxidized there are ***three electrons*** transferred. This partly accounts for the high energies of reaction.

Aluminum hydroxide is **amphoteric** (reacts with acids or bases) so other reactions are possible:

$$Al(OH)_3\ (s) + 3H_3O^{+}\ (aq) \longrightarrow Al[(OH)]_6^{+3}\ (s)$$
$$Al(OH)_3\ (s) + OH^{-}\ (aq) \longrightarrow Al[(OH)]_4^{+}\ (s)$$

Aluminum corrosion in the marine environment and in aircraft are especially problematic and complex mechanistically.

## Types of Corrosion

There are numerous kinds of possible corrosion mechanisms as follows:

***Atmospheric Corrosion:*** This is actually the most common, long term form of corrosion and is defined as corrosion due to exposure the earth's atmosphere, rather than corrosion due to immersion in a liquid. Not surprisingly there is a correlation between the humidity level and rate of corrosion.

***Uniform Corrosion:*** General corrosion in a liquid at either high or low pH; alkaline or acidic, respectively.

***Galvanic Corrosion:*** This occurs when dissimilar metals are connected electrically. The more active metal will preferentially corrode. A useful example here is zinc anodes connected to iron vessels. The zinc preferentially corrodes away because it is the more electronically active metal.

***Crevice Corrosion:*** This requires the presence of a crevice, salt water and oxygen. By way of example a space between a bolt and an underlying structure works well.

***Pitting Corrosion:*** Pitting is a key mechanism of corrosion in aluminum. It occurs especially in the presence of chloride ion and defects in the protective oxide layer.

***Intergranular Corrosion:*** This type of corrosion occurs at the interface of grains of alloys. Often the grains themselves, other than their surfaces, remain unchanged. The exact corrosive effects are alloy dependent. That is, it depends on whether the additive metals are more or less noble that the main metal as well as surface chemistries.

***Stress (Fatigue) Corrosion:*** This occurs when a specimen is placed under tensile stress in the presence of corrosive conditions. Even the smallest crack may propagate rapidly. The exact mode of action will vary depending upon the nature of the grain boundaries and associated chemistries.

***Microbiological Corrosion:*** Microorganisms can affect the acceleration of corrosion. A well-known example involves the growth of fungus at the fuel/water interface in aluminum aircraft fuel lines. The high octane fuel serves as a food source for the fungi which in turn excretes and acid that degrades the aluminum resulting in leaks. Clearly not a good situation.

## 11.5 Corrosion Prevention

An old adage in the field of corrosion is that you can slow corrosion down, but never stop it.

Any time you have an active metal, such as Li, Al, Fe, Mg, Zn and others, it will corrode eventually. In essence what one accomplishes is to slow down the rate of corrosion by using preventive measures. What measures are to be used?

Widely used techniques for protecting active metals from corrosion include the following.

1. ***Noble Metal Coating***

   Here a nobler (less reactive) metal coating is applied to an active metal. Basically, the nobler metal, unreactive under the environmental conditions imposed, insulates the underlying metal to exposure. An example here is a copper cladding on top of aluminum.

2. ***More Active Metal Coating***

   This is the reverse of the number 1 just above. In this case a more active metal is placed on the surface, and is sacrificially corroded rather than the underlying metal. A common example here is galvanized steel where a zinc layer is electroplated on the steel surface.

3. ***Cathodic Protection***

   An extension of number 2 here a sacrificial anode is connected to the metal structure to be protected (Figure 11.10).

FIGURE 11.10. Galvanic Corrosion Protection.

The anode preferentially corrodes away, but must be replaced periodically. A common example here is the sacrificial zinc anodes widely used to protect steel hulls from corrosion in the marine environment. Another example is the use of magnesium anodes to protect iron rebar in bridges.

*4. Protective Oxide Coatings*

Aluminum naturally has a dense protective oxide coating, but many metals do not. Iron forms the oxides, but it flakes away and exposes more virgin metal to the elements. Similar to aluminum; however, protective oxide coatings can be readily applied, if expensive. A familiar example here is ceramic clad iron used for inert countertop applications.

*5. Protective Polymer Coatings*

Coatings and paints based upon organic or metallo-organic coatings are widely employed to protect all kinds of metals. Since dissimilar materials are being combined here it is important that compatibility issues are addressed. Common polymeric examples are epoxies, polyurethanes, and silicones.

## Chapter 11 Summary

Metallurgy is the study of the physical and chemical properties of metals, as well as how they are prepared and purified.

*PURE METALS:* Reactivity series of some common metals is as follows:

***reactive*** ***Li*** > ***Mg*** > ***Al*** > ***Fe*** > ***Zn*** > ***Cu*** > ***Au*** > ***Pt*** ***inert***

*METALLURGY*

*Ores:* Naturally occurring mixtures of minerals.

*Mineral:* A component of an ore that has a specific chemical composition.

*Gangue:* The component of an ore that is not of particular economic value.

Basic steps involved in the preparation of a pure metal from an ore:

*Ore pretreatment:* removal of gangue.

*Refinement to Metals:* an important application of oxidation-reduction. Processes include roasting and electrolysis. In general done form the metal sulfide, oxide or carbonate compounds using carbon as the ultimate electron source.

*Thermal Refinement Processes Described:* iron, titanium, copper and zinc.

*Electrolysis Refinement Processes Described:* aluminum, lithium, silicon

*ALLOYS:* An ***alloy*** is a mixture of two or more relatively pure elements, one being a metal, that which after mixing retains the properties of a metal.

Common alloys of iron and aluminum are described.

*CORROSION:* Iron and aluminum corrosion are discussed.

*Types of Corrosion*

*Atmospheric Corrosion:* corrosion due to exposure the earth's atmosphere.

*Uniform Corrosion:* General corrosion in a liquid at either high or low pH.

*Galvanic Corrosion:* This occurs when dissimilar metals are connected electrically. The more active metal will preferentially corrode.

*Crevice Corrosion:* This requires the presence of a crevice, salt water and oxygen.

*Pitting Corrosion:* Pitting is a key mechanism of corrosion in aluminum. It occurs especially in the presence of chloride ion and defects in the protective oxide layer.

*Intergranular Corrosion:* Occurs at the interface of grains of alloys.

*Stress (Fatigue) Corrosion:* Occurs when a specimen is placed under tensile stress in the presence of corrosive conditions.

*Microbiological Corrosion:* Due to the presence of biological organisms.

*CORROSION PREVENTION:* Techniques include:

1. ***Noble Metal Coating:*** Here a nobler (less reactive) metal coating is applied to an active metal.
2. ***More Active Metal Coating:*** A more active metal is placed on a metal surface, and is sacrificially corroded rather than the underlying metal.
3. ***Cathodic Protection:*** A sacrificial anode is connected to the metal structure to be protected.
4. ***Protective Oxide Coatings:*** Applied to a metal surface.
5. ***Protective Polymer Coatings:*** Coatings and paints based upon organic or metallo-organic coatings are widely employed to protect all kinds of metals.

## CHAPTER 11 EXERCISES

1. What are the two most important types of ores mined to make steel and copper?
2. Smelting is the process by which ores are reduced to base metals. What does the reduction process involve?
3. Why do some metals, such as silver, gold and copper, occur in their elemental form sometimes, but other metals, such as lithium, iron and magnesium never do?
4. What is the chemistry and chemical process involved in the production of iron?
5. What is the chemistry and chemical process involved in the production of aluminum?

6. What is the chemistry and chemical process involved in the production of gold?
7. What is the chemistry and chemical process involved in the production of copper (when present in compounds not as the pure metal)?
8. What is the chemistry and chemical process involved in the production of zinc?
9. Cadmium is in the same periodic table group as zinc. Describe a possible process for preparing pure cadmium metal.
10. Invent and outline a process for preparing pure germanium, an important metalloid widely used in the semiconducting industry.
11. What is an alloy?
12. There are many alloys of iron some of the most important of which contain carbon. What are the key properties, good and bad, of carbon steels?
13. Other important steel alloys contain chromium. What advantages and disadvantages does adding chromium to iron result in?
14. What are the advantages and disadvantages of using aluminum rather than iron in large scale applications?
15. Corrosion processes are ubiquitous, and very costly. Aluminum is a more active metal than iron, and yet aluminum parts do not typically rust away. Why?
16. What is the Hall-Heroult process?
17. Bauxite, the ore principally used to manufacture aluminum, is ~55% aluminum. The production of pure aluminum in 2016 was $6x10^7$ metric tons. How much bauxite was needed to produce this quantity of Al?
18. How can corrosion processes be slowed down?
19. Write out the corrosion chemistry for iron (Fe).
20. Write out the corrosion chemistry for aluminum (Al).
21. Copper slowly oxidizes to yield either the familiar green or a black oxide coating depending partly on particle size, but also basic chemistry. Remembering that copper can form both $Cu^{+1}$ and $Cu^{+2}$ write out corrosion chemistries for copper metal reacting with oxygen and water in the environment.

# Credits

CHAPTER 12

# Sources of Energy (Power)

## Energy: So familiar, so important, so mysterious...

## 12.1 Basics

Energy usage is so fundamental we often take it for granted. The usual definitions relate to a capacity to do work, and certainly the relationship between the two topics is fundamental; however, at a deeper level a definition of energy is elusive. In chemistry and physics we discuss electron transfers, heats of reactions, and chemical transformations within a framework of energy changes, but none of this dispels a feeling of it all being magical. There is nothing wrong with this. Maybe we just accept energy as being in the same category as entropy—mysterious.

The concept of power is hazy as well, but since it is by definition just the expenditure of energy per unit time power is easier to put into concrete, measurable terms. On the other hand it includes the variable of time so now we have two elusive concepts, energy and time, linked together into one.

### Types of Energy

***Internal Energy*** is the total energy associated with a system, That is, the sum of the kinetic and potential energies. Even for the smallest atomic system this represents a phenomenal amount of energy. The two familiar subcategories of energy are ***kinetic energy*** and ***potential energy***. These are well named since the words describe exactly what is involved. Kinetic means the energy of activity or motion while potential is just that, energy waiting to happen. Breaking the types of energy available in common use are familiar at least by name as shown in Table 12.1 below.

This is a loose grouping of energy types. Clearly, there are many subclasses underneath each of these headings, and the interconversions between different energy types are very important.

Half of these forms of energy, the ones highlighted in red, are the ones of particular interest throughout the discussions of chemistry in this book. The next few sections discuss these forms of energy in greater detail.

TABLE 12.1 Basic Forms of Energy

| Type of Energy | Example |
|---|---|
| Mechanical | Turbines (wind, water), machinery |
| Solar | Photosynthesis, solar cells |
| Magnetic | Magnets, motors |
| Acoustic | Sound |
| Chemical | Combustion |
| Electrical | Batteries, fuel cells, capacitors |
| Gravitational | Gravity |
| Thermal | Heat transfer |
| Nuclear | Fission, fusion |

### Units

Another confusing aspect of energy and power are the conventions for units, or lack thereof. The calorie (cal), yes the one on food packaging, is most common to us, but we will stick with the Joule (J) unit of heat energy. There are 4.184J/cal. Note also that in the following descriptions the acronym MJ refers to 1 MegaJoule of heat = 1,000,000 Joules.

## 12.2 Traditional Fossil Fuels

We use the term traditional fuels to refer to naturally derived fuels that have been known and used since antiquity, but are finite resources. They are all biofuels in the sense that they are derived from sunlight. Here we have somewhat arbitrarily put "biofuels" into a separate section that follows.

Power derived from the widespread use of natural gas, oil, and coal underlie the economies of the world. They are all finite resources in the sense that, for example, copper ore is finite. Like copper fossil fuels are being mined and will one day run out.

In spite of the recent focus upon the development of renewable resources there is little doubt that fossil fuels will remain the backbone of energy generation for some time to come.

### Fossil Fuel Basics—Solar Energy Derivation

All three types of conventional fossil fuels are derived from the power of the sun, more specifically photosynthesis. The biochemical processes involved in photosynthesis are complex and are described in the next chapter; however, the fundamental chemistry involved can be distilled down to just one equation:

$$\text{Sunlight} + 6CO_2\,(g) + 6H_2O\,(l) \longrightarrow C_6H_{12}O_6\,(s) + 6O_2(g) \qquad (12.1)$$

This equation is a drastic simplification of the biochemical reality, but it is at the heart of the matter. Ubiquitous carbon dioxide and water are combined to form a carbohydrate and oxygen. This is an energetically demanding process that is ***driven***

***by solar energy***. The fixing of $CO_2$ into a fuel, the carbohydrate, is one key result. The second key result, clearly just as important, is the formation of $O_2$. Oxygen formation as a reaction product is a rare chemical event.

Fossil fuels are derived from the slow settling and decomposition of dead plants and animals in sedimentary rock formations over eons of time. The chemical conditions required are demanding and include intense heat, high pressures and anaerobic conditions. From a chemistry standpoint the organic materials are slowly deoxygenated to form simpler hydrocarbons, and a lot of them. If geological conditions are suitable, a relatively rare event, vast quantities of petroleum fuels may be entrapped. Figure 12.1 shows an example of such conditions:

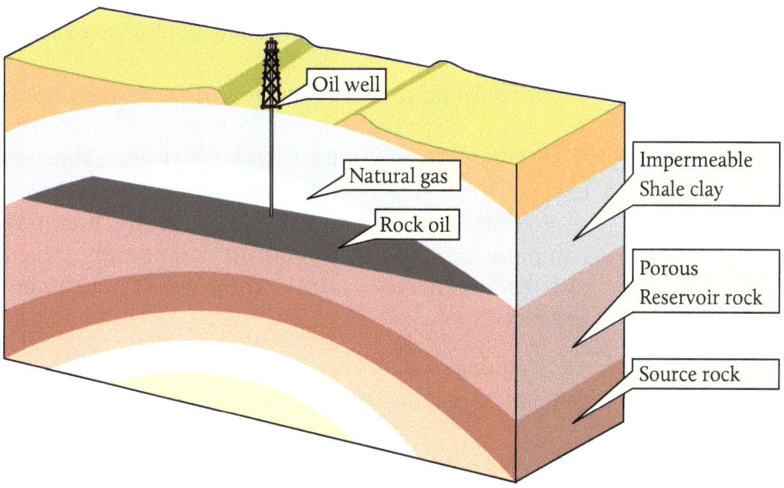


FIGURE 12.1. Anticline Geologic Formation.

This kind of geological formation is called an ***anticline*** and is characterized by layers of water and/or rock that underlie a porous sedimentary layer filled with petroleum on top of which is another impermeable rock layer. On top of the petroleum liquids is generally a layer of gases that include natural gas and helium. A classic oil well involves accessing the fuel layer via the drilling of a vertical shaft. Once the shaft is drilled the complex mixture of fuels is tapped as described in the next sections.

A fascinating fact, which is outside the scope of this chapter, is that nature is left handed. Your right and left hands are not exactly the same. In three dimensions they are mirror images of each other that cannot be superimposed. Nature is the same way, and molecules can have exactly the same chemical formula, but different atomic arrangements in space. This apparently subtle difference turns out to be extraordinarily important in the functioning of biochemical systems.

It turns out that fossil fuels are handed as well; like the rest of nature is left handed. In addition even numbered carbon chain lengths are dominant. ***Lignins*** (see Chapter 17) are chemically complex molecules that form the basis for peat and coal, peat being the chronological predecessor for coal. ***Lipids*** are basically fatty acids (long linked-carbon chain-containing molecules similar to paraffins) that convert to crude oil mixtures over millions of years.

## Natural Gas

Natural gas is the term used for the fossil fuel composed primarily of methane ($CH_4$), but also containing higher molecular weight hydrocarbons, and sometimes other minor components such as hydrogen sulfide and helium. About 33% of the electricity generated in the United States in 2015 was based upon this fuel. It occurs in large quantities alone in gas fields widely dispersed throughout the world, and also as the top layer on oil field anticline sites. It is by far the cleanest burning fossil fuel type, and has a high energy content, but still results in substantial air pollution as shown by equation 12.2:

$$CH_4\,(g) + 2O_2\,(g) \longrightarrow CO_2\,(g) + 2H_2O\,(g) \quad (12.2)$$

Water formation is fine, but carbon dioxide is not since it is a greenhouse gas implicated in the earth's ongoing climate changes. Other products are formed as well such as carbon monoxide, CO.

In addition, recall that nitrogen gas ($N_2$) is abundant as 80% of the earth's atmosphere. Although $N_2$ is a very stable molecule, some of it does react during the combustion process to form a variety of nitrogen oxide compounds as well as nitro-derived organic molecules, all of which are pollutant compounds. Similarly any sulfur compounds present will also react with oxygen and/or methane to yield undesirable products.

***Methane contains ~56 MJ/kg of stored energy.***

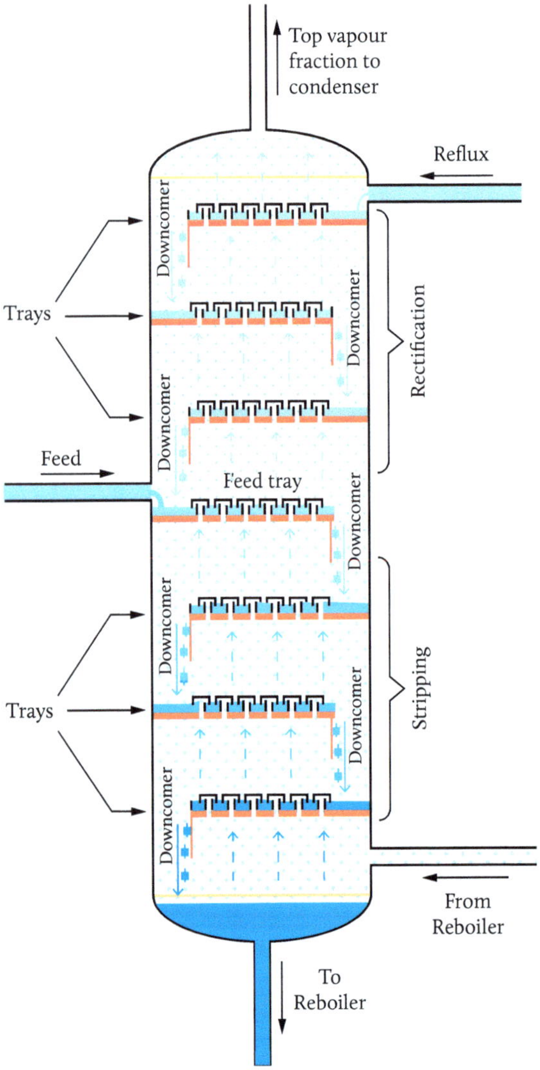


**FIGURE 12.2.** Distillation Tower of a Typical Refinery.

## Petroleum

Petroleum is a complex mixture of organic liquid hydrocarbons. Although the exact chemical compositions vary from place to place, significant familiar components are gasoline, kerosene, oils, and asphalt. Figure 12.2 is a schematic of a typical distillation tower used to separate the mixture into different fractions.

The process is done on a very large scale at oil refineries, but is conceptually simple. In essence the mixture is heated to high temperatures and the various liquid components distill off in the order of their respective boiling points. They then selectively condense at different stages of the tower, and are siphoned off.

The various hydrocarbons collected are usefully combusted and yield more or less the same products as methane does, but not nearly as cleanly. By way of example octane, formula $C_8H_{10}$, if burned cleanly (which it never does) yields just $CO_2$ and $H_2O$:

$$2C_8H_{10}\,(g) + 21O_2\,(g) \longrightarrow 16CO_2\,(g) + 10H_2O\,(g) \quad (12.3)$$

Also, as described above for natural gas, the reaction of nitrogen and sulfur compounds during these combustions produces deleterious pollutant products.

***Petroleum contains ~47 MJ/kg of stored energy.***

## Coal

Over millions of years woody plant detritus can be compressed to release small molecules such as methane and water. If the process continues long enough solid state ***coal*** is formed. A general formula in mole fractions for coal is $CH_{0.8}S_xN_yOz$ where x, y and z are <0.1. So basically coal has a C:H ratio of 1:1. Note that continuing the treatment at very high pressures and temperatures results in the formation of diamond, which is basically pure carbon atoms covalently and tightly bonded to each other in three dimensions.

Another 33% of the electricity generated in the United States in 2015 was based upon this fuel. Depending upon the extent of the pressure-temperature treatment two forms of coal will obtain:

***Bituminous:*** Known as soft or dirty coal this is the commonly mined coal used in steam-fired power plants for electricity generation (Figure 12.3). The basic combustion products are $CO_2$ and CO, but because it is relatively high in sulfur and nitrogen-based impurities substantial air pollution issues result from its use as a fuel. Equally unfortunate is the generation of airborne mercury and organo-mercury compounds upon combustion of bituminous coal.

***Anthracite:*** This coal is much rarer, and has glassy, rock-like properties. It is significantly higher in carbon content (~86–97% C) than bituminous coal (~45–86%). Because it is substantially more expensive than bituminous it is used primarily for specialty heating purposes where a cleaner combustion process is required.

***Coal contains ~30 MJ/kg of stored energy.***

FIGURE 12.3. Bituminous Coal Seam (Nearly River, Alaska).

## Heavy Oil & Bitumens (Tar Sands)

These fossil fuels are high molecular weight compounds that are closely related to petroleum in terms of chemical structure. They too are found dispersed in sands and rock formations, but are difficult to process because they are viscous materials that do not flow easily. Basically the rock is mined, ground up and then the resultant is extracted using high pressure steam at ~500°C in the absence of oxygen to remove the fuel. As might be expected the overall process is very disruptive to the natural environment, causes pollution of many different kinds, and uses tremendous quantities of water to make the steam.

## Hydraulic Fracturing

The straightforward vertical drilling of oil fields has been known since the late 1800s. However, the readily accessible sources of petroleum have been pretty much exploited leading to the search for secondary sites that tend to be difficult and expensive to develop (think deep-ocean and arctic).

A recent technological development involves the use of horizontal drilling, known as hydraulic fracturing or fracking, to access petroleum in rock strata (Figure 12.4).

A vertical shaft is first drilled down to the fuel bearing layer, then the drill bit is turned to follow this layer. Physical cracking via high pressures and explosives of the rock releases the petroleum which is returned to the surface and collected.

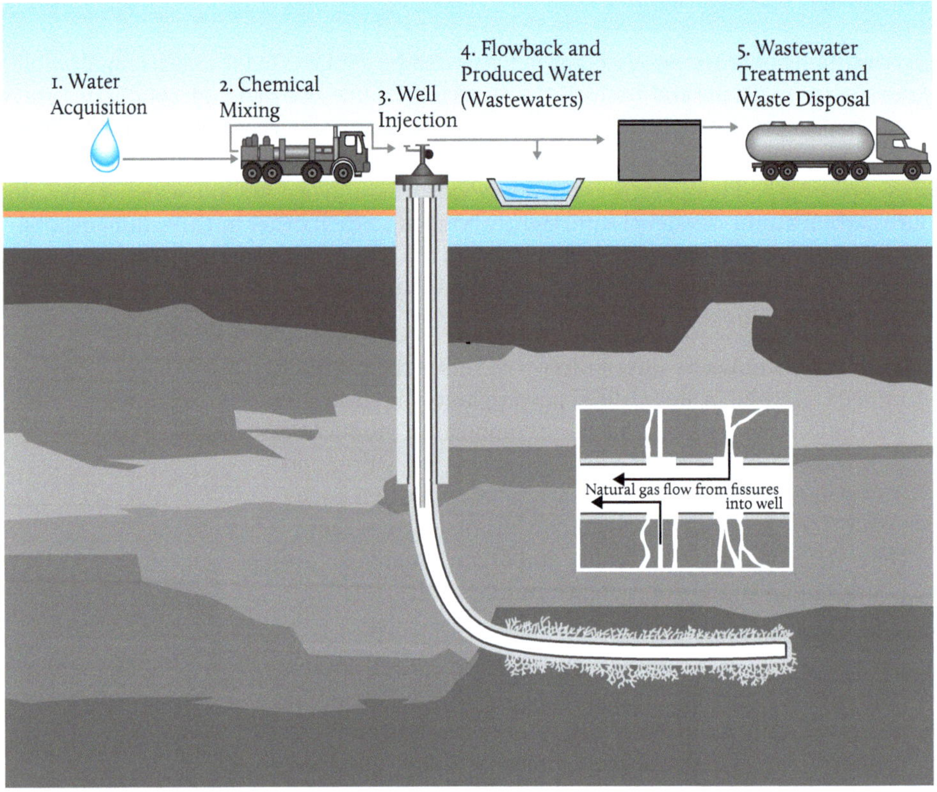


**FIGURE 12.4.** Hydraulic Fracturing.

The fracking process is very controversial since it involves significant land disturbance, uses tremendous amounts of water, and involves a number of different kinds of chemicals. It is difficult to know he long term effects of the fracking processes. In addition, when compared to traditional oil wells indications are that the useful lifetimes of these newer wells might be quite limited. Nevertheless, the economic and political impact of this new technology has been dramatic.

# 12.3 Biofuels

The harnessing of plant derived fuels in a sustainable way is very attractive as a potential renewable energy source replacement for fossil fuels (see Figure 12.5).

## Wood

Wood was the originally harnessed biofuel, and one that is still widely employed worldwide for heating and cooking. Because it is low in energy content relative to the fossil fuels it is not practical for use where energy demands are high, such as industrial plants and large power generation stations.

Growing and harvesting woods takes a lot of space, time and is demanding, dangerous work. The demands for and concerns for preserving open woodlands makes the future

**Biofuel: From Field to Fuel Pump**

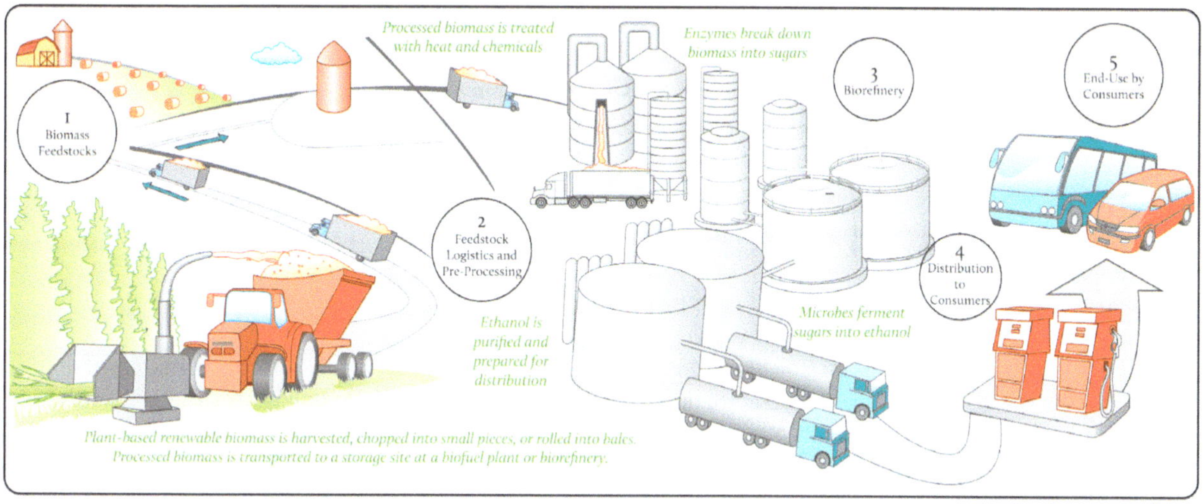


**FIGURE 12.5.** Biofuels Life Cycle.

use of large quantities of wood for power generation less and less attractive. There are far more important uses for wood than simple combustion applications.

***Wood contains ~16 MJ/kg of stored energy.***

## Alcohols: Methanol and Ethanol

The various oil supply crises that have occurred over the past 40+ years made alcohols, especially ethanol ($CH_3CH_2OH$) as derived from corn, attractive prospects as fuels to replace fossil fuels. Although methanol costs less to produce and burns more cleanly, ethanol has dominated the arena for political reasons, especially since ethanol is readily derived from corn grown in the American Midwest.

Taking ethanol as the poster child of the alcohol group some advantages over petroleum are:

1. Derived from domestic starting materials
2. Modern engines will run efficiently.
3. Combust more cleanly to yield fewer pollutants than petroleum.

Some substantial disadvantages are:

1. Dedication of vast acreages of farmland to growing corn for non-food applications.
2. Large amounts of water and fertilizers are needed, so there is a question of whether there is a net gain in energy costs/efficiencies.
3. Farm machinery use fuels to plant, be taken care of and harvest crops.
4. Corn processing to ethanol requires energy, especially the distillation process.
5. Huge amounts of water are consumed.

There is also considerable doubt concerning whether or not corn-derived ethanol is carbon neutral or not, as well as the overall effects on the environment since substantial quantities of herbicides and pesticides are used in the growth of corn.

Ethanol can be derived from corn, sugar cane, algae, cellulosics and lignins. In many ways the latter two starting materials are superior since there is a lot of wood waste products available at very low cost. Unfortunately, cellulosics and lignins are much more difficult to convert to alcohols than corn or sugarcane. There is a lot of research going on in this area.

***Ethanol contains ~26 MJ/kg of stored energy.***
***Methanol contains ~20 MJ/kg of stored energy.***

## 12.4 Synthetic Fuels and Derivatives Therefrom

The distinction between a natural fuel and a synthetic fuel is largely a matter of definitions. By synthetic we mean man-made, but in the final analysis any fuel we synthesize is based upon one naturally occurring substance or another as a reactant.

### Hydrogen

Hydrogen has the distinction of being far and away the most energy dense fuel outside nuclear fuels. $H_2$ is being made naturally occurring processes in great quantities constantly. Unfortunately it does not collect in substantial quantities because it is so light, small and mobile. Even after being synthesized it is difficult to contain long term. It is also extremely flammable so storage and safety are significant concerns.

Hydrogen is synthesized by a number of different procedures in very large quantities throughout the world. The most important current procedures is ***steam reforming***:

$$CH_4\,(g) + H_2O\,(g) \xrightarrow{1000^\circ C} CO\,(g) + 3H_2\,(g) \qquad (12.4)$$

This endothermic reaction requires significant amounts of heat to run. The product combination, CO and $H_2$, is known as ***syngas***. Using a fossil fuel, $CH_4$, to make hydrogen may seem curious, but is based purely upon economics. The instant that natural gas becomes rare, and thus expensive. Other processes will come to the fore.

The CO produced by reaction 12.4 can be reacted with water in a reaction known as the ***water gas shift reaction*** to form more $H_2$:

$$CO\,(g) + H_2O\,(g) \longrightarrow CO_2\,(g) + H_2\,(g) \qquad (12.5)$$

This reaction is exothermic and the heat released from it can be used to help run the overall 12.4–12.5 synthetic sequence. Hydrogen production overlaps with the derivatization of coal *vide infra*.

### Coal Derivatives

It is unfortunate that coal has become the political black sheep of the fossil fuel family. Coal is a very valuable resource, and far more abundant and available worldwide than petroleum and natural gas combined.

There is good news here, however. It is technically well-established that it is possible, and likely soon practical, to convert coal into value added products. Let's look at a couple of important technologies now.

### Coal Gasification

In essence coal is pure carbon, so on paper we can simply add hydrogen to the carbons to form hydrocarbons such as methane and petroleum. As mentioned above though there is no natural source of elemental hydrogen so we need an alternative. Reactions 124 and 12.5 both apply here, but if we start one step further back with pure carbon these reactions are possible.

$$C(s) + H_2O(g) \longrightarrow CO(g) + H_2(g) \quad (12.6)$$

The CO produced in this reaction can be separated from the $H_2$ and reacted with water further per reaction 12.5.

Another reaction is the direct reaction of oxygen with carbon to form carbon monoxide:

$$C(s) + O_2(g) \longrightarrow 2CO(g) \quad (12.7)$$

Once again this reaction can be followed by 12.5. Many different kinds of reaction chemistries are known to converty coal into other useful products. Most of them are endothermic, and require significant energy inputs. One set of reactions is based upon Fischer Tropsch chemistry *vide infra*.

### Fischer-Tropsch (FT) Chemistry

This important set of chemistries is conceptually simple. Basically carbon monoxide and hydrogen are mixed together and reacted to form hydrocarbons of many different kinds, the simplest being methane:

$$CO(g) + 3H_2(g) \longrightarrow CH_4(g) + H_2O(g) \quad (12.8)$$

The astute reader will point out that 12.8 is simply the reverse of equation 12.4. True, we seem to have come full circle, but remember that the only reason reaction 12.4 is carried out is because of economics, that is, natural gas ($CH_4$) is cheap today. Not tomorrow though. If we start further back with coal, per 12.6 and 12.7, we are ahead of the game.

Equation 12.8 is a drastic simplification of the FT scope of importance. It can be used to synthesize a whole family of useful hydrocarbon compounds, and compounds. By way of example gasolines, jet fuels, fuel oils, lubricants, alcohols, and high molecular weight polymers, such as polyethylene and polystyrene can be synthesized (Figure 12.6).

Catalysts are essential components in most of these reactions. FT reactions are run at temperatures of 150–300°C and high pressures. Higher temperature-pressure conditions favor higher molecular weight hydrocarbons as opposed to producing simple methane.

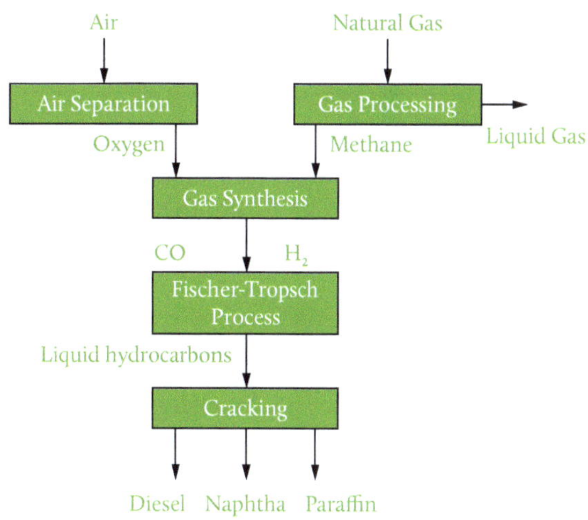


**FIGURE 12.6.** Synthesis Gas Derivatives.

## 12.5 Solar Power Systems

The most attractive source of power is the sun. Clearly, photosynthesis overwhelms everything else in terms of importance, but the solar influx of energy hitting the earth every day tremendous. In a year all of humanity on earth uses just 0.001% of the solar

FIGURE 12.7. Solar Panel Array—Apple Park.

energy received from the sun. Yes, that is a percentage. All we need do then is to harness a very small amount of this generous energy supply (Figure 12.7).

Photosynthesis is referenced in later chapters. The section here contains two subsections, photovoltaics and photolysis.

## Photovoltaics

Solar energy generation via solar cells or photovoltaics is a commercial technology that has grown and continues to grow by leaps and bounds. The market is dominated by crystalline silicon usage, a semiconductor that is described in Chapter 10 (see page **). Solar energy systems have a number of advantages and disadvantages compared with other energy systems as follows;

### Advantages

1. ~20% energy conversion efficiency (sunlight to electricity) and getting better by the day. The maximum efficiency possible with today's cells is ~35%. The theoretical maximum is ~86%.
2. Durability. Very dependable with little change in performance over time, and no moving parts, that is totally solid state.

3. Performance and efficiencies more or less independent of the ambient temperature.
4. Clean power. There are no greenhouse gas emissions.
5. Silent.
6. Can be made into virtually any configuration, and put into any location that receives sunlight.

### Disadvantages

1. Solar cell manufacturing uses toxic chemicals.
2. Manufacturing uses rare elements that are subject to depletion over time due to mining.
3. A variable energy resource that is subject to weather conditions.
4. Expensive relative to fossil fuels.

### Photolysis

This technology uses solar energy to make compounds. In essence the conversion of solar energy into stored chemical energy. Probably the most promising is the simplest, the splitting of water into hydrogen and oxygen:

$$2H_2O\,(g) \xrightarrow{\text{sunlight}} O_2\,(g) + 2H_2(g) \qquad (12.9)$$

This reaction has been known for a long time, and a tremendous amount of work has gone into the development of catalysts that will efficiently promote it. It is still not economical or competitive with the $H_2$ synthetic processes described earlier, but once again, this is so because of low cost fossil fuel availability.

There are a large number of photolytically driven reactions that are known, some of which are reasonable candidates for large scale energy storage. Further discussion is outside the scope of this book.

# 12.6 Wind, Nuclear, Geothermal, Hydroelectric, and Tidal Power Systems

Except for tidal power these types of power are important to very important sources of power. They all deal with chemical topics in a way, but will be covered here only briefly as will become evident right now.

## Wind

Wind turbines have become a very common sight in the United States in the recent past (Figure 12.8).

Wind power has become an important, if small, component of our total energy supply, and one that is growing. It suffers from the same significant disadvantage that solar has in that it requires the wind to be blowing, so it is intermittent. In addition there

FIGURE 12.8. Wind Turbine.

are only a limited number of sites where enough wind blows to make the investment in hardware worthwhile.

Terrestrial sites tend to involve a significant habitat disturbance in tower placements. Marine siting has already proven to be problematic in the United States; for example take the strong resistance in the Northeast to a wind farm placement in Nantucket Sound. Another factor to be considered is that the financial and environmental costs of manufacturing the wind turbine components, as well as maintenance costs need to be included in any benefit/cost analysis. Finally there is some indication that wind turbines affect the survival of birds on migration, as well as bats.

## Nuclear

Nuclear power supplies about 20 of the electricity in the U.S., but this is in decline. The most serious problem with this technology is what to do with the waste products which can remain radioactively dangerous for 40,000 years or much longer. In addition, the nuclear disasters at Chernobyl and Fukishima have awakened everyone to the hazards involved. In both cases substantial areas of land; many, many square miles, have been rendered uninhabitable for many generations of humanity.

## Hydroelectric

Arguably the best source of power in many ways, but this resource is dependent upon the damming of rivers, and quite simply, most such rivers have been dammed already. In essence, this resource is tapped out. In addition, the building of dams is almost always a local environmental disaster, and the effects can effects can much farther. For example, the recently completed *Three Gorges Dam* in China has caused unknown, but surely large, environmental and societal changes. Millions of Chinese people were forcibly moved to make way for this massive project whose outcome is still in doubt.

Another factor with dams is that these resources are dependent upon regular rain and snowfall to maintain water supplies, something which has been notably absent in the American west for many consecutive years now. *Lake Mead*, one of the largest reservoirs in the world is at its lowest water level since the dam was built in the 1930s (Figure 12.9).

FIGURE 12.9. Hoover Dam/Lake Mead.

## Tidal

Also known as hydrokinetic power or water turbines, this is a developing technology, but one that is fraught with issues that wind turbines don't have to deal with (Figure 12.10).

For example, biofouling is a tremendous issue, as well as the effect of turbines on marine life in general. Currently there are no commercial installations using this form of power generation.

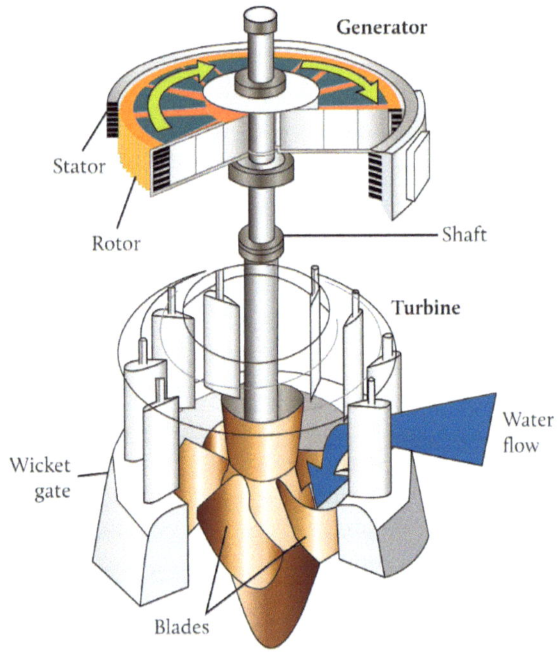


**FIGURE 12.10.** A Hydrokinetic Turbine.

## 12.6 Energy Storage Technologies

As mentioned in the previous section wind and solar power generation are intermittent sources. Two factors immediately become apparent. One is that you want to generate power when possible (sun shining or wind blowing), even if the electricity demand is low. So what do you do with any excess power? The second factor deals with how to access that stored power in a timely manner. A critical factor in both cases is that the grid can require almost instantaneous response, that is, in ***milliseconds***! For example, say the electricity demand in Los Angeles suddenly leaps up due to an overnight heat wave. To avoid outages it is critical to be able to bring extra power on line quickly. There are a couple of ways to do this. One is to have a power station idling off line which can then be almost instantaneously fired up. This is an expensive option, and assumes there is such a plant already just waiting around. Another way is to have significant amounts of power stored that can then be instantaneously accessed. Here are technologies being employed already or seriously considered with their advantages and disadvantages.

### Hydropower

This is one of the most efficient and attractive ways to store energy. In the presence of excess electricity pumps can be used to move water uphill where it is stored as potential energy. When extra electricity is needed the water is simply run downhill through turbines which convert the kinetic energy into electricity. The hardware used here is very robust. Also the response time to electricity demand is very fast. This approach is about 80–90% energetically efficient.

The disadvantages to this technique are that large land areas are required which are difficult to find and can be very disruptive to both existing land uses, people, and the environment. Clearly, geology is important here as well as a substantial elevation differential is required. As a practical matter virtually every geologically attractive spot in North America has been dammed already. In fact, many old dams are being removed. Finally silt buildup behind dams is very substantial, and the practicality of removing the same is dubious. In essence this means that dams too have a lifetime.

### Batteries/Capacitors

Batteries have come a long way over the past ten years, and hold great promise for further advancements. As compared to transportation applications battery weight is not a significant concern since the power plant sites are stationary. Rechargeable batteries store energy efficiently with substantial energy densities, but are relatively slow in terms of response to electricity demand. Capacitors, however respond very fast, but have very limited energy density capacity. Coupling the two closely related technologies together probably makes sense.

Disadvantages here are that the term reversible is a relative one. Anytime chemical reactions are involved there are possibilities for side reactions to occur, so the batteries will have finite lifetimes. Also, many of the newest battery/capacitor technologies require rare and/or expensive elements, so over time availability and cost factors could be significant.

## Power-to-Gas

This technique involves the electrolytic splitting of water into hydrogen and oxygen using electricity generated by solar or wind power systems. The hydrogen thus generated can be compressed and used as is or converted into natural gas ($CH_4$) via the Sabatier Reaction:

$$CO_2 + 4H_2 \longrightarrow CH_4 + 2H_2O \quad \Delta H = -165.0\ kJ / mol$$

Note here that this reaction is exothermic, and the heat thus generated can be claimed as well increasing the overall process efficiency. The following table shows some efficiencies for this technique:

TABLE 12.2 **Power to Gas Conversion Efficiencies**

| Fuel | Efficiency | Conditions |
|---|---|---|
| **Pathway: Electricity→Gas** | | |
| Hydrogen<br>Methane (SNG) | 54–72 %<br>49–64 % | 200 bar compression |
| Hydrogen<br>Methane (SNG) | 57–73 %<br>50–64 % | 80 bar compression (Natural gas pipeline) |
| Hydrogen<br>Methane (SNG) | 64–77 %<br>51–65 % | without compression |
| **Pathway: Electricity→Gas→Electricity** | | |
| Hydrogen<br>Methane (SNG) | 34–44 %<br>30–38 % | 80 bar compression up to 60% back to electricity |
| **Pathway: Electricity→Gas→Electricity & heat (cogeneration)** | | |
| Hydrogen<br>Methane (SNG) | 48–62 %<br>43–54 % | 80 bar compression and electricity/heat for 40/45 % |

Source: Fraunhofer IWES, p. 18, February 2011 (in German)[11]

These are respectable efficiencies, but note the overall conversion numbers from power to gas to power. Using the cogenerated heat dramatically improves the overall processes.

## Compressed Air

Large scale commercial systems using compressed air storage systems have been known since the mid-19th century. The compression of air results in heating so in terms of efficiency it is critical that the heat thus generated is used as well. If so efficiencies of the order of 40–70% can be achieved. Systems based on constant volume or pressure are both well known, although not many large scale commercial systems have been developed.

# Chapter 12: Summary

TABLE 12.3 **Sources of Energy: Potential and Kinetic Energies.**

| Type of Energy | Examples |
| --- | --- |
| Mechanical | Turbines (wind, water), machinery |
| Solar | Photosynthesis, solar cells |
| Magnetic | Magnets, motors |
| Acoustic | Sound |
| Chemical | Combustion |
| Electrical | Batteries, fuel cells, capacitors |
| Gravitational | Gravity |
| Thermal | Heat transfer |
| Nuclear | Fission, fusion |

*Traditional Fossil Fuels:* Ancient deposits of carbonaceous fuels derived from solar energy.

*Natural Gas:* Basically methane, $CH_4$.

*Petroleum:* Liquid hydrocarbons including gasoline, kerosene, diesel and fuel oils.

*Coal:* Solid carbon with impurities. Bituminous and anthracite are the two major forms.

*Hydraulic Fracturing:* A vertical shaft is drilled to fuel bearing rock layers, then the drill bit is turned to follow these layers horizontally. Physical cracking via high pressures and explosives of the rock releases entrained petroleum which is returned to the surface and collected.

*Biofuels:* Fuels, such as wood, cellulosics, ethanol and methanol, derived from solar energy conversions into biomass.

*Synthetic Fuels:* Hydrogen and coal derived hydrocarbons

*Solar Power:* solar cells and photovoltaics.

*Other Energy Sources:* Wind, Nuclear, Hydroelectric, Geothermal and Tidal.

*Power Storage Technologies:* Methods to store power for future usage.

*Electrochemical:* Batteries and capacitors

*Hydraulic:* Primarily using dams.

*Power to gas:* Conversion of electrical power into chemical bonds via chemical reactions.

*Compressed Air:* Conversion of electrical power into high pressure, contained gases.

## CHAPTER 12 EXERCISES

1. List nine different forms of energy.
2. What are the different kinds of fossil fuels?
3. What is an anticline, and how does it relate to fossil fuel locations in the earth?
4. What are the basic biochemical process leading to a vein of coal?
5. There is no large terrestrial source for the essential element hydrogen. What are two chemical synthetic pathways for making hydrogen?
6. When oil is extracted from an oil well it is a mixture of a tremendous number of substances in gaseous, liquid and solid forms. How is this mixture separated into useful products?
7. Draw a diagram of the process for hydraulic fracturing.
8. What chemically comprises a Fischer Tropsch synthesis?
9. What are biofuels and what differentiates them from fossil fuels?
10. What are the most important renewable energy sources?
11. What are the basic forms of electrical energy storage on a large scale? What are the advantages and disadvantages of each?
12. What is steam reforming used for? Write out the chemical equation for this reaction.
13. What is the water gas shift reaction? Write out the chemical equation for this reaction.
14. What is the coal gasification? Write out the chemical equation for this reaction.
15. Which of the following compounds could be made using the Fischer Tropsch synthesis? Show possible synthetic routes!

    a. Cyclohexane

    b. Phenol

    c. Calcium carbonate

    d. Hydrogen
16. List the common fossil fuels in order of energy content.
17. Calculate the energy content of ethanol compared to that of ethane. Which compound has a higher energy content?
18. How is a solar cell constructed? Start with the elements and compounds you would need.
19. How much energy does it take to manufacture a wind turbine blade?
20. What are the environmental consequences from building a dam?

# Credits

Fig. 12.1: Copyright © 2004 by MagentaGreen, (CC BY-SA 3.0) at https://commons.wikimedia.org/wiki/File:Anticline_trap.svg.

Fig. 12.2: Copyright © 2006 by H Padleckas, (CC BY-SA 2.5) at https://commons.wikimedia.org/wiki/File:Tray_Distillation_Tower.PNG.

Fig. 12.3: Copyright © 2007 by Kai Point Coal Mine, (CC BY-SA 3.0) at https://commons.wikimedia.org/wiki/File:Uncovered_Coal_Seams.jpg.

Fig. 12.4: Source: https://upload.wikimedia.org/wikipedia/commons/7/73/Hydraulic_Fracturing-Related_Activities.jpg.

Fig. 12.5: Source: https://www.energy.gov/sites/prod/files/2017/09/f36/bioenergy_basics_infographic.pdf.

Fig. 12.6: Copyright © 2008 by Dexcel, (CC BY-SA 3.0) at https://commons.wikimedia.org/wiki/File:GTL_process.GIF.

Fig. 12.7: Copyright © 2018 by Daniel L. Lu, (CC BY-SA 4.0) at https://commons.wikimedia.org/wiki/File:Aerial_view_of_Apple_Park_dllu.jpg.

Fig. 12.8: Copyright © 2008 by Philip May, (CC BY-SA 3.0) at https://commons.wikimedia.org/wiki/File:Windrad-Wind-Turbine.jpg.

Fig. 12.9: Copyright © 2012 by Adam Kliczek, (CC BY-SA 3.0) at https://commons.wikimedia.org/wiki/File:Hoover_Dam,_02.jpg.

Fig. 12.10: Source: https://commons.wikimedia.org/wiki/File:Water_turbine_(en_2).svg.

CHAPTER 13

# Organic Chemistry

## Organic chemistry is the framework for all life as we know it...

## 13.1 A Different Kind of Chemistry

Students usually take organic chemistry immediately after two general chemistry courses, and usually they expect more of the same. It can be unsettling when this is not the case. Organic chemistry is not as precise as general chemistry, which is a reflection of how rich and interesting the topic of organic chemistry really is.

Organic chemistry underlies all of biochemistry, that is, all of life as we know it in both the plant and animal kingdoms. Certainly inorganic chemistry is also a vast and important field, underlying such important industries as electronics; however, without organic chemistry we would not have been here to it to have been developed in the first place.

Organic chemistry is based upon six elements***; carbon, hydrogen, oxygen, nitrogen, phosphorus*** and ***sulfur***. By definition all organic compounds contain carbon and hydrogen. For example take this essential biochemical reaction:

$$6CO_2 + 6H_2O \longrightarrow C_6H_{12}O_6 + 6O_2 \qquad (13.1)$$

This is the basic reaction that occurs in photosynthesis wherein two inorganic compounds, water and carbon dioxide, are reacted to form a sugar, $C_6H_{12}O_6$, and oxygen. The structure of the organic sugar glucose can be represented as:

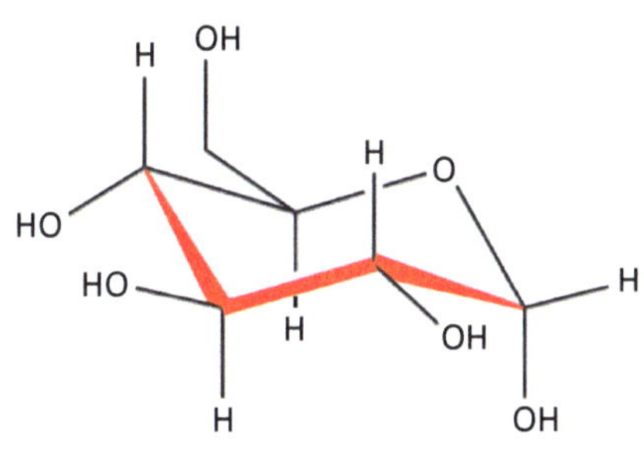


CHEMICAL STRUCTURE 13.1. Organic Sugar Glucose Structure.

The broader red C-C bonds here come out of the page towards the reader. As an aside equation 13.1 drastically simplifies the actual biochemistry involved in the harvesting of solar energy. It is indeed complex.

The span of organic compounds ranges from the very simple, such as methane ($CH_4$), to the marvelously complex, such as DNA. In essence there are an infinite number of possible organic chemical structures. A key point to keep in mind concerning organic chemistry is that the compounds are three dimensional. We generally write out and look at molecular structures in two dimensions for practical reasons, but being able to visualize the structures in 3D is a very useful skill to possess.

## 13.2 Organic Compounds—Bond Strengths, Compound Stabilities, and Reactivities

In Chapter 5 we discussed bond energies and reaction enthalpies (heats of reaction). It is a seamless transition to a discussion of bond strengths and reaction mechanisms in organic chemistry in general. The covalent bonding characteristics of carbon are central to all of this topic. These covalent bonds are strong, but not immortal. For example, a comparison of the reactivity of octane, a primary constituent of gasoline, and your basic rock, mainly composed of silicon dioxide, $SiO_2$, is informative. Octane is composed of just C - H covalent bonds and is stable under ambient conditions, but once ignited reacts rapidly with oxygen. $SiO_2$ on the other hand is very stable, and can only be reacted under harsh and energetically demanding conditions. The basic difference here distills down to the relative bond strengths between the various atoms.

The rest of this chapter introduces the most important organic functionalities, their interconversions from one to another, and their relative reactivities. It is worthwhile noting that the remaining Chapters 14–17 all rely on organic chemistry as a foundation of understanding.

There are two large and closely intertwined classes of organic compounds, the aliphatic group and aromatic group. They are composed of carbon and hydrogen combined in a great many different ways. The difference between the two classes can be subtle, and structures usually include both moieties. We will look at these two groups first followed by an examination of important subgroups of compounds.

## 13.3 Aliphatic Compounds

The simplest members of this class of compounds are composed of just carbon and hydrogen. Along with aromatics (section 13.4) they form the basis for the rest of organic chemistry, biochemistry, and all of the biological sciences. It is important to keep in mind that although the following discussion separates out compounds on the basis of functional groups, in general organic compounds are composed of mixtures of these different moieties.

### Alkanes

Alkane molecules contain covalently bound carbon and hydrogens. Many of these compounds are important to our daily life, for example, propane ($C_3H_8$). Natural gas is methane ($CH_4$) with assorted impurities thrown in. Gasoline is another example.

Molecular formulas like these do not tell us why these compounds have the physical and chemical properties they do, however, so let's explore some of compounds more deeply.

The alkanes are known as fully saturated compounds meaning that there are only single bonds to the carbons, no double or triple bonds. The parent group of this class is basically a homologous series of C-H, C-C bonded atoms. By way of example Figure 13.1 shows ten straight-chain alkanes:

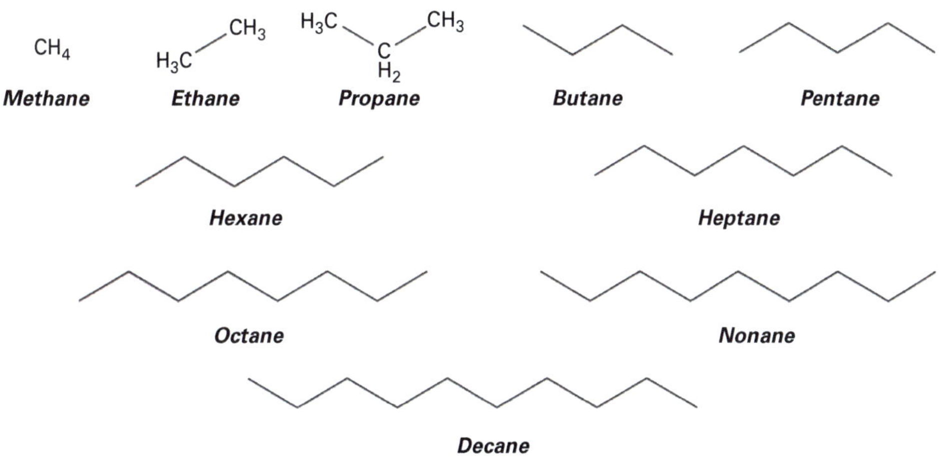


**FIGURE 13.1.** The First Ten Members of the Straight Carbon Chain Series of Alkanes.

They are called straight chain because of the linear sequence of C-C bonds. There is a general formula for this homologous series of $C_nH_{n+2}$ where n is a whole number. Some good news here is that the physical and chemical properties of this subclass of compounds (there are many more members), although not exactly the same, are similar to each other. Some are gases, some liquids and are some solids under typical ambient pressure and temperature conditions. Table 13.1 contains a list of a few of the physical properties of these ten compounds:

TABLE 13.1 **Physical Properties of Alkanes**

| Molecular Name | Formula | Melting Point (°C) | Boiling Point (°C) | Density (20°C)* | Physical State (at 20°C) |
|---|---|---|---|---|---|
| methane | $CH_4$ | –182 | –164 | 0.668 g/L | gas |
| ethane | $C_2H_6$ | –183 | –89 | 1.265 g/L | gas |
| propane | $C_3H_8$ | –190 | –42 | 1.867 g/L | gas |
| butane | $C_4H_{10}$ | –138 | –1 | 2.493 g/L | gas |
| pentane | $C_5H_{12}$ | –130 | 36 | 0.626 g/mL | liquid |
| hexane | $C_6H_{14}$ | –95 | 69 | 0.659 g/mL | liquid |
| octane | $C_8H_{18}$ | –57 | 125 | 0.703 g/mL | liquid |
| decane | $C_{10}H_{22}$ | –30 | 174 | 0.730 g/mL | liquid |

As can be seen the boiling and melting points of this series change in a systematic way for compounds C1-C10. Notice also the gradual transition from the gas phase to volatile liquids to high boiling liquids through waxes and eventually to solids as the molecular weight increases. Notice also the far more gradual increase in density.

Learning the names of organic compounds, known as ***nomenclature***, basically involves learning a new lexicon. Alkanes names end with ***–ane***, hence hexane contains six carbons (hex means 6) with –ane tacked on to the end. Meth- means one C, eth- means 2 carbons, prop- means 3 etc. This convention follows through with the other classes of compounds to be discussed hereafter.

## Isomers

Isomers are compounds with the same molecular formula, but different 3D structures.

Here is where nomenclature gets more complicated. Let's do an example, pentane. Remember that each carbon has four single bonds, but note that we are free to mix and match these bonds as long as we stay within the constraints of the formula $C_2H_{12}$. Also, fortunately for us, each H can only have one single bond. What are the possibilities here? Chemical structure 13.2 shows there are only three shown with their common names.

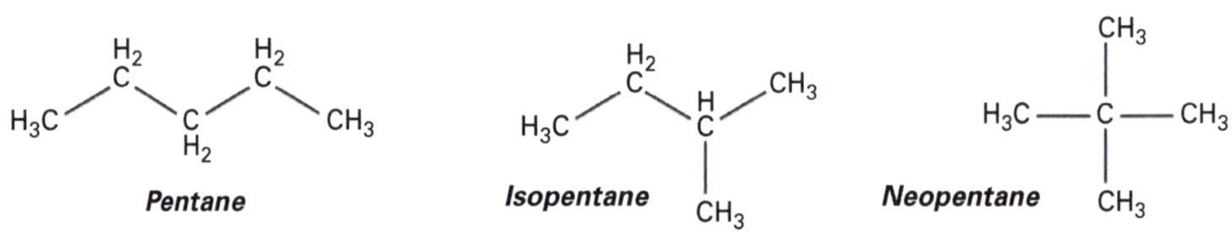


**CHEMICAL STRUCTURE 13.2A.** Isomers of Pentane.

There are just three possibilities here. At first blush it may appear there are more, but consider the following two pentane molecular line structures:

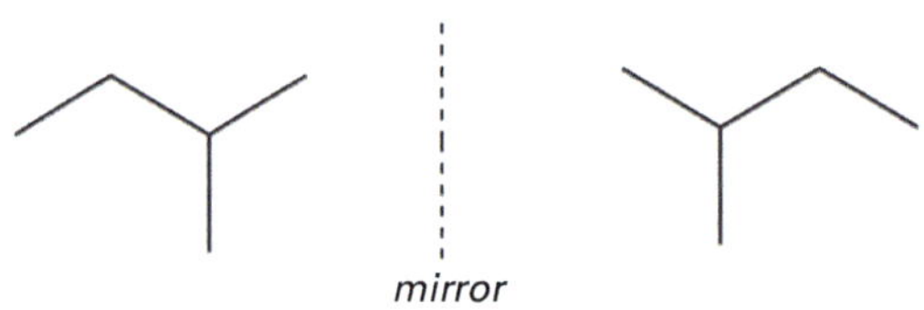


**CHEMICAL STRUCTURE 13.2B.** Mirror.

Each qualifies as a possible isomer since each has a formula of $C_5H_{12}$. However, note that they are mirror images of each other! Flip one over and they become superimposable structures, that is, the same structure.

Pentane is a small molecule. Clearly, as the number of carbons increases arithmetically the number of possible isomers increases geometrically and naming a compound becomes problematic in a hurry. For the simple alkane class of molecules the convention is to determine the longest straight carbon chain and use that as the root name for the structure. By way of example consider this $C_8H_{18}$ alkane molecule.

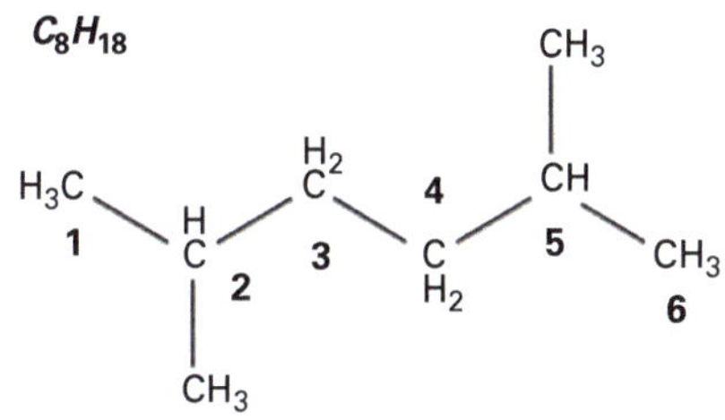


CHEMICAL STRUCTURE 13.3. 2,5-dimethyl hexane.

The longest carbon chain here is C6, so it is a hexane. Note too that the carbons are numbered sequentially from 1 to 6. Note also we have two $CH_3$ groups to account for. These are given the name **methyl = -$CH_3$.** Their positions on the carbon chain are denoted by the C chain numbers. That is this compound is named as:

**2, 5-dimethyl hexane**

**PRACTICE** Draw all of the possible isotopes for hexane, $C_6H_{14}$.

## Cycloalkanes

An important subclass in the alkane family are cycloalkanes. An example is cyclohexane shown in three representations:

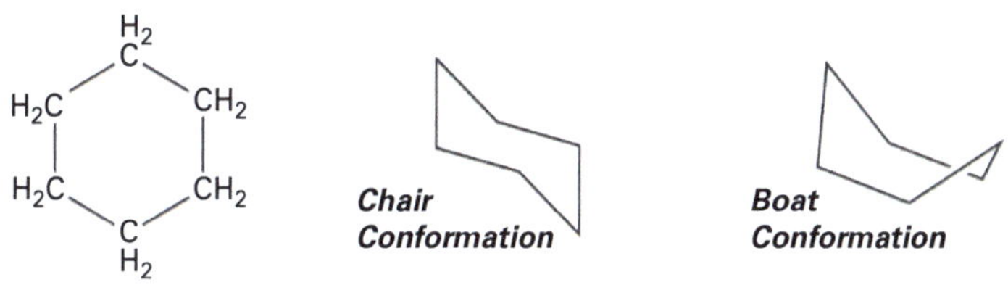


CHEMICAL STRUCTURE 13.4. Cycloalkanes.

These three drawings hopefully give a sense for how flexible alkanes are having free rotations around all of the single bonds. A related point to remember is that, although we use drawings that are static, in reality molecules are very dynamic systems with stretching and twisting motions occurring constantly around the bonds.

Note here that the extra C-C bond to make the ring eliminates two H atoms so the generic formula for cycloalkanes is $C_nH_{2n}$.

**PRACTICE** Draw the structure of cycloheptane.

## Alkenes and Alkynes

Alkenes and alkynes are closely related to alkanes in terms of structure. ***Alkenes*** are named by substituting an *-**ene*** ending for the *-**ane*** of the corresponding alkane with the elimination of two hydrogen atoms with the concomitant formation of a C to ***C double bond. Alkynes*** are named by substituting an *-**yne*** ending for the *-**ane*** of the corresponding alkane with the elimination of four hydrogen atoms with the concomitant formation of a C to C ***triple bond***. For example ethane is transposed into either ethylene or ethyne:

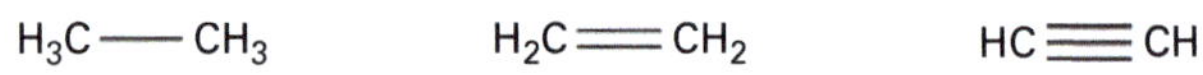

***Ethane*** ***Ethylene*** ***Ethyne***

**CHEMICAL STRUCTURE 13.5.** Alkenes and Alkynes.

Ethyne is far more commonly known as the welding gas **acetylene.**

Although alkenes and alkynes have similar properties to alkanes, they are far more reactive because of the presence of the double and triple bonds. These multiple bonds are energetically strained. We will look at their relative reactivities later in this chapter.

# 13.4 Electron Delocalization and Resonance Energy Stabilization

Underlying half of organic chemistry is the concept of ***resonance energy stabilization***.

Consider two arrangements of the double bounds in linear pentadiene molecules:

***1,3-pentadiene*** ***1,4-pentadiene***

**CHEMICAL STRUCTURE 13.6.** Pentadiene.

These molecules are surely isomers and would be expected to have more or less the same energies. However, the 1,3 isomer has a substantially lower total energy, and is thus more stable, than the 1,4 isomer. By more stable we mean less chemically reactive.

In both cases there are two four electrons involved in double bonds; however in the 1,3 case these double bond electrons ***are delocalized over four carbon atoms***. This concept can be visualized as follows in which the electrons are evenly spread out:

***1,3-pentadiene*** ***delocalized π electrons***

**CHEMICAL STRUCTURE 13.7.** Pentadiene and Delocalized Electrons.

***This type of delocalization induced stability is extremely important in organic chemistry.***

Another example of primary importance as we develop the topic of organic chemistry is the resonance stabilized ***phenyl*** ring:

**CHEMICAL STRUCTURE 13.8.** Phenyl Ring.

This six carbon ring in a formal sense has three double bonds, but the reality is six electrons that are delocalized around the hexagonal carbon ring. This structure possesses substantially more reactive stability than predicted; a stability that is extremely important in all biological systems. The phenyl ring is the basis for the second huge class of organic compounds, the aromatic class.

## 13.5 Aromatic Compounds (Arenes)

There is energy released upon the hydrogenation of cyclohexene via the following reaction:

**CHEMICAL STRUCTURE 13.9.** Hydrogenation of Cyclohexene.

The reaction is exothermic because of a release of double bond strain to form two new single bonds.

By extrapolation the compound cyclohexatriene (known as benzene) should be exothermic to the tune of $\Delta H = -360\ \boldsymbol{kJ / mol}$ upon hydrogenation; however, in reality it is $\Delta H = -208\ \boldsymbol{kJ / mol}$, or substantially lower by **152 kJ/mol.** This difference is due to the resonance stabilization of the six electrons shared by all six carbons. Remember here that there are still six single C-C bonds in the molecule as well.

A point of potential confusion here is nomenclature or naming of structures. ***Benzene*** is the name of the six membered carbon aromatic ring with only hydrogens bound to the ring carbons. The ***phenyl*** name is used when one or more of the hydrogens is replaced with another structural moiety.

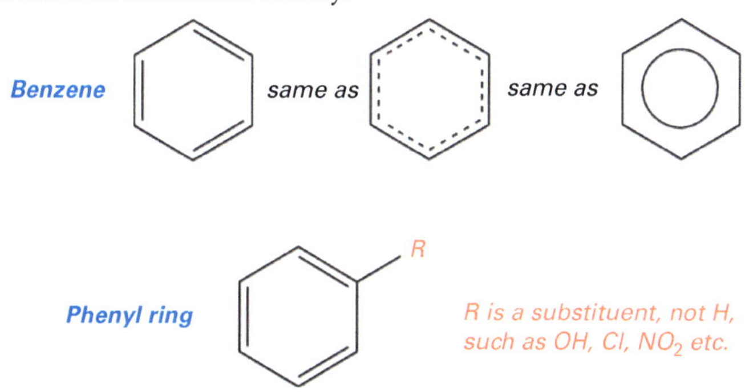


**CHEMICAL STRUCTURE 13.10.** Benzene and Phenyl Ring Structures.

## Fused Ring and Some Important Aromatic Compounds

### Fused Ring Aromatics

Following are a few of the most important and basic fused ring structures based upon the phenyl unit:

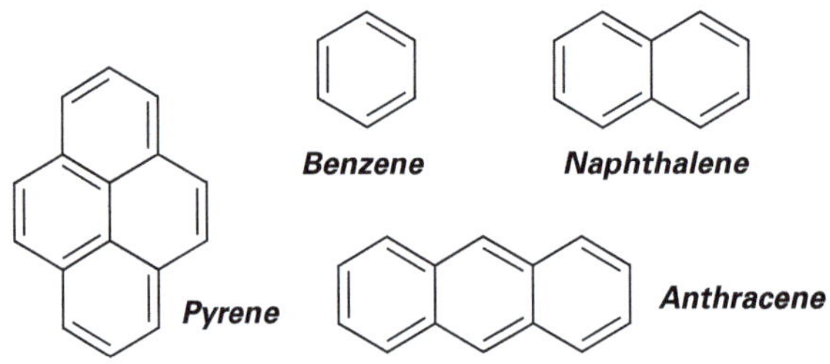


**CHEMICAL STRUCTURE 13.11.** Fused Ring Aromatics Structure.

### Some Other Important Aromatic Compounds

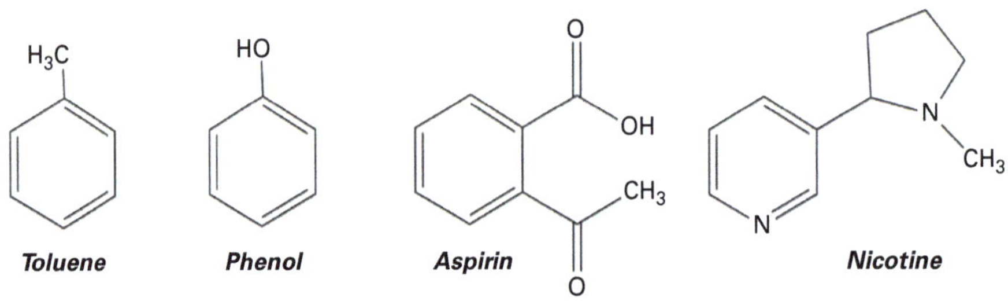


**CHEMICAL STRUCTURE 13.12.** Some Other Important Aromatic Compounds.

The two on the right are certainly familiar by name and are more complex than the others. These are but two examples of a vast number of known organic compounds that are substituted with a wide array of possible groups. We will now look at some of the most important classes of substituted organic molecules.

## Substituents in Organic Chemistry

Organic chemistry is a broad and complex topic. This chapter is but an overview of this essential chemistry. As such, we will touch on the naming of compounds in passing. The interested reader is referred to texts that specialize in organic chemistry; a few of which are discussed here briefly.

The classes of functional groups are discussed now. At the end of this chapter basic reactivity mechanisms are introduced as well.

***Important Note:*** In many structures the letter R is used generically to represent a chemical group. For example ***R*** can stand for a $-CH_3$, $-OH$, $-NH_2$, $-Br$ etc.

# 13.6 Halides

Most of the ***halogens*** (group VII) are important in all kinds of chemistries, in particular Cl, Br and I. To a lesser extent F is also a player in organic chemistry, but only where manmade compounds are involved. Representative structures for both the aliphatic and aromatic classes are shown in Chemical Structure 13.13:

$CH_3Cl$

***Methyl Chloride***

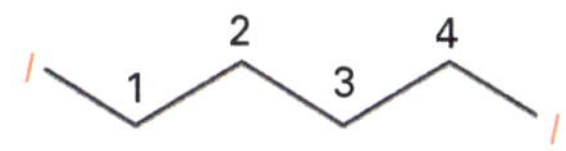

***1,4-diiodobutane***

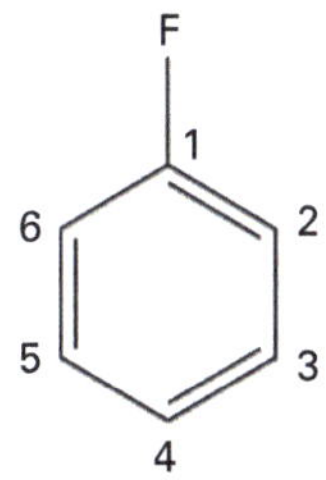

***Fluorobenzene***

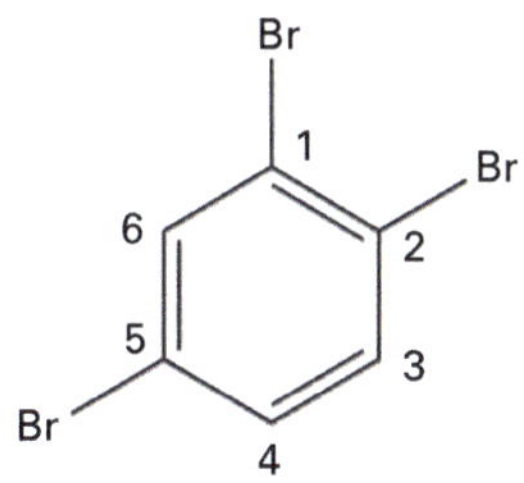

***1,2,5-triibromobenzene***

**CHEMICAL STRUCTURE 13.13.** Representative Organic Halide Compounds.

The halo-organics have a wide range of properties, existing as gases, liquids and solids. Most are biologically active and many have found wide use as insecticides and herbicides, a double edged sword as they often present toxicological and environmental issues (see Chapter 17). The reactive properties of this class are discussed at the end of this chapter.

# 13.7 Alcohols

This large and diverse group of compounds are based upon the C-OH moiety, a group that is ubiquitous in all biological systems. Representative structures for both the aliphatic and aromatic classes are shown in Chemical Structure 13.14:

Here again the chemistry nomenclature gets confusing. In general alcohol names end with **-ol**. For example **$CH_3CH_2OH$** is ethan***ol***; the name derived from the parent alkane compound **ethane** with an ***-ol*** substituted for -e at the end.

Another confusing naming feature shown here is the use of different names for the same compound, for example phenol and hydroxybenzene. Long ago, when there weren't many compounds known anyway, common names were given to compounds as they were identified and characterized. That's the bad news. Better news is that as compounds have gotten more complex even the best organic chemists probably can't name them correctly! Or draw their structures given the name. Fortunately today's world with the internet allows for the rapid identification of compound structures from names and vice versa.

$CH_3OH$

***Methanol***

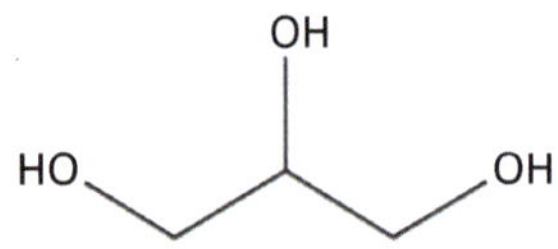


***1,2,3-trihydroxypropane (glycerine)***

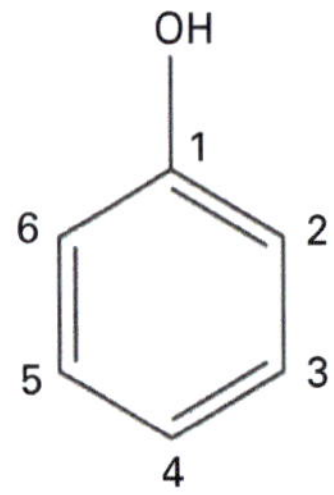


***Hydroxybenzene (Phenol)***

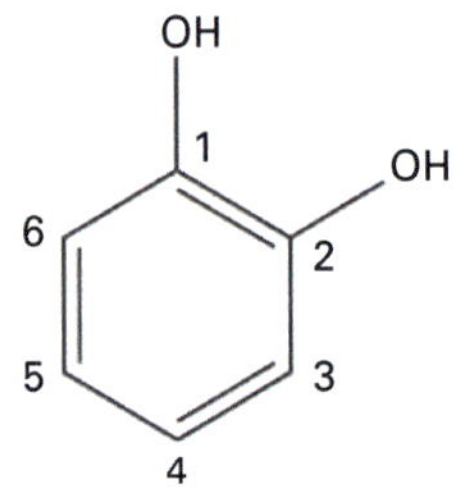


***1,2-dihydroxybenzene (a cresol)***

**CHEMICAL STRUCTURE 13.14.** Representative Alcohol Compounds.

## 13.8 Amines

***Primary*** amines are structurally very similar to alcohols, but with the –OH replaced by an $-NH_2$, ***secondary*** amines –OH replaced by –NHR, and ***tertiary*** amines by $-NR_2$. Examples of amines are shown in Chemical Structure 13.15:

**Representative Amine Compounds**

### Primary Amines

$CH_3NH_2$

***Methyl amine***

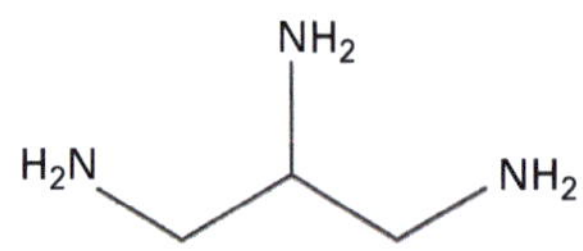


***1,2,3-triaminopropane***

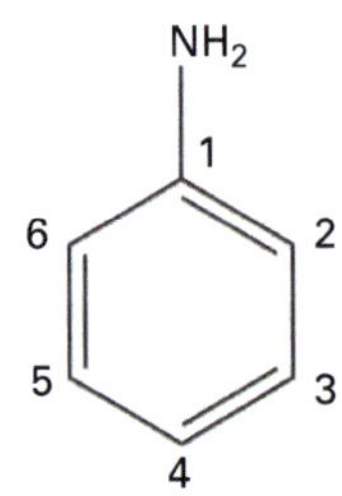


***Aminobenzene (aniline)***

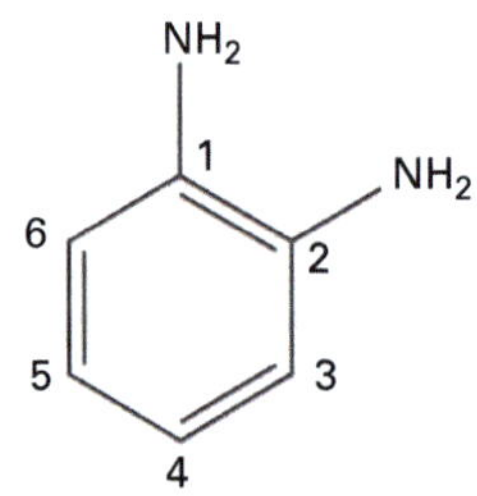


***1,2-diminobenzene***

**CHEMICAL STRUCTURE 13.15.** Primary Amines.

## Secondary and Tertiary Amines

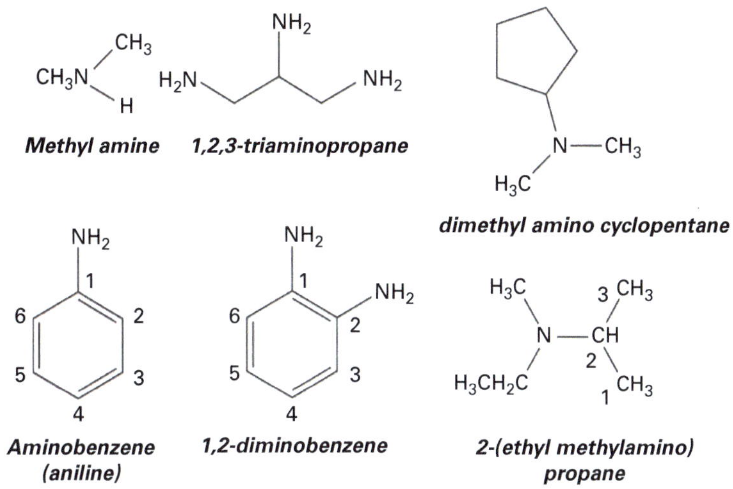


**CHEMICAL STRUCTURE 13.16.** Secondary and Tertiary Amines.

In terms of reactivity amines are also very similar to alcohols, but generally more reactive. Also note that the NH2 group has two N-H linkages whereas the O-H has but one O-H.

The amine group is ubiquitous and critically important in the metabolic function of all biological systems without exception.

# 13.9 Carbonyl (C = O Containing) Compounds

The organic compounds described in this section all have a common structural feature, the carbonyl group, hence their physical properties and chemical reactivities are similar, but by no means exactly the same.

## The Carbonyl Group

This is the carbonyl group:

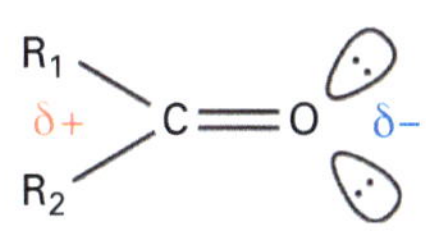

**CHEMICAL STRUCTURE 13.17.** The Carbonyl Group.

Once again $R_1$ and $R_2$ are generic symbols representing other groups bonded to the ***carbonyl carbon***. Also the two lone pairs of electrons on the oxygen atom are shown.

Although these electrons are not typically shown in discussions of carbonyl chemistry, they are critically important in terms of physical and chemical properties. An important point here is that ***a lone pair of electrons occupies as much or more 3D space as does an H atom***.

Another important point here is the distribution of the electrons within the molecule. As the drawing indicates the electrons are pulled towards the very electronegative oxygen atom (indicated by the electron rich symbol δ-) and away from everything else (indicated by the electron poor symbol δ+). ***This electronic distribution underlies much of the rich chemistry that we will briefly discuss in Section 14.11, Common Reaction Types and Mechanisms.***

Now we will look at the families of compounds that are anchored by the carbonyl functional group and its associated chemical properties.

***Note: the symbol***  ***represents a single covalent bond to another carbon***

Table 13.2 shows the names of these families, their corresponding key chemical functionalities, and an example of each class:

TABLE 13.2 **Carbonyl-Based Functionalities and Names**

| Group | Structure | Name | Example |
|---|---|---|---|
| **Aldehyde** | 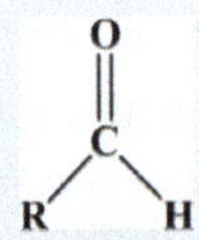 | **-aldehyde** | **Acetaldehyde** |
| **Ketone** | 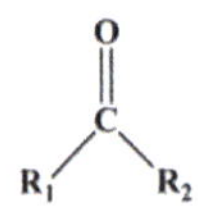 | **-one** | **Acetone** |
| **Carboxylic Acids** | 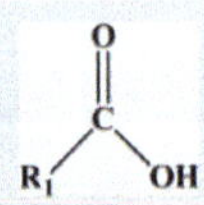 | **-ic acid** | **Acetic Acid** |
| **Esters** | 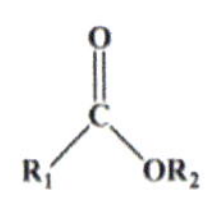 | **-ate** | **Acetate** |
| **Amides** | 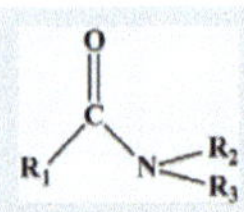 | **-amide** | **Acetamide** |
| **Carbonates & Carbonic Acids** | 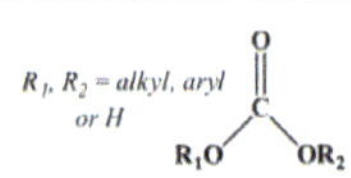  | **carbonate carbonic Acid** | **Dimethylcarbonate Methyl carbonic acid** |
| **Carbamates (Urethanes)** | 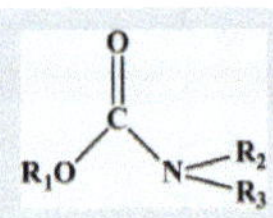 | **carbamate** | **Polyurethane** |
| **Ureas** | 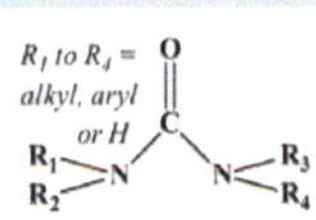  | **urea** | **Methyl urea** |

## Aldehydes and Ketones

Here are the aldehyde and ketone functionalities:

CHEMICAL STRUCTURE 13.18. Aldehydes and Ketones.

There is not much apparent difference in these structures, but as we shall see the substitution of the lone hydrogen (in red) in the aldehyde with a second carbon-based group in the ketone significantly affects a compound's resultant properties.

### Interesting Examples:

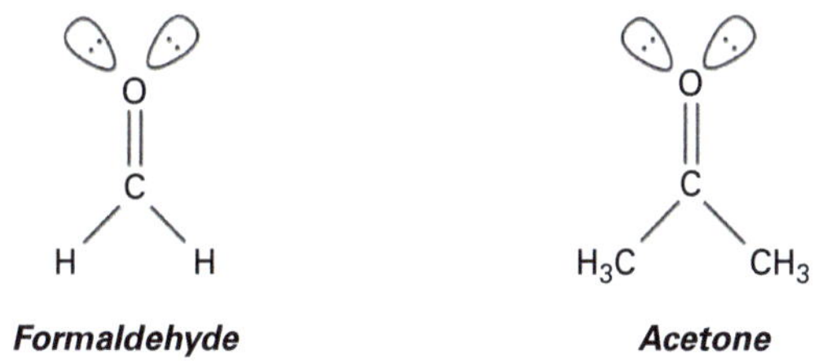


CHEMICAL STRUCTURE 13.19. Formaldehyde and Acetone.

Formaldehyde, where the second group bonded to the carbonyl carbon is another H, was used to embalm and preserve dead organisms; however, another of its properties is being carcinogenic, so it is no longer used for this application in general.

Acetone is an excellent solvent that is widely used to remove nail polish.

## Carboxylic Acids

These compounds are weak acids characterized by the presence of –COOH group:

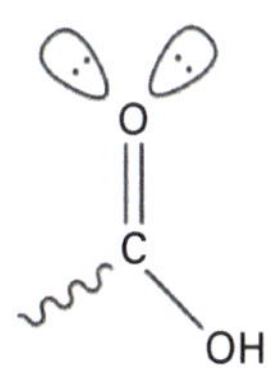


CHEMICAL STRUCTURE 13.20. Carboxylic Acids.

They are ubiquitous in nature, and serve many different biological functions, an important one being to act as buffers in regulating the pH of chemical systems.

Here are two examples of carboxylic acid structures. Acetic acid is a key ingredient in vinegar. Benzoic acid is a common example of an aryl organic acid

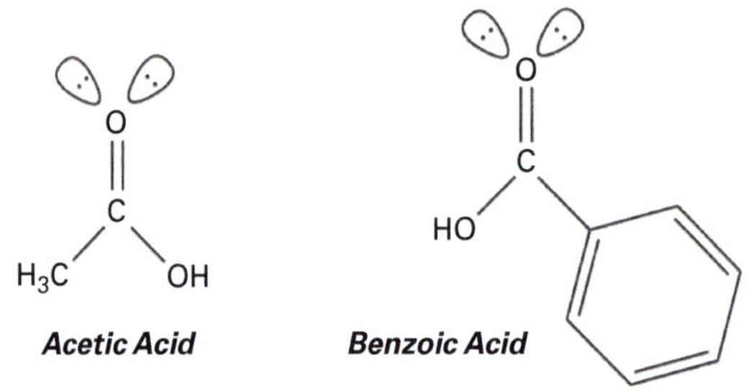


CHEMICAL STRUCTURE 13.21. Acetic Acid and Benzoic Acid.

## Esters

As a group esters are known for their aromas, sometimes those being fragrances, sometimes odors. They all contain the following structural unit:

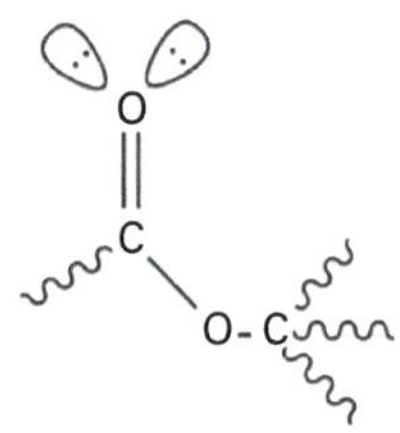


CHEMICAL STRUCTURE 13.22. Esters.

### Interesting Examples:

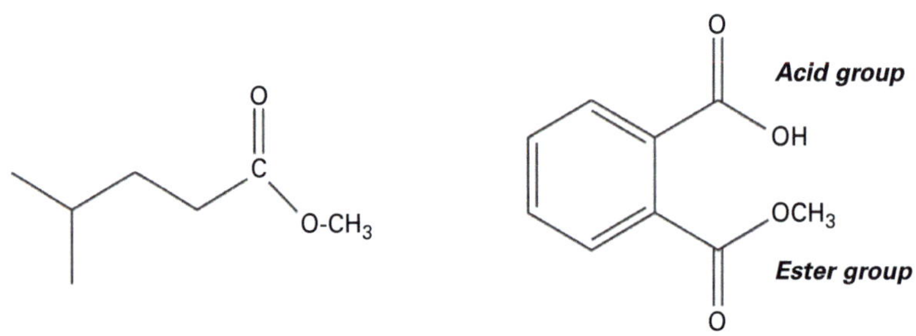


CHEMICAL STRUCTURE 13.23. Essence of Banana and Aspirin.

Note above that in the Aspirin structure, besides the ester group, a carboxylic acid moiety is also present. Both are critical to Aspirin's well-known medicinal properties.

## Amides

The amide functionality is closely related to esters wherein the –OR group is replaced by an $-NR_2$. This similarity qualitatively parallels the pairing of the alcohol (–OH) and amine ($NH_2$) functionalities described above.

Amides contain the following moiety:

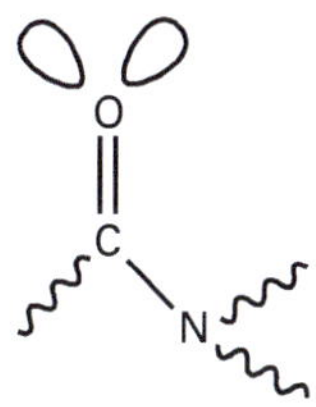


CHEMICAL STRUCTURE 13.24. Amides.

A key difference here from the corresponding ester structure is that there are two groups bonded to the nitrogen atom, versus only one to the ester's singly bonded O.

### An Example Well Known by Name, Nylon:

Polymers are high molecular weight compounds, many families of which are characterized by being long chains of repeat structural units. The Nylon group of polymers is one of these families. One classic example of this group is Nylon 6,6:

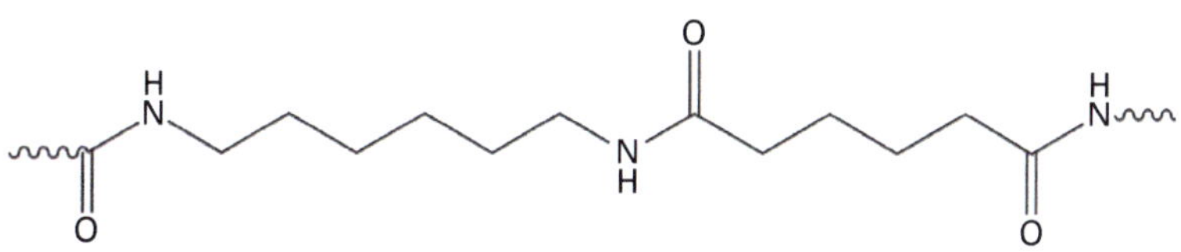


CHEMICAL STRUCTURE 13.25. Nylon.

The 6,6 stands for the number of carbons, including C=O carbons, that are between each pair of Ns in the chain. Varying the number of these spacer carbons changes the molecule synthesized.

The amide functionality is fundamental to all of protein chemistry, and proteins are basic building blocks in all life forms on earth, both plant and animal. A subgroup of compounds therein are the polypeptides, a basic example of which is shown below with the amide linkage colored red:

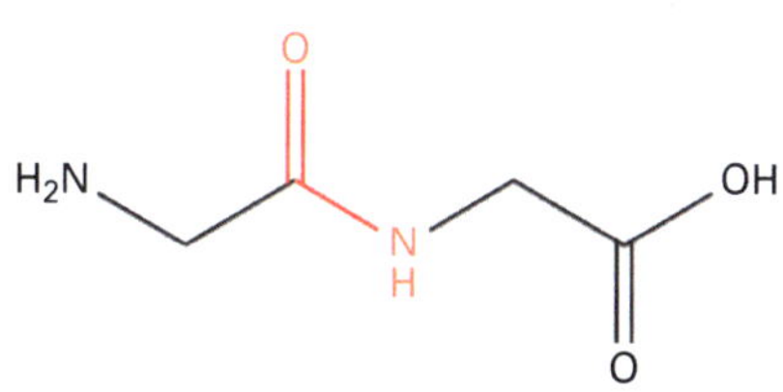


CHEMICAL STRUCTURE 13.26. Polypeptides.

Present too are the amino and carboxylic acid functionalities. Chapter 15, Biochemistry, contains describes protein chemistry in detail.

## Carbonic Acids and Ureas

These groups of carbonyl-based compounds and the next set are not as well-known as those groups described above; however, they are just as important. The parent member, though not an organic compound, is carbonic acid, $H_2CO_3$.

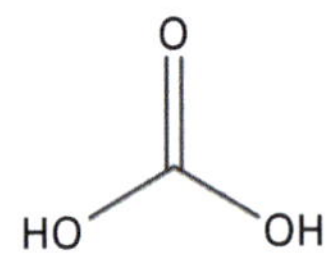


**CHEMICAL STRUCTURE 13.27.** Carbonic Acid.

It is difficult to argue there is a more important compound in chemistry. It is key in an untold number of biochemical reactions, and as a weak acid, acts as the buffer that regulates the acidity of the world's oceans. Enough said.

Substituting for one of the two acidic H atoms on this molecule can result in an organic carbonic acid, such as methyl carbonic acid or phenyl carbonic acid:

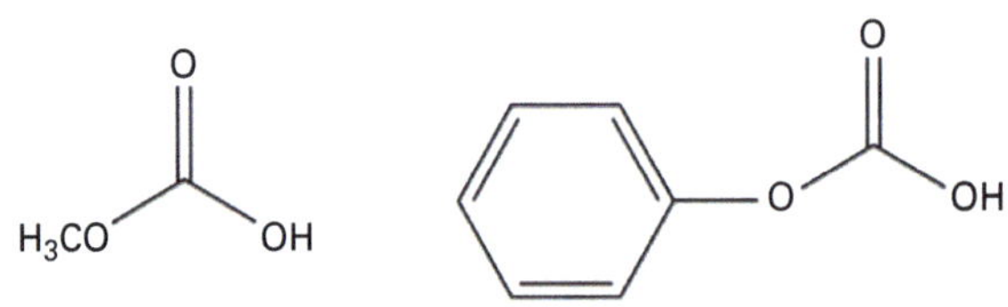


***Methyl Carbonic Acid*** ***Phenyl Carbonic Acid***

**CHEMICAL STRUCTURE 13.28.** Methyl Carbonic Acid and Phenyl Carbonic Acid.

Here note that there is both an ester group and a carboxylic acid group present in the molecules!

***Ureas*** are the corresponding structures wherein the –OH groups are replaced by –NH2 moieties:

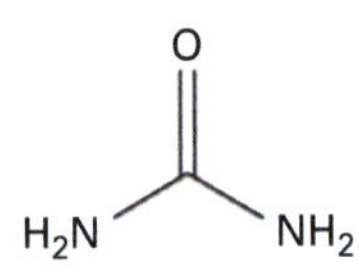


***Urea***

**CHEMICAL STRUCTURE 13.29.** Ureas.

The parent compound, urea, is manufactured in very large quantities from $CO_2$ and $NH_3$ and is used as a fertilizer as well as a starting material for many other useful chemical products. As might be imagined the amide structures corresponding to the esters shown above are readily obtainable.

## Carbonates and Carbamates (Urethanes)

The last set of functional groups to be discussed are the next logical iteration of the carbonic acid category. Carbonates that are fully esterified can be represented by dimethylcarbonate:

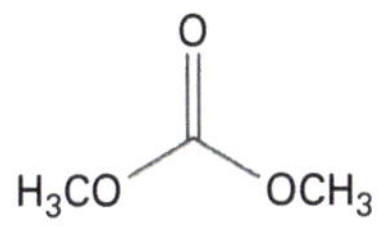


**CHEMICAL STRUCTURE 13.30.** Dimethylcarbonate.

This compound is used as the basis for the electrolyte in virtually all rechargeable lithium batteries of commercial importance. The structure is characterized by in essence having two ester linkages, one on either side of the carbonyl group.

***Carbamates*** are especially important as components of synthetic herbicide formulations. Here is a structure of one of these herbicides:

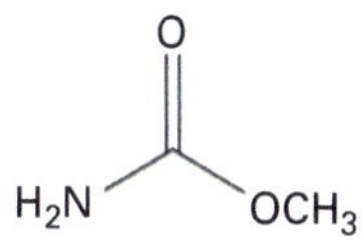


**CHEMICAL STRUCTURE 13.31.** Carbamates.

This structure essentially combines ester and amide groups to the central carbonyl. The biochemistry and toxicology of this and related compounds are discussed in chapters that follow in this book.

# 13.10 Ethers

Although the ether linkage, shown in Chemical Structure 13.32, is widespread in nature and occurs in conjunction with the chemical moieties described above, in many ways it has a chemistry all of its own.

This figure is informative on a number of levels. First is that the ether linkage is C - O - C. Second, the top left hand structure, diethyl ether, is an interesting compound. It was one of the first anesthetics used, shortly after chloroform (an organic halide, section 11.1, trichloromethane, $CHCl_3$). It is also remarkably dangerous since it is highly volatile (boiling point 34.6°C) and explosively flammable. It is notoriously reactive with oxygen via radical chemistry (see Section 12.11) in which peroxides can build up to form shock sensitive explosive products.

A second point dealing with Chemical Structure 13.32 is nomenclature, once again we are dealing with a new and complex language here. Compound B is named by the groups on either side of the ether oxygen, in this case cyclopentyl and methyl. C is logically methyl phenyl ether, but the methoxy group ($-OCH_3$) name is also used. In naming D the substituents are placed in alphabetical order, hence the amino substituent comes first. E's name reinforces yet another convention, using the longest carbon chain as the root and then naming the substituents alphabetically. Notice here also that this pair of conventions results in the carbon chain numbering beginning from the right side rather than the left.

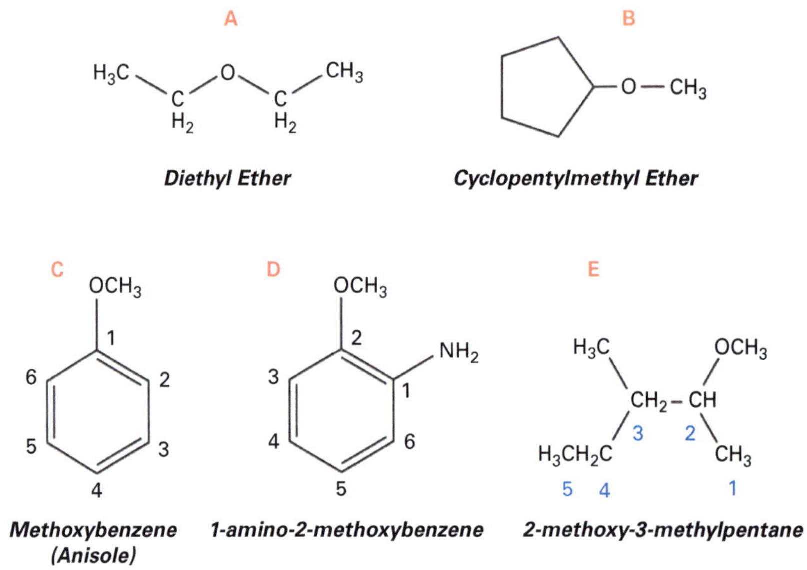


**CHEMICAL STRUCTURE 13.32.** Representative Ether Compounds.

# 13.11 Common Reaction Types and Mechanisms

Now that we have discussed what underlies organic chemistry via the basic functional groups, we will look at what happens in the basic reaction mechanisms of the same. In essence this involves the ability to look at the structure of a compound and predict what kinds of reactions will occur under a specific set of conditions. Let's do an example. Here is a compound that has a number of functional moieties incorporated into it:

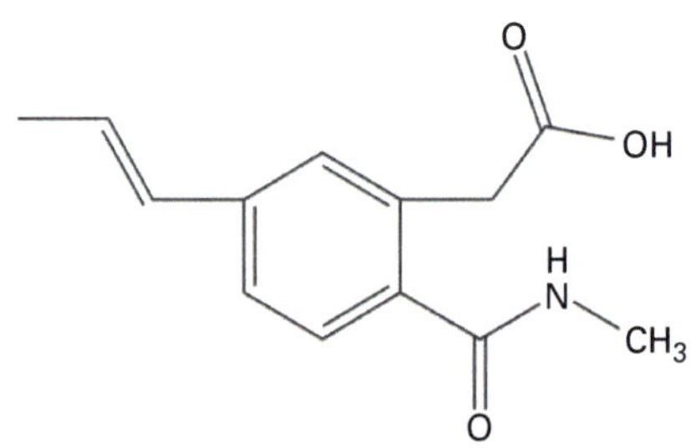

**CHEMICAL STRUCTURE 13.33.** Multiple Functional Moieties Structure.

We will not even attempt to name this compound. In the final analysis who cares about the name anyway? What is really important is how it will behave in terms of reactivity.

There is a lot going on structurally in this molecule. Present are alkane, alkene, aromatic, acid and amide functionalities. Now suppose that the reaction conditions are typical for what we experience every day. Room temperature and pressure, and the presence of ubiquitous oxygen and water as reactants. Notice here that we have drastically simplified real conditions. Reality includes the presence of atmospheric $N_2$, $CO_2$, argon, etc. as well as a multitude of sulfur and amine based compounds, alcohols........... it goes on and on, but we have to start somewhere. Somewhere is what follows in this section. We will look individually at the most important possible types of reaction, and apply what we learn to the random molecule shown above.

## Types of Organic Chemistry Reactions

In this chapter we have established the structural basics of organic molecular chemistry. Especially important to take away from this chapter are the functional group structures and the electron distributions, that is, where do the electrons want to spend most of their time?

In the next Chapter, 14, we will be discussing two things. First is the types or general classes of organic reactions. This will be followed by a look at what of reaction mechanisms. Here are the subtopics of Chapter 14.

## Reaction Types

Substitutions
Eliminations
Additions
Decompositions

## Reaction Mechanisms

Nucleophilic
Electrophilic
Hydrolyses
Oxidations
Free Radical

# Chapter 13 Summary

*Organic Chemistry:* The chemistry of carbon and hydrogen combined into covalently bonded compounds.

*A Different Kind of Chemistry:* Organic chemistry is basically the chemistry of six elements, carbon, hydrogen, oxygen, nitrogen, phosphorous and sulfur with a smattering of the rest of the periodic table. This chemistry revolves around the concept of electronegativity.

## Compound Classes

*Alkanes:* Contain carbons and hydrogens covalently bound together in a wide variety of molecular arrangements.

1. *Isomers: Compounds with the same molecular formula, but different 3D structures*
2. *Cycloalkanes: Alkanes that contain a ring structure.*

*Alkenes and Alkynes:* Hydrocarbons that contain a double bond (C=C) or a triple bond, respectively.

*Electron Delocalization or Resonance and Energy Stabilization:* Where electrons are capable of occupying larger volumes a lowering of the total energy is observed. This results in a stabilization of molecules and a lessening of reactivity.

*Aromatic Compounds:* A large class of compounds that possess electron delocalized, stable structures.

## Substituents in Organic Molecular Structures

***Halides:*** Molecules that contain F, Cl, Br or I as substituents._
***Alcohols:*** Molecules that contain F, the –OH group as substituents.
***Amines:*** Molecules that contain an –NH or $-NH_2$ group as substituents.
***Ethers:*** Molecules that contain an (C– O – C) moiety.
***Carbonyl Containing Compounds:*** Ones that contain the C=O group.

| Group | Structure | Name | Example |
|---|---|---|---|
| ***Aldehyde*** | 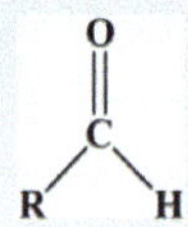 | ***-aldehyde*** | ***Acetaldehyde*** |
| ***Ketone*** | 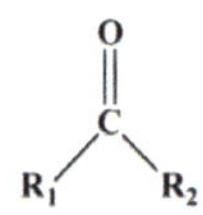 | ***-one*** | ***Acetone*** |
| ***Carboxylic Acids*** | 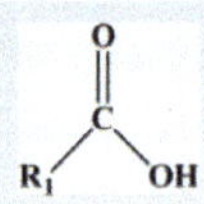 | ***-ic acid*** | ***Acetic Acid*** |
| ***Esters*** | 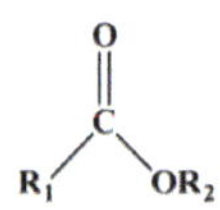 | ***-ate*** | ***Acetate*** |
| ***Amides*** | 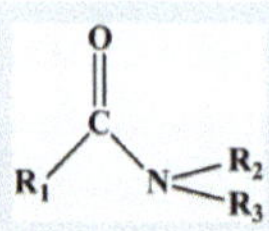 | ***-amide*** | ***Acetamide*** |
| ***Carbonates & Carbonic Acids*** | 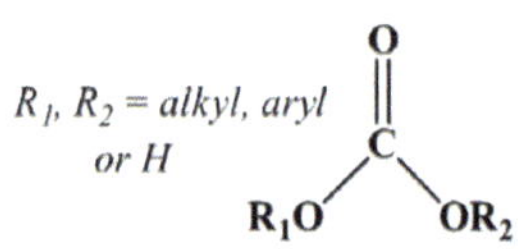 | ***carbonate carbonic Acid*** | ***Dimethylcarbonate Methyl carbonic acid*** |
| ***Carbamates (Urethanes)*** | 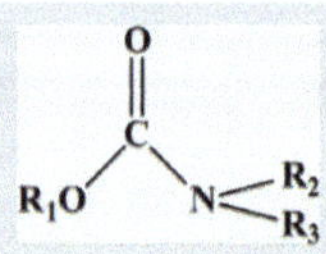 | ***carbamate*** | ***Polyurethane*** |
| ***Ureas*** | 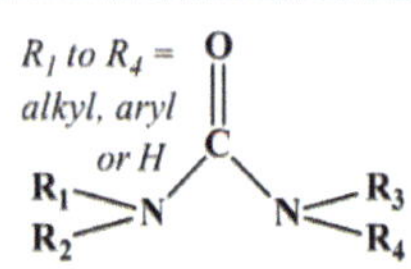 | ***urea*** | ***Methyl urea*** |

# CHAPTER 13 EXERCISES

1. Why is organic chemistry so important?
2. What are the six elements that form the foundation of organic chemistry?
3. One of the two or three most important chemical reactions is photosynthesis. Although the whole process is exceptionally complex, the fundamental result can be distilled down to one that is very simple. What is it? (Hint: at its simplest just two reactants and two products).
4. Which of the following formulas could be an alkane?

   a. $CH_2O$

   b. $CH_3CH_2CH_2CH_2CH_3$

   c. $CH_3CH_2CH_2NHCH_2CH_3$

   d. $NaHCO_3$
5. Which of the following formulas could be an alkene?

   a. $CH_2CH_2$

   b. $CH_3CH_2CH_2CH_2CH_3$

   c. $CH_3CH = CHNHCH_2CH_3$

   d. $NaCHCH_2$
6. Given the data in Table 13.1, what is a likely boiling point for dodecane ($C_{12}H_{26}$): 72°C, 216 °C, 318 °C?
7. Draw the possible isomers of butane, $C_4H_{10}$.
8. What is meant by the term resonance?
9. What makes an aromatic compound special?
10. Which of the following compounds has a lower total energy? Why?

***2,4-heptadiene*** ***2,5-heptadiene***

CHEMICAL STRUCTURE 13.34. Heptadienes.

11. Which of the following compounds has a lower total energy? Why?

***1,3-cyclohexadiene*** ***1,3,5-cyclohexadiene***

CHEMICAL STRUCTURE 13.35. Cyclohexadienes.

12. The hydrogenation of 2-heptene to yield heptan results in a decrease in energy of the molecule. Why is this so?

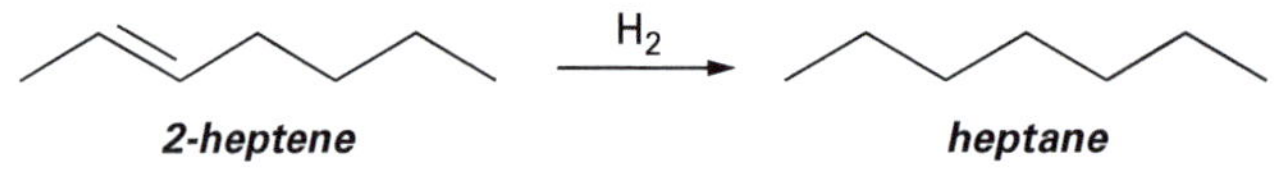


CHEMICAL STRUCTURE 13.36. Heptene and Heptane.

13. Draw the structure of the following molecules:

a. 1,2-diiodobutane

b. 1-chloro propyne

c. 1,3,6-tribromocycloheptane

d. 1,2,3,4-tetrahydroxy benzene

14. Draw the structure of the following molecules:

a. 1-bromo-6-nitrooctane

b. 4-nonene

c. methyl amine

d. propylene

15. What is the carbonyl group?

16. Draw the structure of the following molecules:

a. Dimethylketone

b. Butaldehyde

c. Ethanoic acid

d. 1-propyl ethyl ether

17. Draw the structure of the following molecules:

a. Diphenyl ketone

b. Benzaldehyde

c. Trimethylamine

d. Urea

18. Draw the structure of the following molecules:

a. Acetamide

b. Benzoic acid

c. 1,3-dichloro benzoic acid

d. the methyl ester of benzoic acid

19. What is the difference in structure between a carbonate and a carbamate?

20. What is the name of the following compound?

CHEMICAL STRUCTURE 13.37. Compound.

21. What is the name of the following compound?

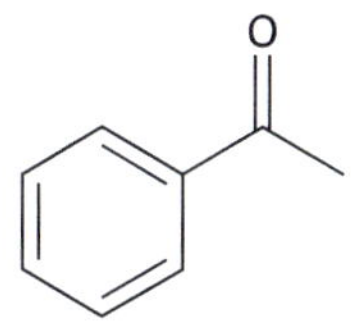


CHEMICAL STRUCTURE 13.38. Compound 2.

22. What is the name of the following compound?

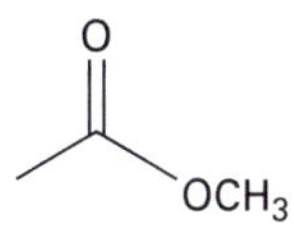


CHEMICAL STRUCTURE 13.39. Compound 3.

23. What is the name of the following compound?

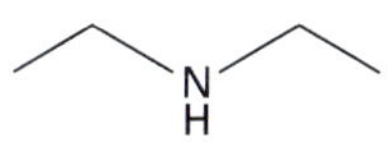


CHEMICAL STRUCTURE 13.40. Compound 4.

24. What is the name of the following compound?

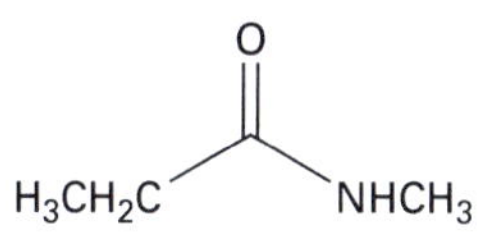


CHEMICAL STRUCTURE 13.41. Compound 5.

25. What is the name of the following compound?

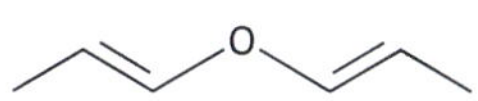


CHEMICAL STRUCTURE 13.42. Compound 6.

# Credits

Table 13.1: Source: http://www.wou.edu/chemistry/files/2017/01/table_7.3.png.

CHAPTER 14

# Organic Chemistry—Reactions and Mechanisms

## Now that we know the functional groups, let's look at the really interesting stuff, what goes on ...

## 14.1 Starting Material Functional Groups—Reprise

Recall the following functional group table from Chapter 13. The functional groups described were not random choices, rather they represent the key chemical moieties involved in the biochemistry of biological systems.

TABLE 14.1 Carbonyl-Based Functionalities and Names

| Group | Structure | Name | Example |
|---|---|---|---|
| **Aldehyde** | 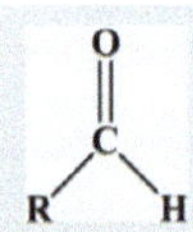 | **-aldehyde** | **Acetaldehyde** |
| **Ketone** | 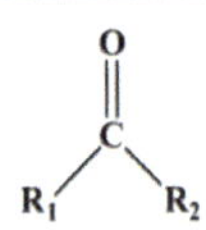 | **-one** | **Acetone** |
| **Carboxylic Acids** | 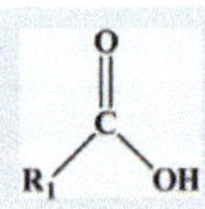 | **-ic acid** | **Acetic Acid** |

| | | | |
|---|---|---|---|
| ***Esters*** | 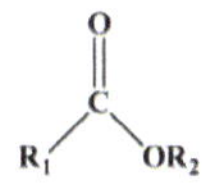 | ***-ate*** | ***Acetate*** |
| ***Amides*** | 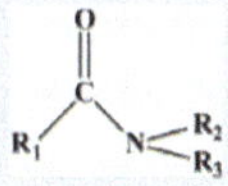 | ***-amide*** | ***Acetamide*** |
| ***Carbonates & Carbonic Acids*** | 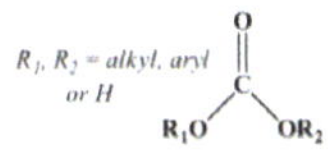 | ***carbonate carbonic Acid*** | ***Dimethylcarbonate Methyl carbonic acid*** |
| ***Carbamates (Urethanes)*** | 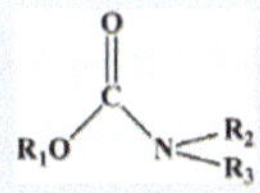 | ***carbamate*** | ***Polyurethane*** |
| ***Ureas*** | 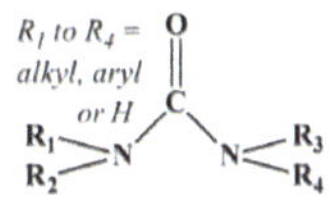 | ***urea*** | ***Methyl urea*** |

The left hand two columns of this table are reproduced from Table 13.2 in Chapter 13. The right hand column gives a common molecular example of one of these functional groups.

## 14.2 Broad Classes of Reactions

In the examples that follow A and B are generic symbols for elemental or molecular components. **C**s are reactive carbon centers, bold lines are bonds. Specific examples of reactions are given in the sections that follow. Also the colored arrows represent electron movements to form new bonds and hence new compounds.

### Substitutions

The class name is very descriptive. Basically one element or group on a molecule is replaced by another, although the mechanism of this replacement varies over a wide range. Here is a generic example where **A** is replaced by **B**:

Here is one example, the nucleophilic displacement of iodine by hydroxide ion:

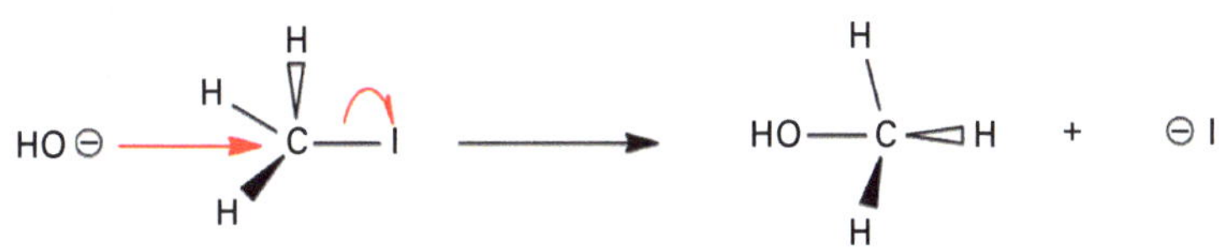


**CHEMICAL STRUCTURE 14.1.** Nucleophilic Displacement of Iodine by Hydroxide Ion.

## Eliminations

Also a descriptive name, basically one or more moieties are removed from a molecular structure. In this example **A** and **B** are removed from adjacent carbons via covalent bond breaking:

Here is an example. The elimination of water from 2-hydroxyethyl benzene to yield 1-ethenylbenzene:

**CHEMICAL STRUCTURE 14.2.** Elimination of Water.

## Additions

Mechanistically these can be considered as the reverse of eliminations. Here is one example, the addition of hydrochloric acid to 2-butene:

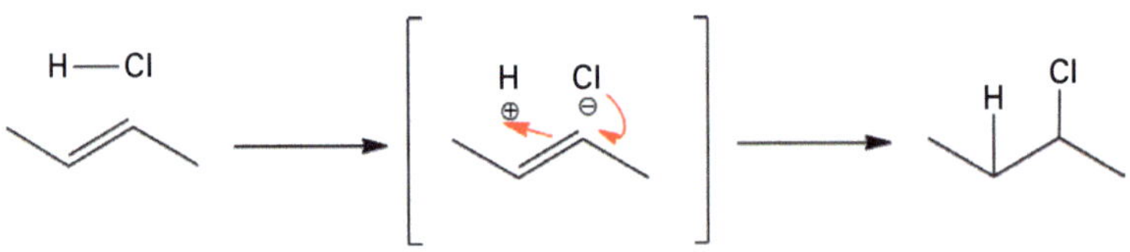


**CHEMICAL STRUCTURE 14.3.** Addition.

## Decompositions

This is a wide-ranging class of reactions that can involve many different mechanisms as individual steps, for example an elimination and combined. As will now be described a wide range of combinations of ionically neutral, cationic, anionic, and radical mechanisms of reaction are possible.

An important yet simple example of this broad class is the thermally induced decomposition if hydrogen peroxide:

**CHEMICAL STRUCTURE 14.4.** Hydrogen Peroxide.

Depending upon the concentration of the $H_2O_2$, and reaction conditions, this decomposition can be modest and well-behaved or explosively destructive.

# 14.3 Some Basic Mechanisms of Organic Reactions

We now look at some fundamental reaction mechanisms of organic functional groups. These kinds of reactions are very common in both biological and inorganic systems. There are an almost infinite number of possible variations on all of these mechanisms; however, the basics will follow through each group. Two examples of each class will be presented. There are so many more.

## Nucleophilic Reactions

These reactions are characterized by a nucleophile, or nucleus seeking species, attacking an electron deficient center shown generically by the following examples:

### Example 1

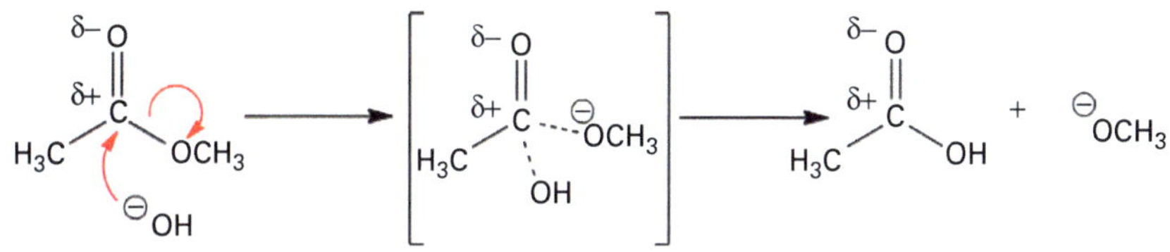


**CHEMICAL STRUCTURE 14.5.** Nucleophilic Example 1.

Here the red arrows represent electron movements. The electrons on the OH group attack the cationic carbon center on the ketone coupled with the concomitant movement of an electron pair to the $OCH_3$ group, The intermediate transition state, which is fleeting, is shown in brackets. This nucleophilic displacement results in new products, a carboxylic acid, and the methoxide anion. The dotted lines represent bonds that are in the process of breaking or forming.

### Example 2

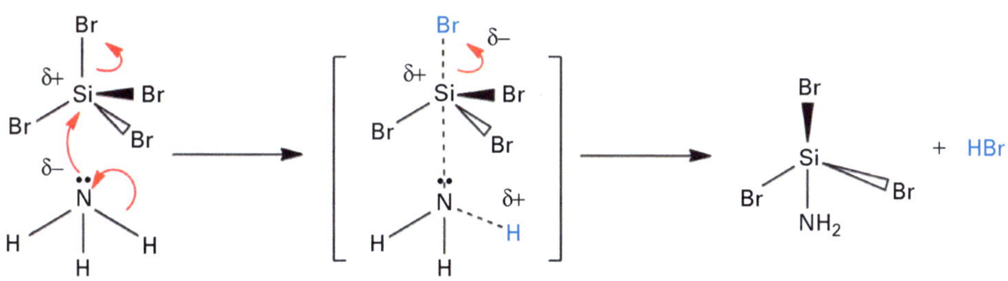


**CHEMICAL STRUCTURE 14.6.** Nucleophilic Example 2.

This example does not involve ions rather it is a neutrally charged reaction wherein the electron rich nitrogen attacks the electron poor silicon. Again the red arrows represent electron shifts to break and form new bonds. Note too that the $Br^-$ and $H^+$ displaced here (shown in blue) then can combine to form hydrobromic acid as a second product. Note also how the remaining Br atoms rearrange themselves to minimize steric interactions.

These are but two examples. The reaction specifics are not important here. What is important is being able to recognize where the reactions are likely to occur on the molecules which all has to do with electronegativity and electron distributions.

## Electrophilic Reactions

Electrophilic reactions are the reverse of nucleophilic reactions wherein an electron seeking entity (the electrophile) finds an electron-rich center and reaction proceeds apace. Here are two examples.

### Example 1

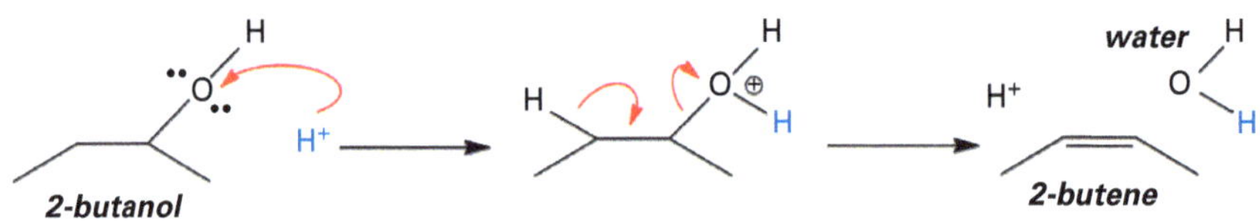


**CHEMICAL STRUCTURE 14.7.** Electrophil #1.

Here the electrophile is **$H^+$** attacking the electron-rich oxygen on the 2-butanol reactant substrate. A clean electron rearrangement ensues resulting in the formation of the products 2-butene and water with the release back into solution of an $H^+$. The latter can then go on to cause reaction of another 2-butanol molecules. This is a catalytic elimination process. Note here too that by reversing the process one performs an addition. Both reaction processes are well known and widespread in practice.

### Example 2

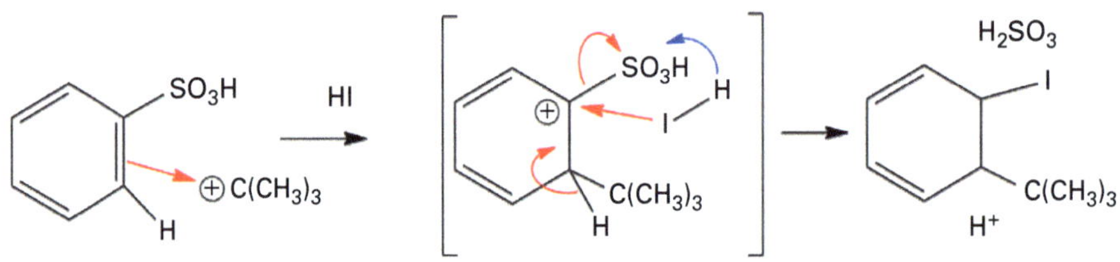


**CHEMICAL STRUCTURE 14.8.** Electrophil #2.

This more complicated example is an aromatic substitution reaction wherein the $+C(CH_3)_3$ cation is serving as the electrophile in the presence of HI. Sulfurous acid is formed as a leaving group product. Here the red arrows represent electron movements and the blue arrow the transfer of the H+ from iodine to the sulfurous acid leaving group.

## Hydrolyses

Reaction with water. One of the three most important overall reaction mechanisms because hydrolyses are ubiquitous in the function of biological organisms regardless of the kingdom under discussion. Catalysts are also very commonly used or observed in hydrolyses, both inorganic and enzymatic kinds of catalysts.

### Example 1: Making Soap

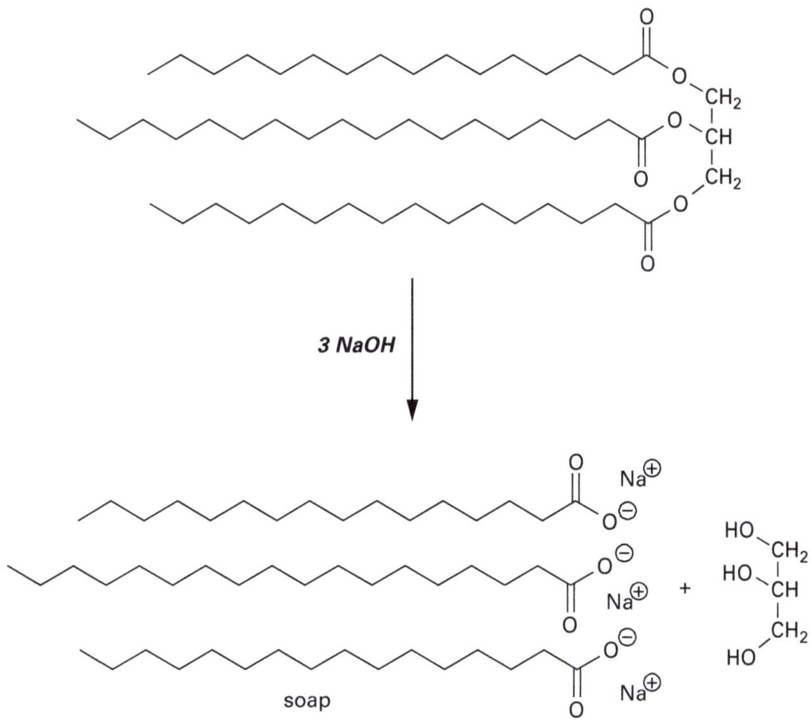


**CHEMICAL STRUCTURE 14.9.** Hydrolyses #1.

Since antiquity soaps have been synthesized using the same basic process as shown above. This simple reaction begins with a fat, in this case a triglyceride, which is then reacted with an aqueous solution of lye, sodium hydroxide to form the soap products, the sodium salts of long chain alkyl fatty acids to form the sodium salt carboxylates and glycerine.

### Example 2:

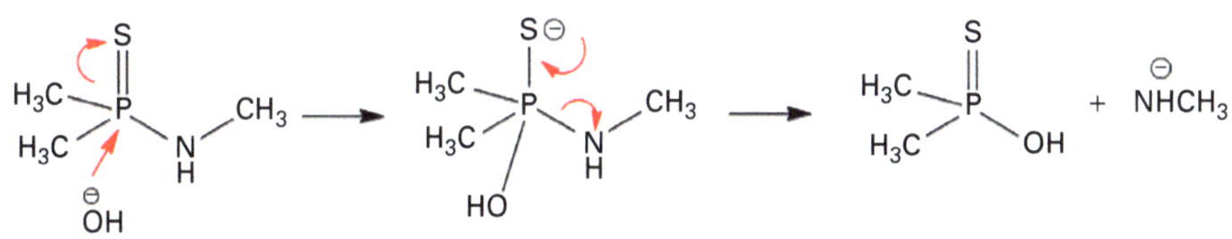

**CHEMICAL STRUCTURE 14.10.** Hydrolyses #2.

The thiophosphorous carbonyl compounds shown here are very interesting classes of compounds that are very active biologically (the term thio- refers to the sulfur atom substituted for oxygen in the carbonyl group). Familiar names for a couple of the compounds in these classes are the herbicide *Roundup* and *Sarin*, a potent war gas. And yes, this is a nucleophilic displacement reaction.

## Oxidations

The second of the three most important reaction mechanisms, oxygen activation is clearly essential to all life as we know it, as well as essential reactions such as combustion:

### Example 1:

$$CH_4 + 2O_2 \qquad CO_2 + 2H_2O$$

**CHEMICAL STRUCTURE 14.11.** Oxidation #1.

$$CH_4 + 2O_2 \longrightarrow CO_2 + 2H_2O$$

Here methane (natural gas) reacts with oxygen to form carbon dioxide and water. This is an ideal reaction representation as the actual reaction is not nearly this clean; other products are also formed such as carbon monoxide, ketones, aldehydes, alcohols, ethers.......

### Example 2:

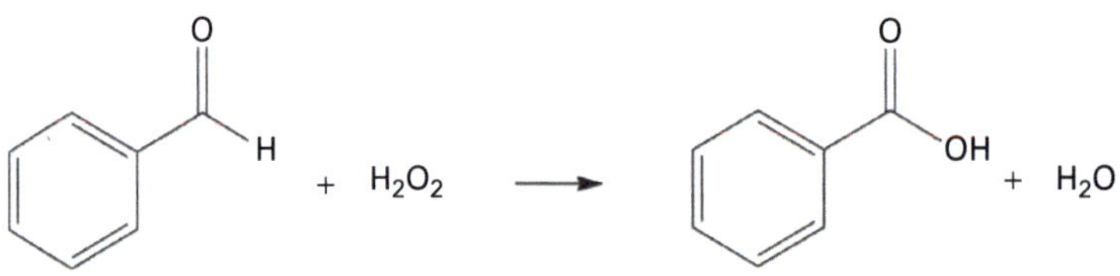


**CHEMICAL STRUCTURE 14.12.** Oxidation #2.

Here again the reaction occurs at the electron poor carbonyl carbon and involves the oxidation of benzaldehyde to benzoic acid.

### Example 3: A Related Reaction

There is a lot going on in this sequence of reactions. The alkane-arene compound ethyl benzene is reacted with hydrogen peroxide to form 1-hydroperoxyethyl benzene. This relatively unstable compound then easily rearranges to release a molecule of water and the ketone acetophenone as products.

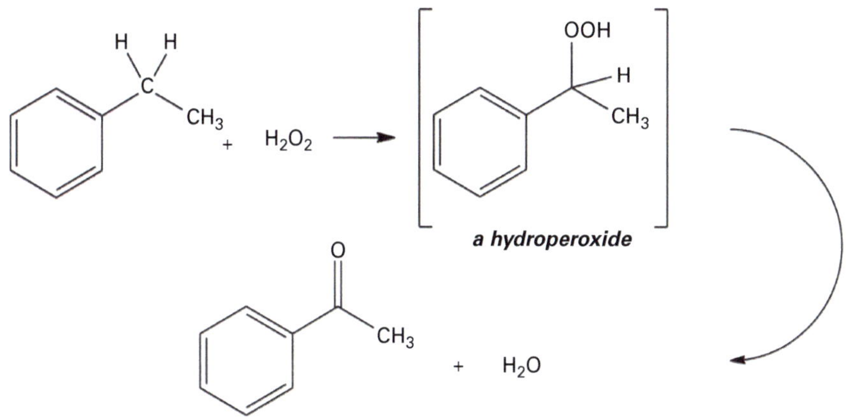


**CHEMICAL STRUCTURE 14.13.** Oxidation #3.

## Free Radical

Yet another broad class of important reactions. A ***radical*** species is by definition one with an unpaired electron which are invariably quite reactive. The reaction types described thus far are generally predictable in terms of what kinds of reactions and compounds are likely. Free radical reactions are different. They are notorious because of their unpredictability, and all too often can result in explosively dangerous product situations. A classic example is the case of tetrahydrofuran (THF). THF is a flammable, low boiling solvent to begin with and by allowing access to oxygen the THF will slowly react and oxidatively form peroxides, one of which is shown as the product of the first step in the following:

### Example 1

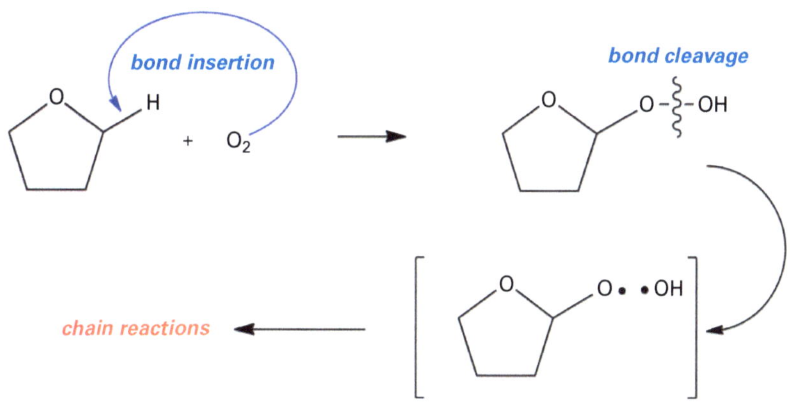


**CHEMICAL STRUCTURE 14.14.** Free Radical #1.

The peroxide bond is weak and cleaves easily in a homogeneous manner to give two radical entities either or both of which can go on to react with other substrates. This is the start of a chain reaction which can proceed very rapidly generating a lot of heat and or gases, the essence of a good explosive. For example suppose the highly reactive OH radical runs into another molecule of THF:

### Example 2

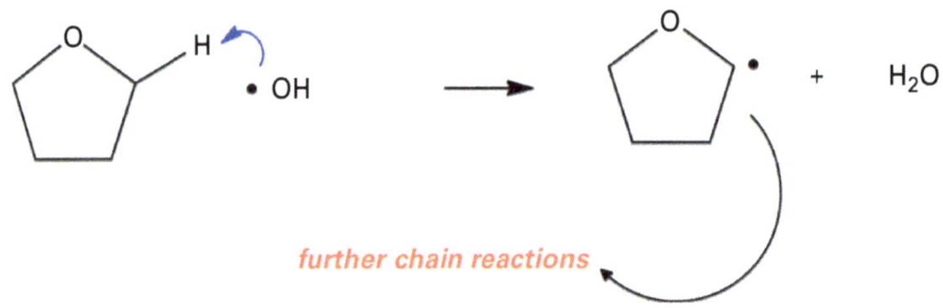


CHEMICAL STRUCTURE 14.15. Free Radical #2.

Here the OH radical abstracts a hydrogen atom from second THF molecule forming water via a new covalent bond. THF is now the radical which can then react with another molecule and so on, and on.....

### Example 3

Here is another free radical reaction spelled out in a good deal more detail:

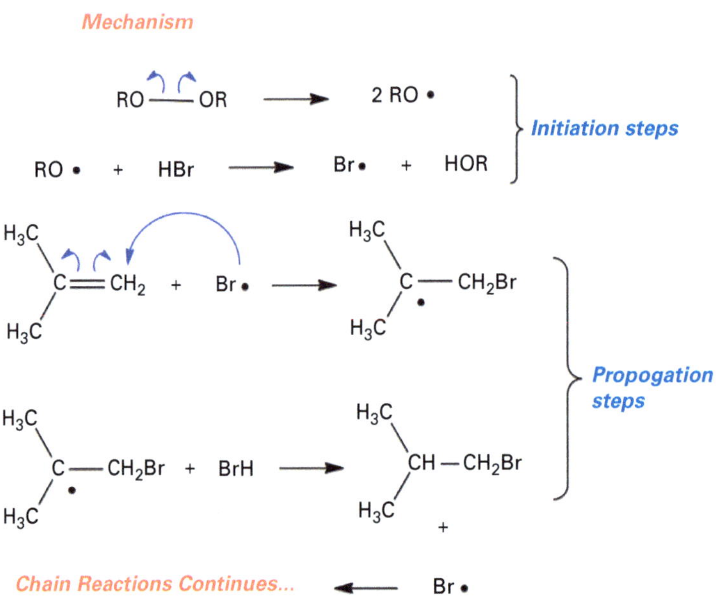


CHEMICAL STRUCTURE 14.16. Free Radical #3.

## Molecular Rearrangements

In these reactions atoms and electrons move around a molecule to form new compounds. By way of example is the following:

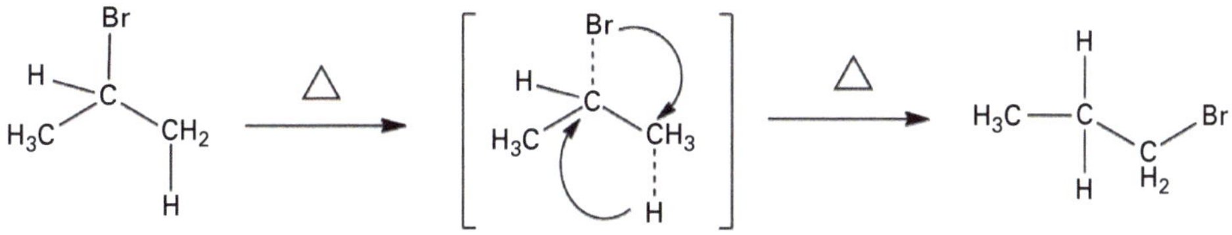


CHEMICAL STRUCTURE 14.17. Rearrangement.

In this rearrangement the bromine on 2-bromopropane migrates or shifts to make 1-brompropane, a similar but distinctly different compound.

## Chapter 14 Summary

*Organic Chemistry: Reactions:* the study of reaction mechanisms in organic chemistry.

### Classes of Reactions

*Substitutions:* Where one chemical moiety replaces another in a compound.

*Eliminations:* When one or more moieties are removed from a molecular structure.

*Additions:* When one or more moieties are added to a molecular structure.

*Decompositions:* When molecules come apart and are rearranged or reacted to form other compounds.

### Basic Mechanisms of Reactions

*Nucleophilic Reactions:* These reactions are characterized by a nucleophile, or nucleus seeking species, attacking an electron deficient center.

*Electrophilic Reactions:* An electron seeking entity (the electrophile) finds an electron-rich center.

*Hydrolyses:* Reactions with water.

*Oxidations:* Reactions with oxygen or reactions involving net loss of electrons.

*Free Radical:* Reactions involving unpaired electrons.

*Rearrangements:* reactions that involve movements of atoms and electrons within molecules to yield new compounds.

## CHAPTER 14 EXERCISES

1. Give an example of a substitution reaction in organic chemistry, not one given in the text.
2. Give an example of an elimination reaction in organic chemistry, not one given in the text.

3. Give an example of an addition reaction in organic chemistry, not one given in the text.
4. Give an example of a decomposition reaction in organic chemistry, not one given in the text.
5. Draw the mechanism of a nucleophilic reaction between the following molecules:

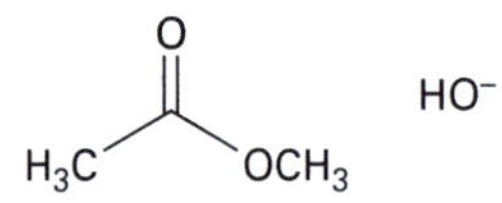


**CHEMICAL STRUCTURE 14.18.** Nucleophilic Reaction.

6. Draw the mechanism of a nucleophilic reaction between the following molecules:

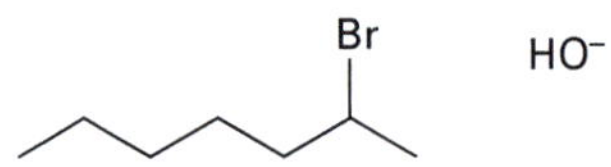


**CHEMICAL STRUCTURE 14.19.** Nucleophilic Reaction 2.

7. Draw the mechanism of a electrophilic reaction between the following molecules:

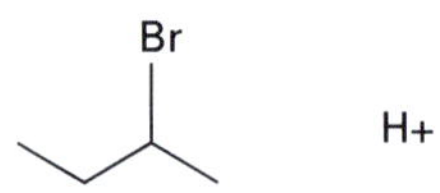


**CHEMICAL STRUCTURE 14.20.** Electrophilic Reaction 1.

8. Draw the mechanism of a electrophilic reaction between the following molecules:

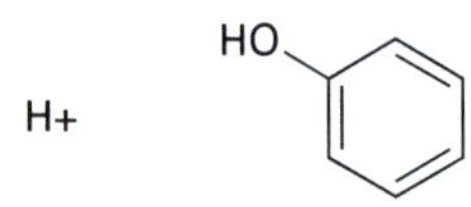


**CHEMICAL STRUCTURE 14.21.** Electrophilic Reaction 2.

9. Draw the mechanism of a hydrolysis reaction between the following molecules:

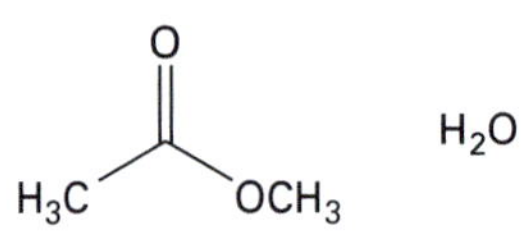


**CHEMICAL STRUCTURE 14.22.** Hydrolysis Reaction 1.

10. Draw the mechanism of a hydrolysis reaction between the following molecules:

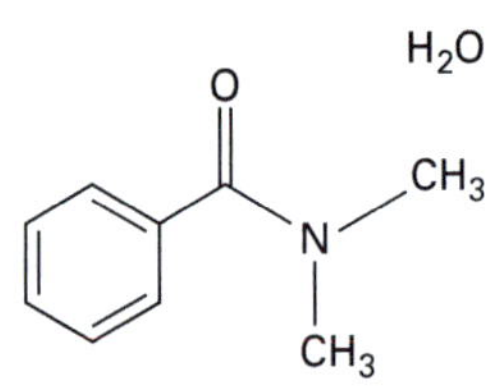


**CHEMICAL STRUCTURE 14.23.** Hydrolysis Reaction 2.

11. Draw the mechanism of an oxidation reaction between the following molecules (many possibilities):

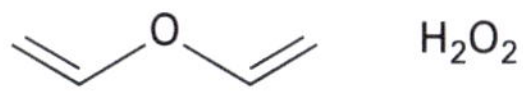


CHEMICAL STRUCTURE 14.24. Oxidation Reaction 1.

12. Draw the mechanism of an oxidation reaction between the following molecules (many possibilities):

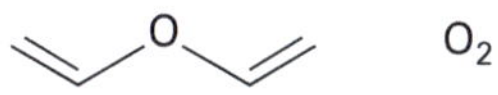


CHEMICAL STRUCTURE 14.25. Oxidation Reaction 2.

13. Draw the mechanism of a free radical reaction between the following molecule and hydrogen peroxide, $H_2O_2$. The first step is the cleavage of the O-O bond to yield two OH radicals that then react with the second molecule.

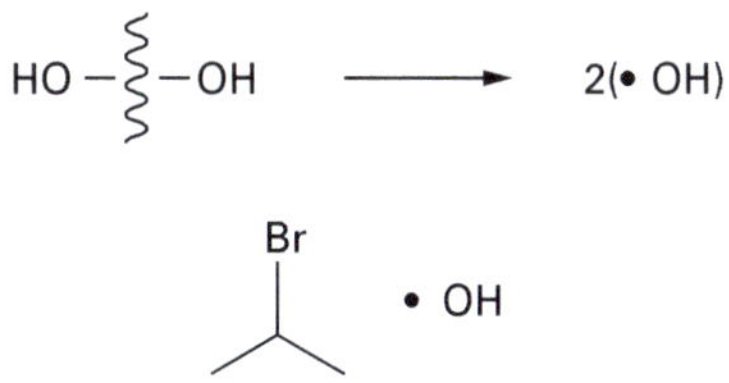


CHEMICAL STRUCTURE 14.26. Free Radical Reaction.

14. Draw the mechanism of a free radical reaction between the following molecule and hydrogen peroxide, $H_2O_2$. The first step is the cleavage of the O-O bond to yield two OH radicals that then react with the second molecule.

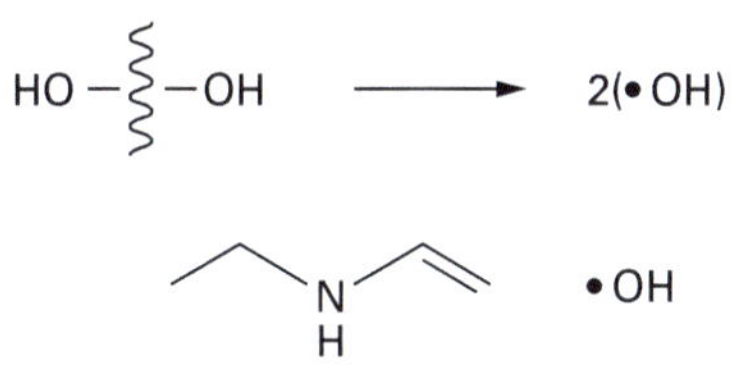


CHEMICAL STRUCTURE 14.27. Free Radical Reaction 2.

15. Draw the mechanism and products of a substitution reaction between the following molecules:

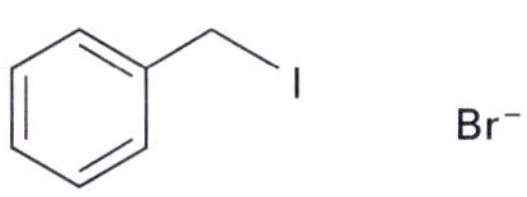


CHEMICAL STRUCTURE 14.28. Substitution Reaction 1.

16. Draw the mechanism and products of a substitution reaction between the following molecules:

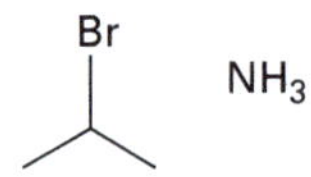


CHEMICAL STRUCTURE 14.29. Substitution Reaction 2.

17. Draw the mechanism of an elimination reaction and products of the following molecule (there is more than one possibility):

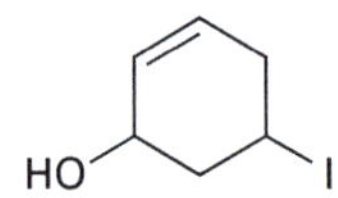


CHEMICAL STRUCTURE 14.30. Elimination Reaction 1.

18. Draw the mechanism of an elimination reaction and products of the following molecule:

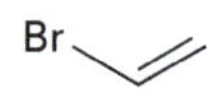


CHEMICAL STRUCTURE 14.31. Elimination Reaction 2.

19. Draw the mechanism of an addition reaction and products of the following molecules:

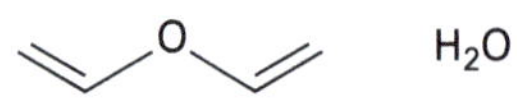


CHEMICAL STRUCTURE 14.32. Addition Reaction.

20. Draw the mechanism of the decomposition reaction and products of the following molecule where $CO_2$ is evolved:

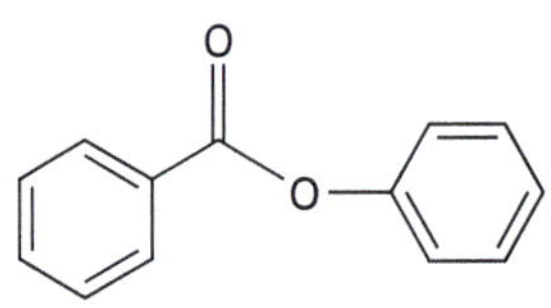


CHEMICAL STRUCTURE 14.33. Decomposition Reaction 1.

21. Draw the mechanism of the decomposition reaction and products of the following molecule where $NH_3$ is evolved:

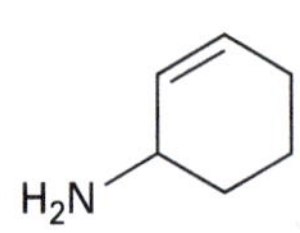


CHEMICAL STRUCTURE 14.34. Decomposition Reaction 2.

22. Where in the following molecule might a free radical, where are very reactive, be most likely to attack? Many possibilities—justify your answer(s).

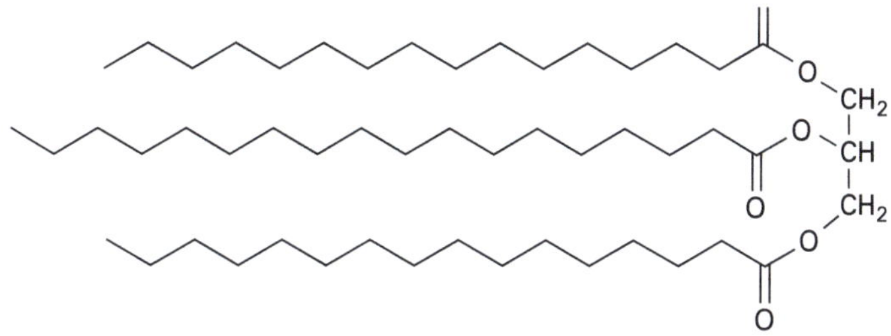


**CHEMICAL STRUCTURE 14.35.** Where in the following molecule might a free radical, where are very reactive, be most likely to attack?

23. Where in the following molecule might hydrolysis occur? Justify your answer.

**CHEMICAL STRUCTURE 14.36.** Hydrolysis.

24. Where in the following molecule might elimination occur? Justify your answer.

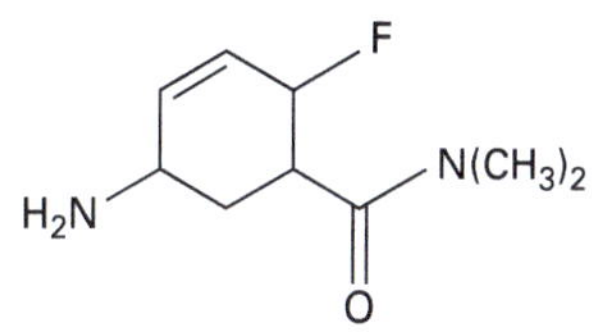


**CHEMICAL STRUCTURE 14.37.** Elimination.

25. Where in the following molecule might addition with HCl occur? Justify your answer.

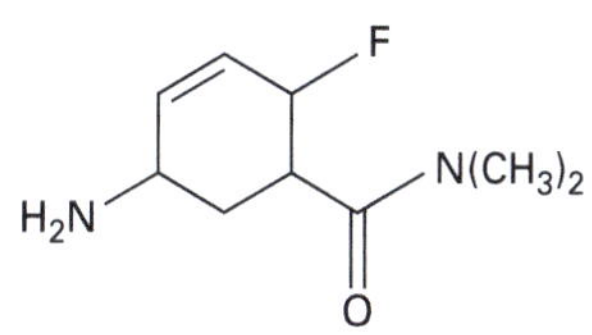


**CHEMICAL STRUCTURE 14.38.** Addition with HCl.

CHAPTER 15

# Biochemistry, Toxicology, and Pharmacology

There are many specialized disciplines in the field of chemistry. One of these is biochemistry. This field focuses on the study of chemistry as it relates to living organism. It is this field of chemistry that explains how sugar is processed by the human body or how plants use sunlight to chemically synthesize sugar. This field can also explain how drugs elicit their effect on the human body and how oxygen is distributed to the body. This chapter will provide a basic introduction to biochemistry and different chemicals and compounds.

## 15.1 DNA and RNA

### Nucleotides

Nucleotides are comprised of a cyclic amine component (referred to as a nitrogenous bases), a five carbon sugar and up to three phosphate groups. Figure 15.1 shows the generic structure of the nucleotides. There are a total of 5 nucleotides which differ by the base component. The nitrogenous bases are placed into two categories the ***purines*** have a double ring structure with a six member ring fused to a five member ring. ***Adenine (A)*** and ***guanine (G)*** are both purine bases. The ***pyrimidines*** have a single, six member ring structure and includes ***thymine (T), cytosine (C)*** and ***uracil (U).***

As mentioned earlier, attached to each of the bases, by a covalent bond, is a sugar component. There are two possible sugars that are attached to a base either a ***ribose*** or a ***deoxyribose.*** Both sugars are comprised of a five member ring. The difference between the two sugars involved the 2' carbon in the ring structure. In ribose, this carbon is attached to a hydroxyl group. In deoxyribose this carbon is attached to a hydrogen, hence the "deoxy" in the name. The structures for both sugars is presented in Figure 15.2.

The final component of the nucleotides are the phosphoryl groups. Using adenine as the example, the bases can either have one, two or three phosphoryl groups attached. This creates three different versions of each nucleotides either a mono-, di- or triphosphate version. The structures of these three versions of the nucleotide are presented in figure 15.3 and the same three versions exist for all five nucleotides. The different versions of the nucleotides have different functions that will be discussed later.

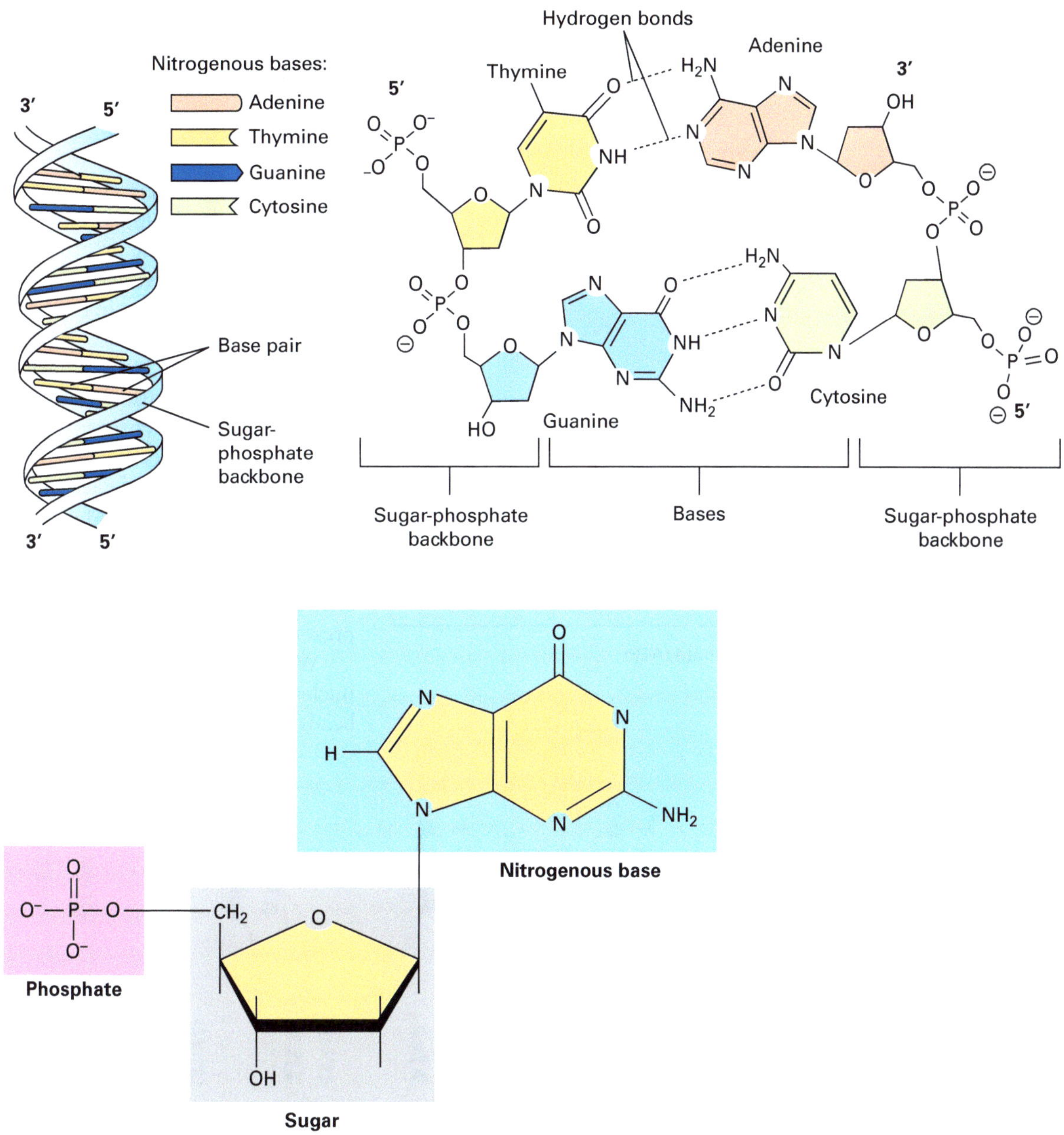


FIGURE 15.1. DNA Nucleotides.

The structure of the nucleotides makes it so that they form specific hydrogen bond interactions (see Figure 15.1). Adenine will form two hydrogen bonds with either thymine or uracil. Guanine can for three hydrogen bonds with cytosine. We will learn later that it is through these interactions, or ***base pairs,*** that creates the genetic code through which genetic information of living organisms is carried.

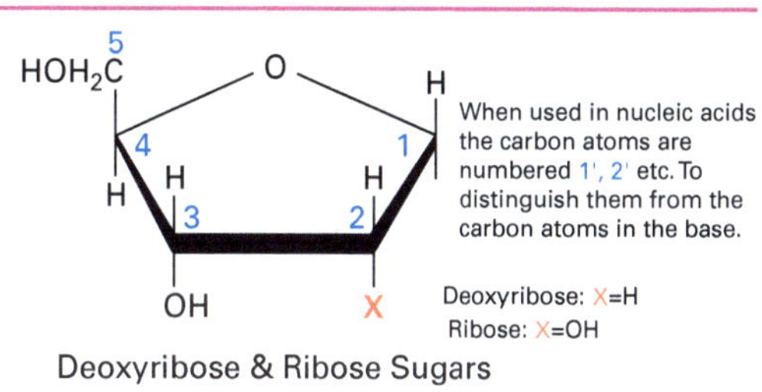


FIGURE 15.2. Ribose Deoxyribose.

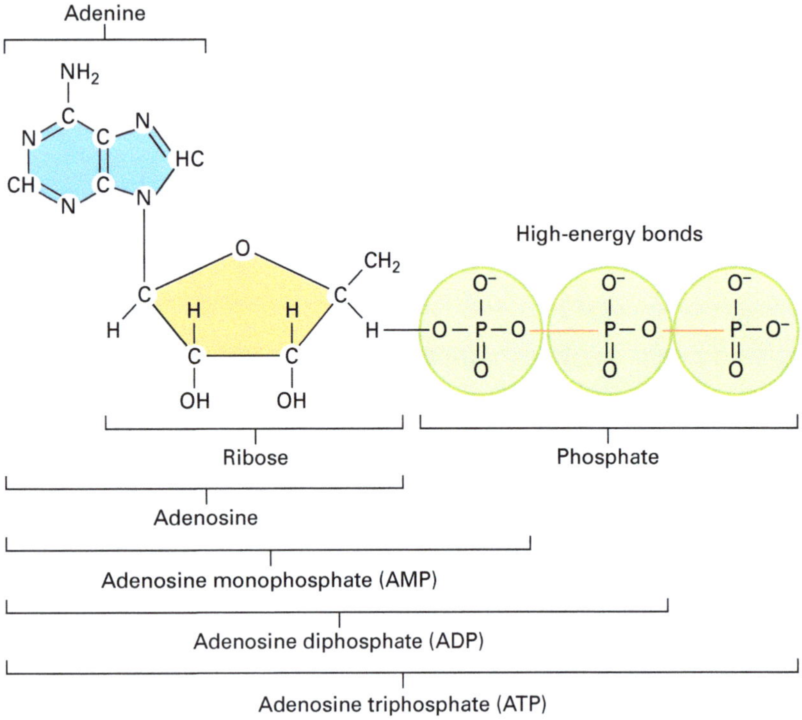


FIGURE 15.3. Phosphoryl Groups.

## Deoxyribonucleic Acid

You may already know that the genetic information for all living organisms comes in the form of ***deoxyribonucleic acid*** or ***DNA. Chromosomes*** are made up of DNA and carry the genetic blue print in the form of ***genes*** for a living organism. The nucleotides present in DNA are adenine, guanine, cytosine and thymine. Focusing on humans, each human being should have two copies of each of 23 chromosomes. Some anomalies occur that stray from the two copies of each chromosome (Figure 15.4). The most common example occurring in the case of Down Syndrome where an extra copy of chromosome 21 is present.

Within the chromosomes, nucleotides are strung together like beads on a string. This is creates the

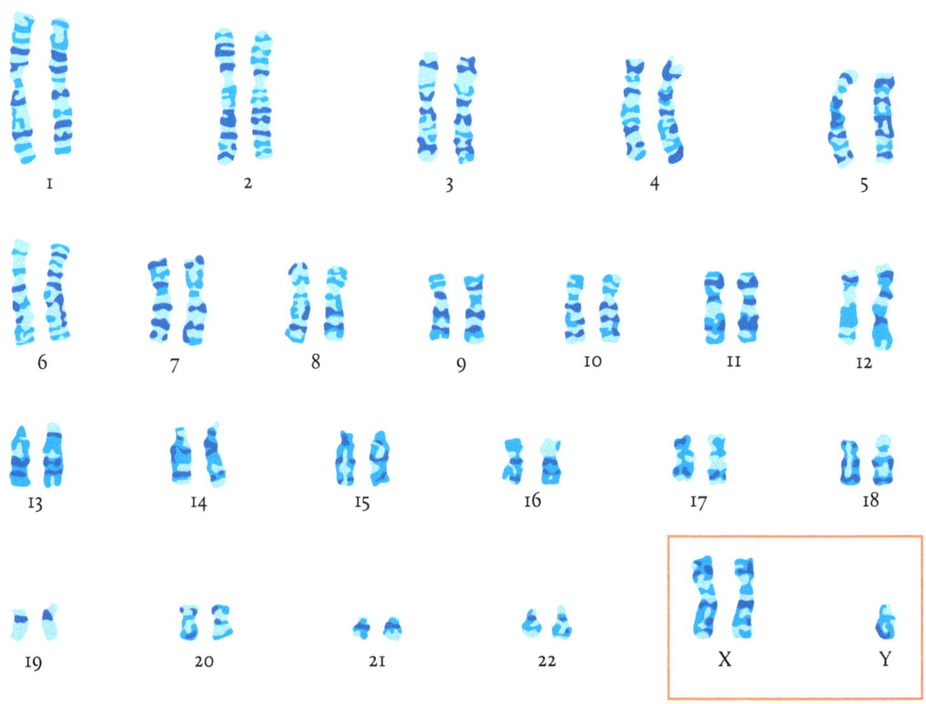


FIGURE 15.4. Human Chromosomes.

primary structure of DNA (Figure 15.5). The nucleotides are attached to each other via the phosphate. The ribose sugar bridges phosphate from 5' to 3', referring to the carbon positions in the ribose ring. The nucleotides are connected by a phosphodiester linkage, referring to the type of bond that forms (Figure 15.6).

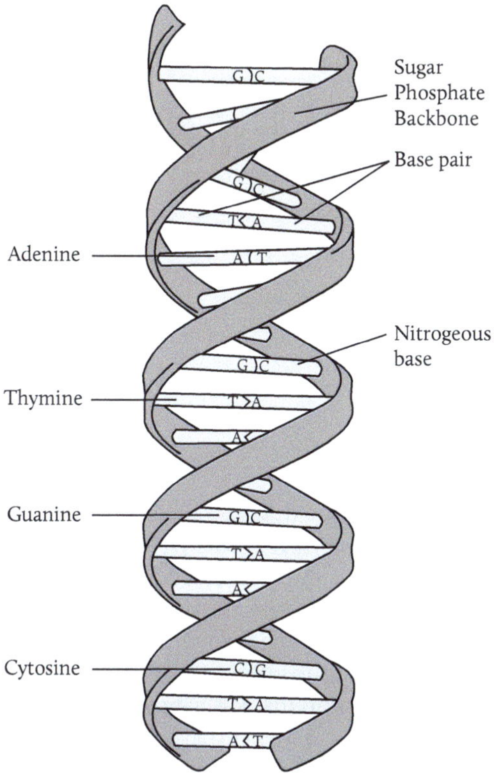


FIGURE 15.5. DNA Structure and Bases.

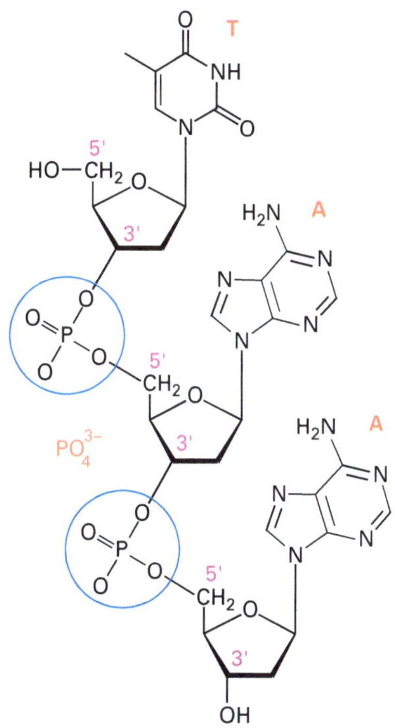


FIGURE 15.6. Phosphodiester Bond.

In DNA, as already discussed, adenine can base pair with thymine, guanine can base pair with cytosine. In the secondary structure of DNA, two strands come together and base pair with each other. One strand is in the direction 5' to 3' and forms base pairs with a second strand in the 3' to 5' direction (Figure 15.7). The two strands eventually form a ***double helix*** structure as part of its tertiary structure.

Through the process of ***replication,*** one strand of DNA is "copied" by separating the double helix and base pairing A to T and G to C (Figure 15.7). This is the process by which new cells are formed and genetic information is passed on from one generation to the next. Replication is a very complicated and intricate process executed by ***DNA polymerase.*** If replication is interfered with or compromised this can lead to the death of cell, can cause cancer, disrupt the proper function of an organism as well as many other problems. There are actually "check points" in the replication process to help maintain the integrity of the process but problems can still arise.

## Ribonucleic Acid

Ribonucleic Acid or RNA is similar to DNA. The difference lies in the nucleotides involved (A, U, G and C) and the sugar (ribose). DNA not only holds the genetic

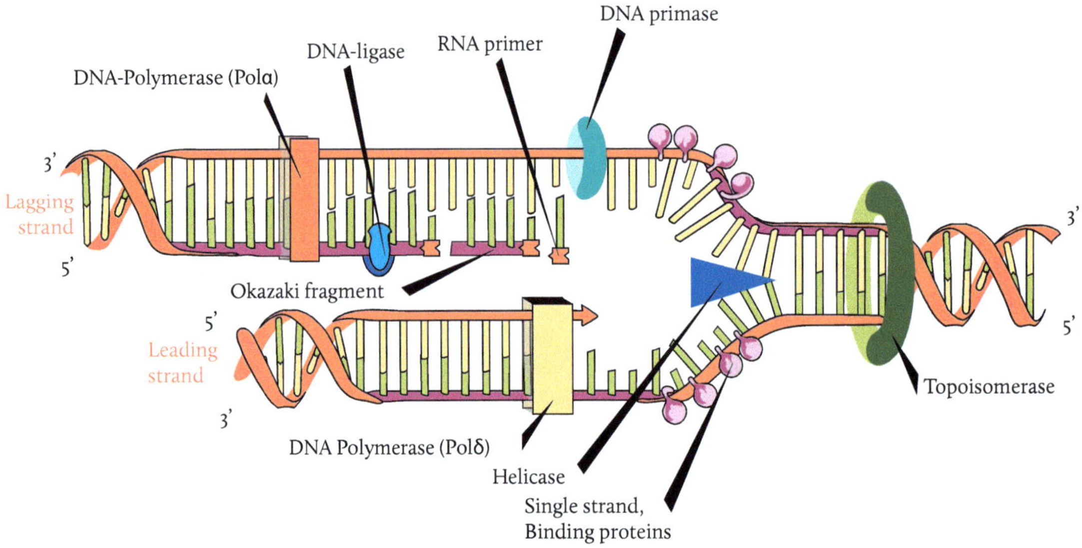


FIGURE 15.7. DNA Replication.

information passed down from generation to generation but it also holds the codes for the production of RNA. Through a process called ***transcription,*** the DNA code is used to create RNA. ***RNA polymerase*** manages this process. Three different forms of RNA are created, they are listed and described below.

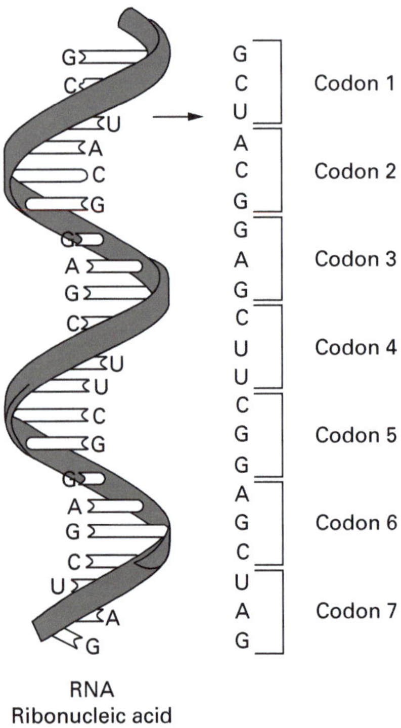


FIGURE 15.8. RNA Codon.

***Messenger RNA*** (mRNA) carries the code needed to create ***proteins.*** In a process known as translation, the code on the mRNA is used to create the ordered arrangement of ***amino acids*** in proteins. ***Ribosomal RNA*** (rRNA) is used to create the Ribosome which is used to synthesize proteins. The ribosome reads the code on the ***Transfer RNA*** (tRNA) is usually 76–90 nucleotides in length. It functions in the formation of proteins from mRNA. tRNA translates the genetic code from mRNA. Attached to the 3' end of each tRNA molecule is a specific amino acid. A specific 3 nucleotide "anti-codon" component of the tRNA attaches to the sequence on the mRNA (fig. 15.8). Based on the pairing of the anti-codons to the mRNA based on the genetic code leads to the formation of the primary sequence of a protein.

## 15.2 Proteins

Proteins can be considered the "workers" in living organisms. They hold many different functions. There are millions of different types of proteins and these functions are well beyond the context of this book. This section will provide an overview of proteins. Different categories will be discussed and reviewed to provide a general background.

### Protein Composition

Proteins are comprised of different amino acids joined together like beads on a string. There are 20 different amino acids. Each type of amino acid provides a specific function to the protein. If the order of the amino acids is altered or one amino acid is replaced with another, the results can be catastrophic to the organism. Different proteins have a different order of amino acids, making each protein unique in both structure and function.

As mentioned in the prior section, the order of amino acids in a protein is dictated by the code in mRNA. The code is translated by the Ribosome which then uses the code to determine the order of amino acids and link them together. The order of amino acids is referred to as the primary structure. The string of amino acids then forms bonding interactions with each other forming the protein's secondary structure. Next the protein "folds" onto itself to form the tertiary structure. For many proteins the tertiary structure forms a complete and functional protein. For other proteins, however, several, folded, strings of amino acids combine together to form quaternary structure in order to yield a functional protein. Usual images of the different levels of protein structure are depicted in Figure 15.9.

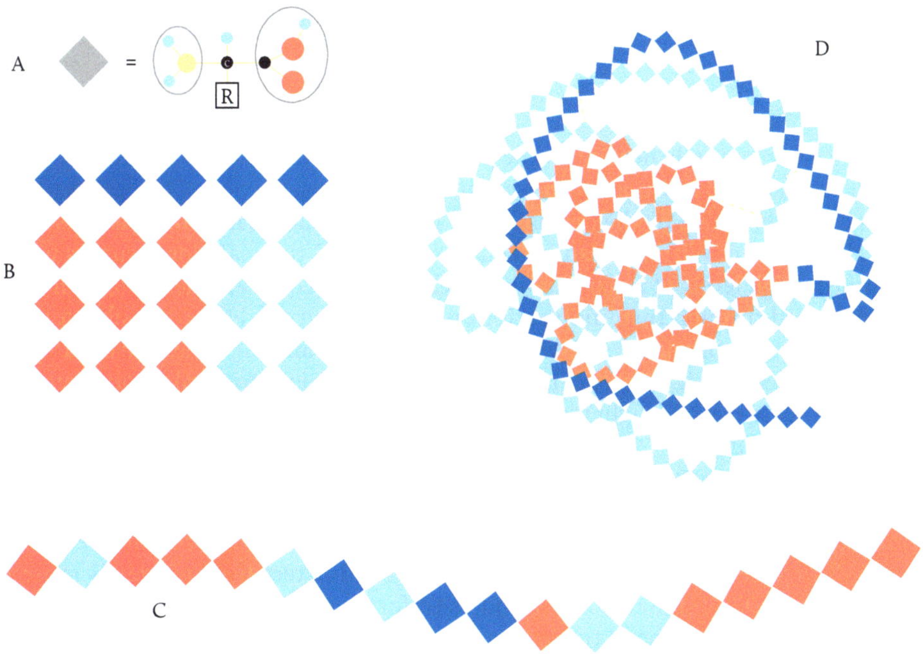


**FIGURE 15.9.** Amino Acids and Proteins.

Once formed, proteins then are capable of execting their specific functions. These are of a great variety but can mostly fall into specific categories. For the scope of this textbook, the discussion of protein function will fall into generalized categories. In actuality each protein has a very specific function. In many cases, if a protein is incapable of executing its task, the result can cause death of an organism.

## Structural Proteins

Proteins under this category, as the name implies, provide structure to cells and other components of living organisms. ***Within most living cells, proteins act as a scaffolding***

***providing shape and stability to proteins. Actin and myosin are proteins also the proteins that are present in muscles and function*** in the movement of limbs and other body parts. ***Kearatin (Keratin)*** found in skin, hair and nails provide important protective coverings in living organisms.

## Storage and Transport Proteins

Proteins are critical for getting various components where they are needed within an organism and other components are stored by proteins until they are needed by an organism. Transport and storage proteins function in these capacities. Hemoglobin, for example, stores iron and acts as its transport protein in red blood cells. Transport proteins are also found in cell membranes. These proteins govern the movement of molecules into and out of cells. Proper function of these proteins are critical to the survival of organisms.

## Hormones and Antibodies

All living organisms rely heavily on the proper signaling provided by hormones. Over or under signaling of hormones can have disasterous consequences. The best example of this can be provided by the discussion of Type I Diabetes. This illness is caused by the lack of insulin. Without insulin, cells are not “told” to break down sugars when consumed. High levels of sugar can ultimately result in coma and death. This illness can be treated through insulin injections when necessary. It is also possible for too much insulin to be provided. This can also lead to death. Similar issues can occur with other hormones as well. Improper hormone function can lead to sterility, obesity, even cancer. For these reasons, tight regulation of hormones is critical for the overall health of living organisms.

Similar to hormones, antibodies provide a signal to the body that something foreign is present and must be destroyed. Antibodies detect a foreign object, called an antigen, they then bind to the antigen. This allows for white blood cells in the organism to destroy the antigen. When vaccines are administered, an antigen is present in the vaccine. This signals for antibodies to form against the antigen. Once vaccinated, the body is prepped to destroy the antigen if it is ever present in the body again.

## Enzymes

In regards to chemistry, enzymes are the most relevant proteins. The proteins are critical to practically every chemical reaction that occurs in the cell. Enzymes catalyze or speed up) chemical reactions (see Section 8.4). Each reaction has a specific enzyme designed to catalyze it effectively. Small changes to these enzymes can cause critical reactions fail in living organisms. The simplest way to explain how enzymes catalyze reactions is by making them “easier” to complete. In this process the compound that needs undergo a chemical change binds to the enzyme in its active site. This is the site where the chemical reaction occurs. After the reaction is complete, the substrate is released and the enzyme returns to its original form and can catyalyze the reaction again.

# 15.3 Carbohydrates

Carbohydrates are another important molecule in living organisms. These compounds are composed of carbon, hydrogen and oxygen and include simple single saccharides with the generic formula of $(CH_2O)n$. Single saccharides are combine to form a polymer to create larger carbohydrates. These saccharides are comprised of anywhere from 3–9 carbons with a varying number of oxygens or hydrogens attached. Larger carbohydrates are formed when monosaccharides a joined creating disaccharides (two monosaccharides) or polysaccharides of up to 10 saccharides. Saccharides with 5 and 6 carbons are the most common and can form cyclic structures resembling furan and pyran rings. Below is the structure for glucose forming a ring structure (Figure 15.10).

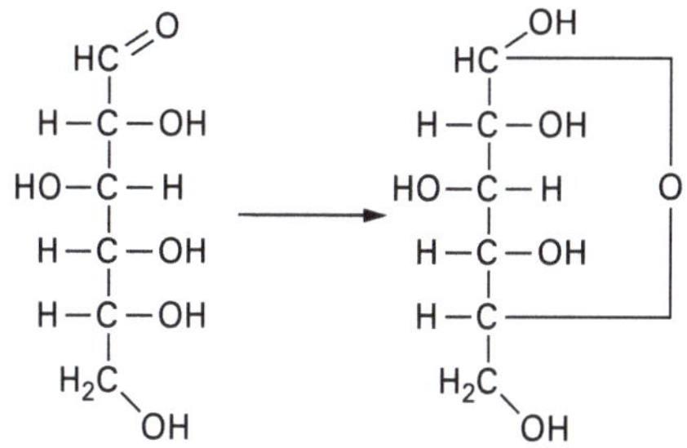

FIGURE 15.10. D-glucose D-glucopyranoside TOLLENS.

Monosaccharides are linked by a covalent bond referred to as a glycosidic linkage in order to form disaccharides (2 monosaccharides), oligosaccharides (3–10 monosaccharides) or polysaccharides (>10 monosaccharides). Starch and cellulose are both examples of polysaccharides.

## Monosaccharides

Monosaccharides fall into the category of an aldose or a ketose. This refers to the carbonyl functional groups present in the saccharide. An aldehyde group is present for aldose sugars and a ketone group is present in the ketose sugars (Figures 15.11a and 15.11b).

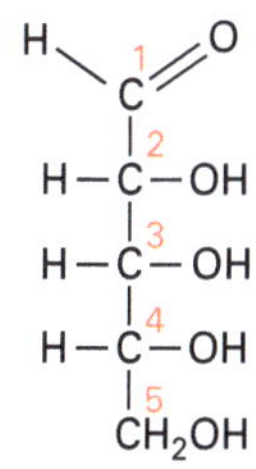

FIGURE 15.11A. Aldose.

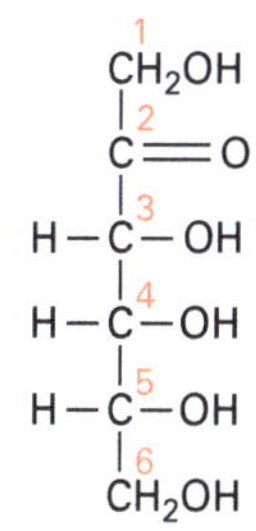

FIGURE 15.11B. Skeletal Structure of Ketose.

Monosaccharides contain anywhere from 3–7 carbon atoms. Each category has a unique names indicating the number of carbons. A triose is a 3-carbon saccharide, tetrose is 4-carbon saccharide, pentose is a 5-carbon saccharide, hexose is a 6-carbon saccharide and heptose is a 7 carbon saccharide.

Glucose is an aldose and likely the most well-known of the monosaccharides. It is also of critical importance as it is the main energy source in most organisms. It is certainly the most important sugar in the human body. In plants, glucose is produced through the process of photosynthesis (Figure 15.12). Blood glucose levels must be tightly controlled and in the body, the hormone, insulin, is key to maintaining these levels. Target levels for glucose in the blood are around 100 mg of glucose per 100 mL of blood. When do these levels peak? You guessed it, right after a meal. Insulin is responsible for telling the body to break down glucose in the blood. What happens when these levels get very low? Often a person will feel shaky and lightheaded when this occurs, within the body, glucagon, another hormone stimulates the release of glucose into the blood to help bring those levels back into the normal range. Whenever considering a diet, it is important to consider processing of sugars in the body as well as the overall well-being of the person consuming the sugars.

$$6CO_2 + 6H_2O \xrightarrow{\text{Light}} C_6H_{12}O_6 + 6O_2$$

Carbon dioxide + Water → Sugar + Oxygen

FIGURE 15.12. Photosynthesis Equation.

Fructose is a ketose and another familiar monosaccharide found in fruits. Not surprisingly, it is the "sweetest" of all the sugars and its presence is what makes fruits so delicious! The sweeter the fruit, the more fructose present. It is also the sugar found

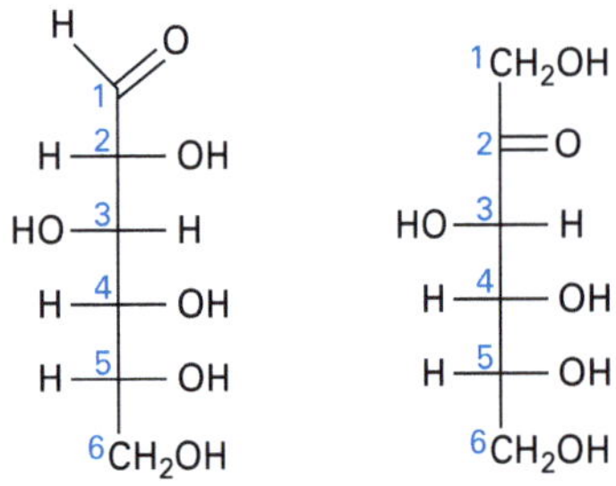

**FIGURE 15.13.** D-Glucose vs. D-Fructose Structural Formulae.

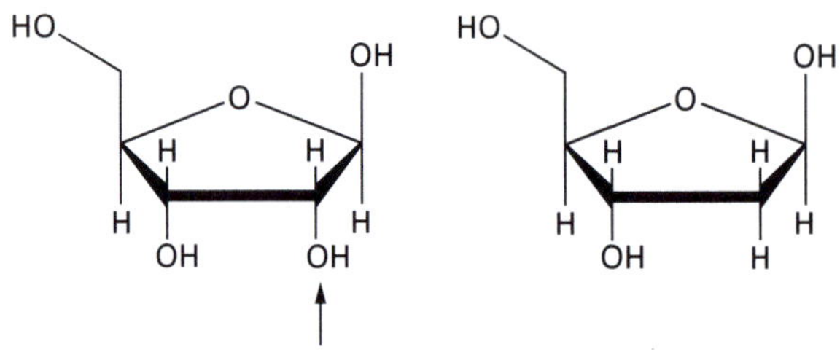

**FIGURE 15.14.** Ribose and Deoxy.

in corn syrup and the honey you add to your tea. Fructose is very similar to glucose, differing only in the arrangements of the atoms around the first and second carbons in these sugars (Figure 15.13).

Two more monosaccharides we have already discussed. These 5 carbon aldose saccharides. Ribose and Deoxyribose are important to the synthesis of DNA and RNA. These sugars differ by the absence of a hydroxyl group on the second carbon in the compound (Figure 15.14).

## Disaccharides

Disaccharides are joined together by a "glyosidic linkage" formed when the hydroxyl groups from the two sugars react. After the reaction the two monosaccharides are linked by an "oxygen bridge" between them.

These bonds can occur between monosscharides can be non-specific, occurring between any of the hydroxyl groups on the two sugars. However, in biological organisms, the linkages tend to be specific. This is most likely due to enzymes directing the formation of the bonds at specific carbon atoms. Sucrose for example always has a bond between the first carbon of glucose and the second carbon of fructose.

Most people have heard of lactose or are familiar with it due to issues with lactose intolerance. ***Lactose*** is a disaccharide found in milk. The body can use this as an energy source. Lactase is an enzyme critical for the proper break down of lactose in the body. This enzyme is responsible for breaking the bond between the first carbon in glucose and the fourth carbon in galactose, another monosaccharide (Fig 15.15). Without this enzyme functioning properly, lacrosse intolerance will occur. The cramping, gas and diarrhea that is associated with lactose intolerance occurs when bacteria in the gut breakdown lactose, producing $CO_2$ gas in the process. Treatments for this ailment included milk products with the glycosidic

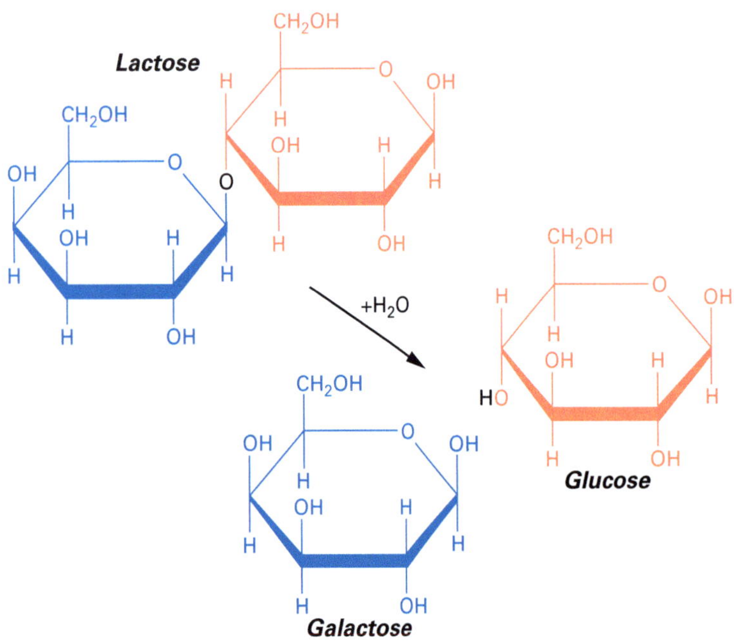


**FIGURE 15.15.** Lactose Color.

linkage in lactose already broken or over the counter tablets contained the necessary lactase enzyme to break the sugar bond.

***Sucrose*** is likely the most widely used dissacharide and it is known by most as table sugar. It is estimated that the average American consumes 130 lbs of sugar in a year. Sucrose is formed when glucose and fructose are linked via the first carbon in glucose to the second carbon in fructose. This is referred to as a 1→4 linkage (Figure 15.16).

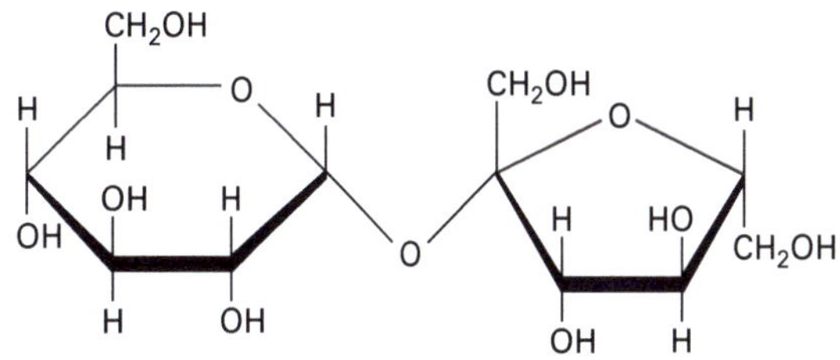

FIGURE 15.16. Sucrose Structure.

## Polysaccharides

Starch is a homopolysaccharide used to store carbohysrates in plants. It is comprise of only one monosaccharide, glucose. After plants produce glucose via photosynthesis, long chains of glucose are formed and starch is produced. Two glucose polymers are present in starch, one is amylose, and is a linear polymer (Figure 15.17A). The other, amylopectin, is a branched version of amylose (Figure 15.17B). Enzymes are then used to breakdown the amylose and amylopectin chains whenever glucose is needed to provide energy to cells.

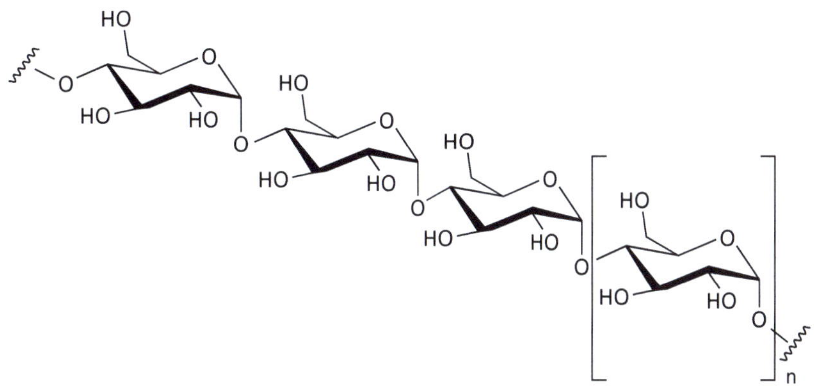

FIGURE 15.17A. Amylose.

Glycogen is the polysaccharide animals use to store glucose in their cells. This is another homopolysaccharide with glucose linked in long chains. Like amylopectin, glycogen also has branches coming of the main chain of glucose saccharides linked together (Figure 15.18).

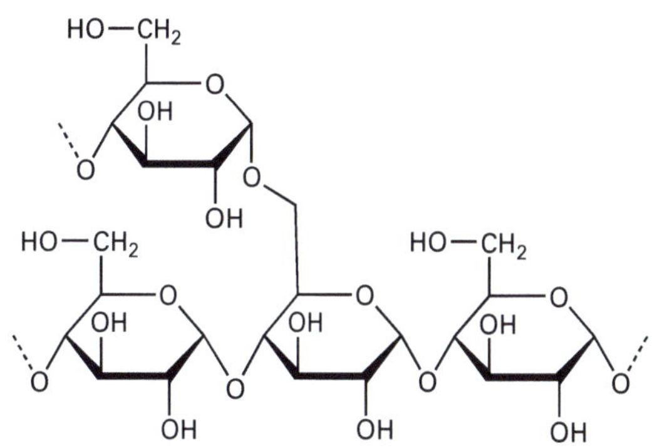

FIGURE 15.17B. Amylopectin.

# 15.4 Lipids

The last group of biological compounds we will discuss are lipids. This group of compounds hold a wide variety of functions. The function of lipids depends on their structure. In this section, we will review lipids and discuss them based on their functions in living organisms.

## Structure

Fatty acids are comprised of a long chain of carbons, with the last carbon a part of a carboxylic acid functional group (hence the name fatty *acid*). Fatty acids can be saturated or unsaturated depending on whether or not any double bonds are present in the fatty acid. A saturated fatty acid has no double bonds between the carbons in its chain. The term "saturated" refers to the fact that the molecule is "saturated" with hydrogen. An unsaturated fatty acid contains one or more double bonds between its carbons (Figure 15.19A). The placement of the double bonds is not random and is dictated by enzymes. The

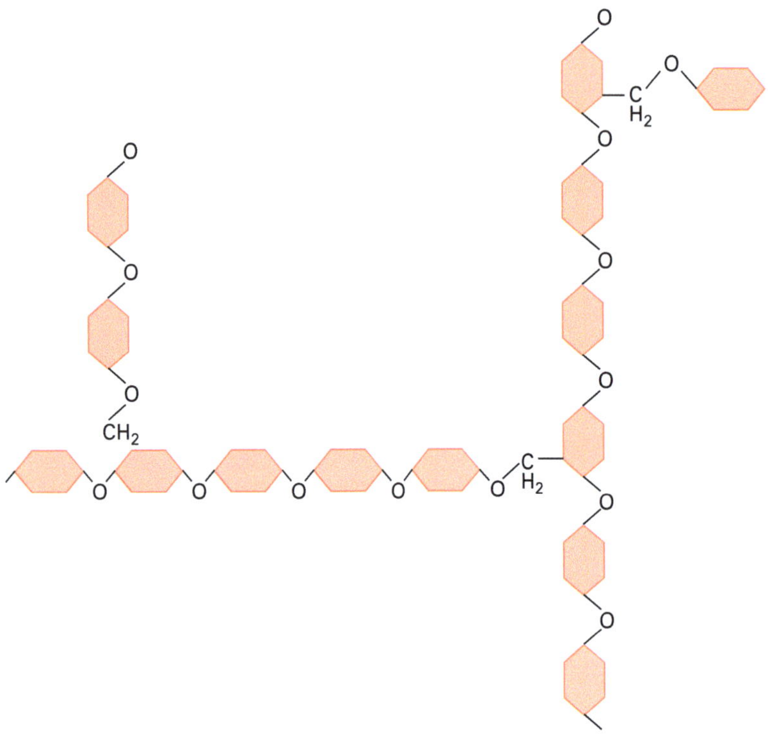


FIGURE 15.18. Glycogen Chain.

placement of the double bonds and the number of the carbons in a fatty acid chain dictates the identity of the fatty acid and contributes to its functionality.

Fatty acids are presents in oils and butter used in cooking. The longer the carbon chains, the higher the melting point of the fatty acid. Also, the ability for fatty acid chains to "pack" together is more difficult if double bonds are present as these creates bends, and curves in the fatty acid chain. Saturated fatty acids are linear and can easily pack together. When these fatty acid chains pack together, a solid can be formed. Generally speaking these solids are quite soft. Take for example butter versus vegetable oil. Butter is comprised of saturated fatty acids, making it easy to pack the chains together and form a solid at higher temperatures. Oil on the other hand contains mostly unsaturated fatty acids. This difference in the structures of the fatty acids explains why oil is a liquid at room temperature while butter is a solid, albeit a soft one.

## Function

Fatty acids serve many different functions in living organisms. One function is providing energy to cells. Actually, when metabolized, lipids can provide more energy per gram than an equal amount of carbohydrates. Generally speaking, the primary energy store in cells is found in the form of lipids and they are stored in fat cells, known as adipocytes. Some vitamins are known as "fat soluble" such as vitamin A. Without lipids, these vitamins would not be accessible. Eicosanoids, including prostaglandins, act like hormones and are important for many biological functions (Figure 15.19B). These include reproductive activity, immune response and blood clotting as well as many other functions.

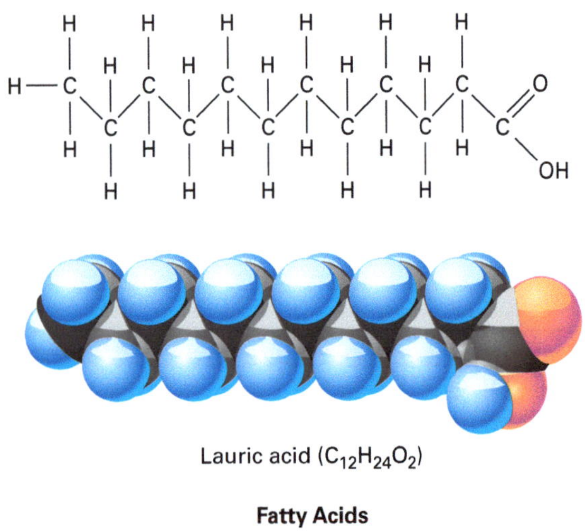


FIGURE 15.19A. Fatty Acid.

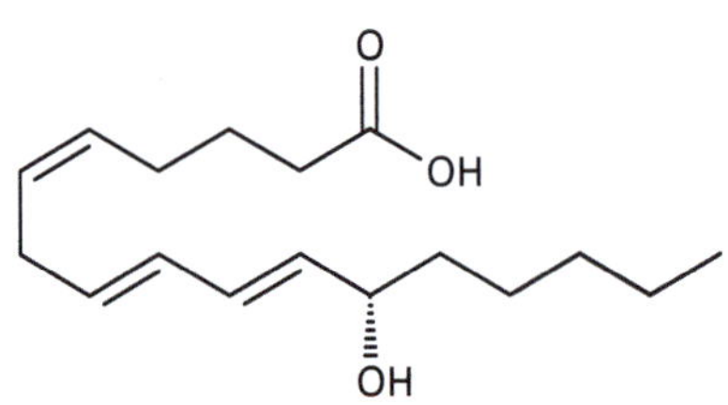


FIGURE 15.19B. 12-Hydroxyheptadecatrienoic Acid.

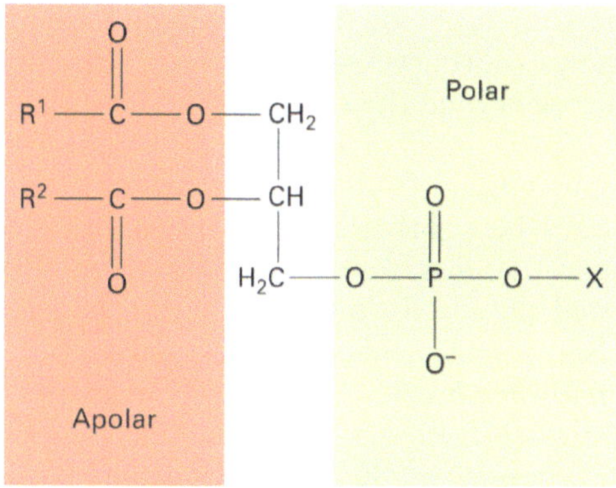


**FIGURE 15.20.** Phospholipid.

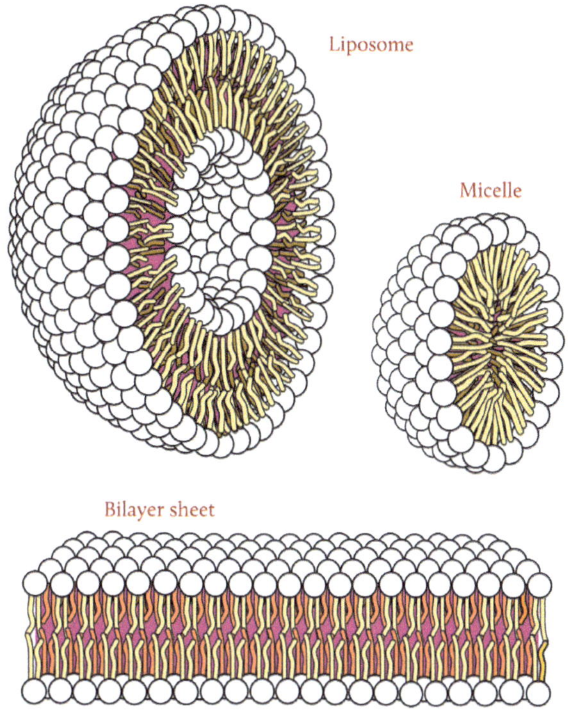


**FIGURE 15.21.** Phospholipids Aqueous Solution Structures.

Phospholipids are also very important to cells. They contain a phosphate group that provides a nonpolar lipid with a polar group on its end (Figure 15.20). Biochemists refer the the polar group as the lipids "head" and the long nonpolar chain as the phospholipids "tail". This polar/nonpolar components of these phospholipids create an amphipathic molecule meaning it can interact with both polar and nonpolar solvents. Of particular importance with this category of lipids is the ability to form lipid "bilayers". These bilayers is what encases cells in living organisms (Figure 15.21).

The bilayer ability of amphipathic lipids are also critical to our nervous systems. Sphingolipids are similar to phospholipids and also contain a polar group attached to the nonpolar chain of a lipid (Figure 15.22). These are part of the protective casing around our nerves. The myelin sheath, as this casing is called, surrounds all nerves. Damage to the sheath can cause severe neurological damage. In fact, damage to myelin sheath is associated with many neurological disorders and illnesses.

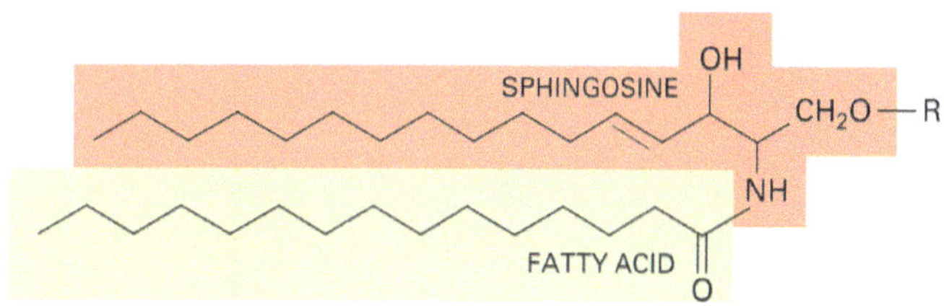


**FIGURE 15.22.** Sphingolipid.

# 15.5 Metabolism

Metabolism refers to the way living organism uses chemical processes to sustain life. It is impossible to cover this topic completely in a single section of this text but an overview of the topic can offer some understanding of the importance of the proper functionality of metabolism for all living things.

Proteins, carbohydrates and fats can all be broken down and ultimately used to provide energy to cells. The most common way cells produce energy is via the breakdown of glucose. Let's consider this process by starting from stored glucose in the form of glycogen. First glycogen enters into glycogenolysis during which it is broken down into its glucose subunits. This is a source of glucose beyond the source that comes from one's diet.

Insulin provides a single to cells to take up glucose from the blood after a meal. Glucose can then enter the glycolysis pathway where glucose gets converted into the compound ***pyruvate*** in a multiple step process that involves several different, vitally important enzymes. Next, pyruvate is converted into acetyl-CoA. It is important to note that acetyl CoA can also be produced from the metabolism of proteins or lipids as well.

The production of acetyl CoA occurs in ***mitochondria,*** football shaped "organelles" found in cells. Acetyl CoA enters into the Citric Acid Cycle (TCA or Krebs Cylce) where it is broken down and the process of producing energy begins. This process also involves several different enzymes with very specific chemical reactions to metabolize.

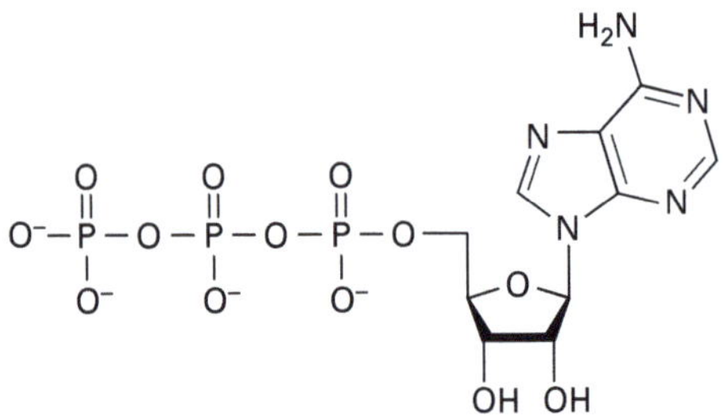


**FIGURE 15.23.** ATP Chemical Structure.

During the TCA Cycle, molecules of $FADH_2$, NADH and ATP are produced. ATP, or adenosine triphosphate, we discussed earlier as it pertains to DNA and RNA. It is also used to power chemical reactions and various processes in the cell. ATP contains an adenine base bound to three phosphate groups (Figure 15.23). Cleavage of the bond between the second and third bond in ATP results in a large release of energy that cells can use to catalyze reactions, move contents of cells from one place or another. Its cleavage is also responsible for the beating of our heart and the movement of our muscles.

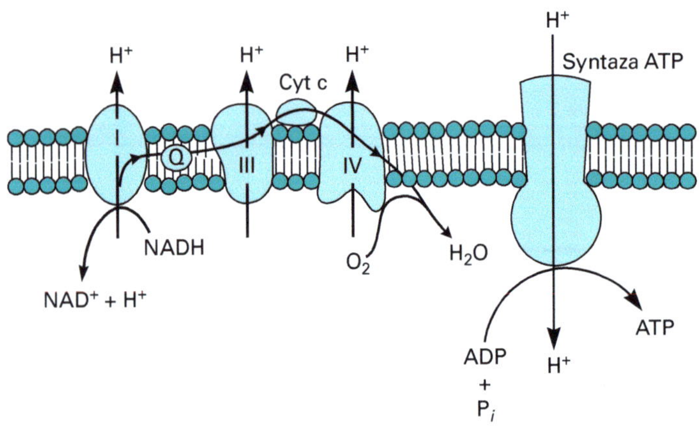


**FIGURE 15.24.** Mitochondrial Electron Transport Chain.

While some ATP is released during the TCA cycle, the largest production of ATP comes from oxidative phosphorylation that occurs in the Electron Transport Pathway (Figure 15.24). The proper function of this cell is critical to the vitality and health of cells. The process by which this pathway produces ATP is quite complex and beyond the scope of this text book. Steps do involve the use of $O_2$. This process and the product of ATP is which we require oxygen to survive. Without oxygen, we cannot product ATP and without ATP our cells would lose the energy they require for some many processes.

# 15.6 Toxicology

## Introduction to Toxicology

Toxicology is the study of the adverse effects chemicals have on living organisms. Pharmacology studies the use, modes of action and effects of drugs on living organisms. Considering these two fields together is useful as it is a balance of effective use of medicines while keeping in mind their toxicity that creates the therapeutic balance needed to treat various ailments.

Toxicologists study of the potential harmful effects a substance, compound or material can have on any living organism or the environment. This field is incredibly complex with a variety of factors requiring consideration. A substance can be highly toxic to one organism while it is considered safe for another species. Depending on the health of an individual, a substance can be highly toxic while it is perfectly safe for someone else. Age is also a critical factor in determining the toxicity of chemicals. Certain chemicals can be safe for a pregnant woman but could cause grave harm to the baby developing in the womb. It is well beyond the scope of this chapter to discuss toxicology in depth, we will however consider it at its most basic level to provide a general understanding of this field as it relates to chemistry.

The first topic relevant to discuss is what does toxic mean? Any chemical is toxic if it can cause illness or death in an organism. So what is a chemical? You, hopefully, by this point in the text realize that EVERYTHING is made up of chemicals. Water, is the chemical dihydrogen monoxide. Your body is comprised of millions of complex chemicals. The compounds produced naturally by plants are also chemicals. If you look up the definition of "chemical" you will get a wide variety of answers. Ultimately, where it is produced in a laboratory or by a plant grown on an organic farm, every compound is formed via chemical processes. So when you read that a product is "chemical free" be aware that this is very misleading and a deceptive form of advertisement.

Next, what is also important to know is all compounds are toxic when presented at a high enough level. We require water to survive. Did you know that people have died from water toxicity? Sounds unbelievable right? However, it is possible for a person to consume a large enough volume of water in a certain period of time a death. What does that mean? It means that water is toxic. In fact, everything is toxic! So how do we deal with the difference between the compounds water versus cyanide? The difference is the level of toxicity. Water has a very low level of toxicity as it requires a very large amount of it to lead to its toxic effects. Cyanide on the other hand requires a miniscule amount in order for it illicit its lethal effects.

So when talking about the toxicity of substances, it is best to discuss the level of toxicity. Does it have a high, mild or low level of toxicity? So next time you read the words "nontoxic" on a product ask yourself what that means (Figure 15.25). Generally speaking, a substance is considered "toxic" if it causes harm in very small amounts

| Toxicity Category | $LD_{50}$ (mg/kg) | Probably Lethal Dose for a 150-Pound Human Adult | Example Compounds |
|---|---|---|---|
| Super Toxic | < 5 | < 0.35 grams | Botulin<br>Aflatoxin |
| Extremely Toxic | 5–50 | 0.35–3.5 grams | Cyanide<br>Vitamin D |
| Very Toxic | 50–500 | 3.5–35 grams | Nicotine<br>Caffeine |
| Moderately Toxic | 500–5,000 | 35–350 grams | Aspirin<br>Salt |
| Slightly Toxic | 5,000–15,000 | 350–1,050 grams | Ethanol<br>Trichloroethylene |
| Practically nontoxic | > 15,000 | > 1050 grams | Sugar |

**FIGURE 15.25.** Toxicity Categories Used for Human Poisons.

$H_3C-Hg^+ \ X^-$

FIGURE 15.26.
Methylmercury.

(parts per million or less). Table 13.1 includes a list of compounds with their toxicity rated based on the amount needed to yield toxic effects. Many crayons are "nontoxic" but do you think it would be safe for a child to consume a box of crayons? Not likely, but nibbling a small piece off will probably be fine. Generally speaking, nontoxic crayons are free of mercury, lead and asbestos, substances that have a high toxicity especially to small children (Figure 15.26).

## Routes of Exposure

Routes of exposure to toxic substances can vary. It does not just come from ingestion of a substance. The main routes are usually dermal, oral, intravenous or via inhalation. Dermal indicates the material is able to pass through the skin and result in harmful effects. Inhalation indicates the material was taken in via the lungs. Oral (or percutaneous) exposure refers to eating or drinking the substance. Finally, intravenous exposure refers to exposure that occurs via a needle into a vein.

When considering each route of exposure, the toxicity level of a given substance will vary. For inhalation to be a concern, the substance must be a gas (carbon monoxide) or a very small particle (asbestos). Dermal exposure can vary from species to species in terms of toxicity as well. When considering the toxicity of a substance or if one is concerned about exposure to a substance, it is important to consider the route by which the individual was exposed.

Exposure can also be acute or chronic. Acute exposure refers to exposure to a substance for a short period of time. For example, if someone were to consume antifreeze one time and it were to result in harmful effects of the toxin, this would be considered an acute exposure to the toxin. Chronic exposure refers to repeated exposure over a long period of time. If someone's well was contamінated with arsenic and they consumed that water over a long period of time, they would be experiencing chronic exposure to arsenic.

## $LD_{50}$

In order to understand the toxicity of substances, studies are performed to determine the lethal dose that kills 50% of the population. Obviously, these tests are not performed on human subjects but rather test animals, usually rats. The population of these animals are fed increasing amounts of the compound of interest and the effect is monitored. The $LD_{50}$ is the dose at which 50% of the population of test animals dies.

Values for the $LD_{50}$ are reported in units of milligrams or micrograms per kilogram. The numerator value is the amount of the compound the denominator refers to the weight of the individual exposed to the compound. This an important reference point as obviously the amount of the compound that would be lethal for an infant would be very different for a full grown adult. In the case of accidental poisoning it is not uncommon for children to be affected first as opposed to adults. Children's tolerance is lower due to their size and they will exhibit symptoms of toxicity first.

## Mode of Action

When considering the toxicity of a compound, understanding of the mode of action is of critical importance. Knowledge of the mode of action can help in terms of treatment, antidotes and prevention of toxicity as well as determine the possible harmful effects of the toxic material. Knowledge of a chemicals composition and structure can also aid in predicting, to a certain extent, the safety of a compound. For example, carbon monoxide's toxicity is related to its ability to mimic oxygen and bind to hemoglobin. Many drugs are created by pharmaceutical companies using other compounds as a guide and making modifications in the generation of a new compound. There is no guarantee that the new compound created will be a "safer" drug and a perfect example of this is heroin. Heroin was marketed as a safe solution to morphine addiction. Its structure is very similar to morphine.

Compounds cause toxicity in a variety of ways they can inhibit the function of an enzyme, accumulate in the brain, cause mutations in DNA and the list can go on. If we understand the mode of action of a compound, its lethal effects maybe be treated if administered in enough time. We will highlight just a couple of compounds.

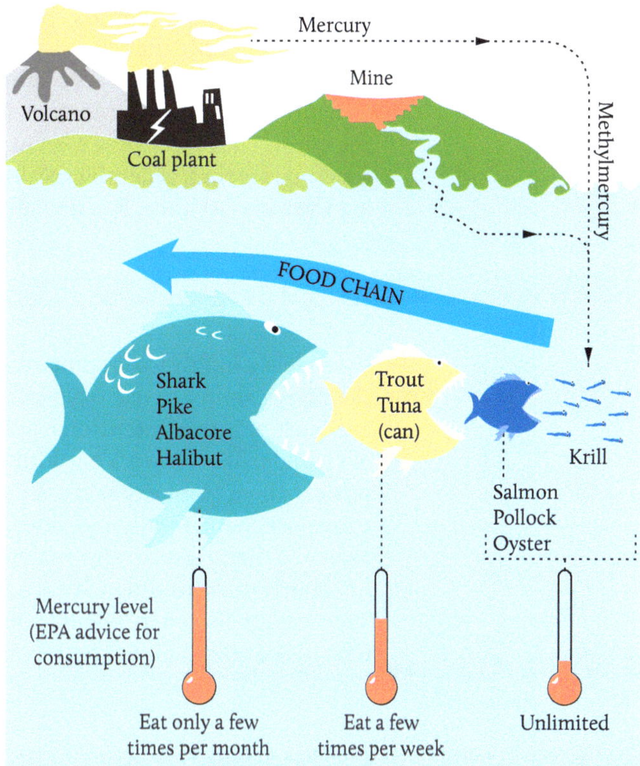


**FIGURE 15.27.** Mercury Food Chain.

## Mercury

Mercury is an element and it can combine with other elements to create different compounds. Exposure to mercury can come in many forms including inhalation, ingest, injection or passage through the skin. Different forms of mercury have different toxicities. Methylmercury is the form that in present in large fish and the form we are exposed to in our diet (Figure 15.27). Methylmercury is formed when mercury in the environment is metabolized by aquatic organisms into methylmercury (Figure 15.27).

Methyl mercury can react with the sulfur atoms on cysteine amino acids found in proteins. As a result, it is easily transported throughout the body, across the blood brain barrier and also across the placenta. This compound is most toxic to babies and growing children, so the diets of young children and pregnant women should be careful to limit their intake of foods that are high in levels of methylmercury. Because methyl mercury can accumulate in the body, it is difficult to remove. If exposure to methyl mercury is reduced (via eating less tuna for example), the levels of methyl mercury will decrease over time.

Even more toxic than methyl mercury is dimethyl mercury. This compound can pass through skin, even latex gloves. Just a small amount can be lethal. This compound is used in chemical laboratories, though its use has been mostly phased out due to its high level of toxicity. It is not likely one would be exposed to this compound unless they worked with or handled it directly.

Thimerosal is another form of mercury that has become commonly known. Thiomersal is a form of mercury that is used as a preservative in vaccines and it is known to chemists as ethyl mercury. Its LD50 is 75 mg/kg. If you are curious about the amounts in a vaccine, 25 micrograms are used in a vaccine well below the toxicity level. It is also important to note that at this time, thiomersal is mostly limited to flu vaccines.

In the case of mercury containing compounds, their mode of action generally relates to neurological effects. Mercury will accumulate in fatty tissues include the tissue in the brain. The presence of mercury in the brain can have all sorts of consequences. The good news is that, generally speaking, we can control our intake of mercury by monitoring the foods we take. It is not advised that one eats tuna or swordfish on a daily basis.

### Cyanide

Cyanide is a highly toxic compound classified as a chemical asphyxiant in that it blocks the body's ability to use oxygen. Exposure to cyanide is usually limited to accidental exposure or acts of intentional poisoning. Hydrogen cyanide gas is the most toxic form of cyanide. Anyone working with this compound must wear appropriate respirators to protect them from exposure. Cyanide inhibits the formation of ATP by the electron transport chain. Orally, cyanide has a toxicity of 1.52 mg/kg. There are some antidotes available that must be administered immediately for them to be effective. In the 1980s in suburbs surrounding Chicago, there were several deaths due to cyanide poisoning. This was attributed to Tylenol bottles that had been tampered with in order to add cyanide to the pills. This led to the tamper resistant packaging we are familiar with today.

### Carbon Monoxide

The main source of carbon monoxide is through the incomplete combustion of carbon fuels. Carbon monoxide is particularly dangerous as it is a colorless odorless gas and the symptoms of poisoning can be missed or attributed to other ailments until it is too late. Signs of CO poisoning include nausea, headache, dizziness as well as other symptoms. CO binds to hemoglobin in the place of oxygen, effectively "out-competing" oxygen. This prevents oxygen from being transported throughout the body.

You might think of CO as a trojan horse. It presents itself as very similar to oxygen. However, it not only binds to hemoglobin in place of oxygen, it binds more tightly. This not only blocks oxygen from binding it also makes it incredibly difficult to remove. Treatment of CO poisoning includes providing patients with oxygen. In severe cases, oxygen is provided in a hyperbaric chamber. This helps oxygen to bind to hemoglobin.

In concluding our discussion regarding toxicity of compounds, it is important to emphasize that anything can cause harm and/or death if present in right dose. In general the term, toxic is associated with compounds that exhibit lethality when consumed in very small amounts. While water may not be labeled as "toxic" it can cause death if consumed in a large enough quantity.

## 15.7 Removal of Xenobiotic Compounds

Anything considered a foreign substance to an organism is considered a xenobiotic compound. An organism actually has a system designed to identify and remove these substances. The way our bodies excretes substances is via urination, exhalation, defecation or perspiration. The easiest way to remove substances from the body is to make them water soluble. This allows them to be circulated in blood and to then be removed via one of these pathways.

## Permeability

The first method by which the body protects itself form the effects of foreign materials is by prevents cells from taking in these substances. As discussed earlier, cell membranes are comprised of lipids. If a compound is polar, it cannot readily pass across a cell membrane. These polar compounds will remain in the blood and eventually be excreted in urine. If a cell requires polar compounds for any reason, the cell membrane contains very specific transport proteins that allow passage of these compounds into cells.

## Bile

Bile from the liver is important in that is aids the digestion of fat soluble vitamins and nutrients an also aids in the elimination of nonpolar waste products in feces. It is produced in the liver and not only aids in the removal of foreign substances but is also important for the removal of metabolites produced in the body. It is also responsible for the removal of broken down red blood cells. If a liver is malfunctioning, one prominent sign is jaundice. The yellowing of skin and eyes it due to the bodies in ability to remove red blood cells which then build up in the body.

## Biotransformation

The liver is critical for detoxifying the blood of xenobiotic material. Liver failure can lead to death as it prevents the body from removing toxic materials. Toxins are removed from the body by making them more polar and water soluble so that they may then be excreted in urine. This is achieved in the liver primarily by a two-step process. In some cases, a third step can be involved. The process is often referred to as the Biotransformation Pathway.

In the first phase of xenobiotic metabolism, nonpolar compounds are "prepared" for the next phase. Compounds are modified so as to make them more reactive. The modification occurs via oxidation, reduction or hydrolysis. This converts the compounds into electrophiles or nucleophiles that can then enter phase II.

After phase I is complete, there are a few possible outcomes. The first is the intended outcome, that the compound is ready to enter phase II. Another possible outcome is the modified compound can be excreted. It is also possible that the toxicity of the compound could be diminished. It is also possible that the compound created could have an increased toxicity. In fact, some drugs are manufactured with the knowledge that they are modified in the liver. When they are first consumed they are inactive (prodrugs) but after they are modified in the liver, they are then elicit their desired effect.

The second phase of xenobiotic metabolism involves a conjugation step. At this point, the compound modified in phase I is now "chemical active". Enzymes in phase II then add a large polar group to the compound via a conjugation or binding step. By adding a highly polar functional group, the compound's water solubility will increase thus increasing the ability for it to be excreted. Modifications include methylation, sulphation, glucouronidation, and acetyalation. Transferase enzymes are used to conjugation the functional groups to the compounds.

At this point the solubility and rate of excretion for a compound should be dramactically increased. In this case, due to the incorporation of these larger, polar groups, increasing the potency or toxicity of a compound is unlikely.

Phase I and Phase II are the most critical phases for xenobiotic removal. However, some compounds can undergo further modification to increase their rate of excretion.

## 15.8 Pharmacology

While toxicology investigates the negative effect compounds have on the body, pharmacology studies their therapeutic effects. Pharmacologists study the effect a compound, refered to as a drug, has on the body. Studies delve into its mode of action, desired effects as well as side effects. In many ways, this field is similar to toxicology.

### Drug Design

The design of drugs begins with a desired outcome whether it be to minimize pain, kill bacteria or one of a multitude of other possibilities. With this in mind, drug companies, have to consider how to achieve this outcome. This requires an understanding of the biology and biochemistry of the target. If designing an antibacterial drug, it is important for that drug to target the bacteria specifically and do so with few if any side effects. Viral drugs are much more difficult to design as viruses share similarities with human cells. This makes it difficult to design treatments that will not have side effects on humans. Cancer drugs are particular difficult to design without harmful side effects. Cancers occur when human cells grow too quickly. To kill off cancer cells, human cells are affected as well. Most affected are fast growing cells, like hair follicles which leads to hair loss in cancer patients.

Drugs are often modeled after natural products. Drug companies will synthesize the same or similar compounds to those found in plants or other organisms. When synthesized in a laboratory, compounds can be made with a higher purity as the synthesis can be directed towards a specific product. When isolating compounds from organisms, the desired compound as well as others can be collected all at once. The other compounds collected can have an undesirable effect, so the ability to synthesize a highly pure compound is desirable.

### Mode of Action

As discussed with toxic compounds, understanding the mode of action of a drug is very important. When drugs are administered, it is best that they are specific. When drugs are nonspecific, side effects can result. It is not always possible to predict the specificity of a drug which is why new drugs undergo a very rigorous approval process. Purity again becomes important as one compound might be very specific and produce the desired effect, while another compound that is contaminating the sample could have serious or undesirable side effects.

Targets of drugs vary widely. They could target and enzyme, cell receptor. They could act as a hormone to trigger a specific response. Some drugs target the nervous system in treatment of mental illness or nervous disorders like multiple sclerosis.

## Effect on Liver

The liver can be greatly affected by drugs used to treat diseases and ailments. When drugs are taken they are in fact xenobiotic compounds. As a result, the enzymes in the liver will work remove them via the process discussed earlier. Some drugs have a greater effect on the liver than others and as a result liver function is closely monitored to ensure it is not damaged during treatment. Alcohol is also metabolized in the liver. It is for this reason it is never a good idea to consume alcohol while taking certain medications as your liver could be overburdened.

## Determining Therapeutic Dosage

We have already learned that all compounds can be toxic, and so it is the case that drugs taken to treat illnesses have a toxicity as well. It is of critical importance that a proper balance be in place when determining the dosage of a drug to consume. The $LD_{50}$ for drugs is determined in the manner discussed previously. Also determined for drugs is the Effective Dose ($ED_{50}$). This values is the concentration of the drug at which 50% of a given population displays the desired effect. Obviously, the desired effect will vary based on what the drug is used to treat. The same units are used for the $ED_{50}$ (mg/kg) again considering the dose as it relates to the weight of the person taking the drug.

Once both the $ED_{50}$ and $LD_{50}$ are determined, the ***therapeutic index*** is calculated and proper dosing for the drug can be determined. It is ideal for the therapeutic index to be very wide). This indicates that the $ED_{50}$ and $LD_{50}$ are far apart. For a drug to be safe and effective, the $ED_{50}$ should be very low and $LD_{50}$ needs to be very high. Basically, one was to feel the desired effect of the drug and not die in the process! The TI is calculated using the formula $\frac{LD_{50}}{ED_{50}}$. The best drugs will have a very high TI indicating that a person can feel the therapeutic effects of the drug with experiencing minimal if any toxic effects. With the knowledge of the TI in hand, the proper dosing for a drug can be determined.

Entire college careers are spent studying Biochemistry, Toxicology and Pharmacology. Hopefully, this chapter exposed you to each of these fields. This chapter is meant to be a survey of these topics. Hopefully you are left with a better understanding of how chemistry relates to the way your body functions.

# Chapter 15 Summary

*Nucleotides* are compounds comprised of a cyclic amine component and a five carbon sugar and up to three phosphate molecules.

*Purine bases* are amine compounds with a double-ring structure and include adenine (A) and guanine (G).

*Pyrimidine bases* are amine compounds with a single-ring structure and include cytosine (C), thymine (T) and uracil (U).

*Base pairs* form between the nucleotides (A to T or U) and (G to C). This forms the double helix structure of DNA is also leads to the genetic code.

*Deoxyribonucleic acid (DNA)* is made of 4 nucleotides adenine, thymine, guanine or cytosine. It also contains the sugar ribose instead of deoxyribose like DNA.

*Ribonucleic acid (RNA)* is made of 4 nucleotides adenine, uracil, guanine or cytosine. It also contains the sugar ribose instead of deoxyribose like DNA.

During *replication* DNA is copied and is used to create new generations of cells.

*DNA polymerase* is a complex enzyme molecule that is the machine that replicates DNA by readings and copying the code given by the nucleotides in the DNA strands.

*RNA polymerase* is a similar complex enzyme molecule to DNA polymerase. It is is the machine that transcribes DNA by reading and copying the code given by the nucleotides in the DNA strand. In this case the code is used to create RNA.

There are three categories of RNA. *Messenger RNA (mRNA)* contains the code for protein synthesis. *Transfer RNA (tRNA)* are used to decipher the code in mRNA and carries the amino acids needed to make proteins. *Ribosomal RNA (rRNA)* is used to create the ribosome which is the large enzyme machine that makes proteins.

*Proteins* have a wide variety of functions in living organisms and can be considered the "workers" of organisms. Each protein contains a unique sequences of amino acids.

*Hormones* are signaling proteins that bind to receptors to elicit a certain response.

*Enzymes* are catalysts that increase the rates of chemical reactions required for organisms to survive.

*Antibodies* are proteins used to signal for immune responses to infections within and organism.

*Storage proteins* function to store and hold proteins until they are needed for cell/organism functions.

*Transport proteins* are carries that bring various molecules in and out of cells or move molecules throughout the body.

*Structural proteins* can be considered the infrastructure component of cells and organisms. They give cells shape and also help muscles move.

*Amino acids* are the subunits that make up proteins. As the name implies it contains an amine component and a carboxylic acid component. There are 20 standard amino acids as well as some specialized amino acids and some that are unique to specific organisms.

*Carbohydrates* are often called "sugars". They are comprised of carbon, hydrogen and oxygen. These are the energy producers for cells and organisms.

*Monosaccharides* are "single" sugars and the smallest type of carbohydrate.

*Disaccharides* contain to sugar subunits.

*Polysaccharides* are a polymer of monosaccharides with more than two monosachharides linked together.

*Lipids* are also known as fats. They are made up of fatty acids with a long chained carbon molecule attached to a carboxylic acids. These make up cell membranes, provide energy and even are types of vitamins.

*Metabolism* refers to the chemical processes that occur within organisms. This includes synthesis and breakdown of various molecules used by the body.

*Toxicology* is the study of the adverse effects chemicals have on living organisms.

***Pharmacology*** is the study of the therapeutic effects compounds can have on a living organism.

***$LD_{50}$*** refers to the dose of a compound at which 50% of a population is killed by the compound.

***$ED_{50}$*** is a measure of the dose at which 50% of a population experiences the therapeutic effects of a compound/drug.

The ***therapeutic index*** is a measure of the safety of a drug. The TI is calculated using the formula $\frac{LD_{50}}{ED_{50}}$ and ideally, the $LD_{50}$ should be much larger than the $ED_{50}$.

***Biotransformation*** is the process by which organisms alter foreign compounds so they can be effectively removed from the body.

# CHAPTER 15 EXERCISES

1. Draw the structures of the two purine bases.
2. Draw the structures of the three pyrimidine bases.
3. Which of the bases are used to make DNA and which to make RNA?
4. How do the nucleotides base pair with each other?
5. List and describe 3 categories of RNA.
6. List and describe 3 categories of proteins.
7. You are tasked with designing and effect drug treatment for migraines. Explain what type of analysis you would perform in order to ensure its safety.
8. You are told by a friend that they made a new cleaning product that is both chemical free and 100% non-toxic. How might you respond to this statement?
9. Define and explain the purpose of the study of pharmacology.
10. How do you determine the effective dose of a drug?
11. Explain the process by which the effects of a toxicant are determined. Draw a simple curve to help explain the procedure by which the LD50 is determined. Also explain what types of subjects are used for the studies and the limitations or inaccuracies this may cause.
12. How do you determine the lethal dose of a compound?
13. What is the ultimate goal of biotransformation?
14. What happens to compounds during phase I biotransformation?
15. What is the major class of enzymes involved in biotransformation?
16. What are three possible fates of compounds after phase I of biotransformation?
17. Describe the chemistry that occurs during phase II biotransformation? What does phase II accomplish for the compound?

18. There are 3 major locations for biotransformation in the body, where are these locations? Why do these locations make sense?

# Credits

Fig. 15.1: Copyright © 2016 by OpenStax, (CC BY 4.0) at https://commons.wikimedia.org/wiki/File:0322_DNA_Nucleotides.jpg.

Fig. 15.2: Copyright © 2009 by Boumphreyfr, (CC BY-SA 3.0) at https://commons.wikimedia.org/wiki/File:Ribose_deoxyribose.png.

Fig. 15.3: Copyright © 2013 by OpenStax College, (CC BY 3.0) at https://commons.wikimedia.org/wiki/File:230_Structure_of_Adenosine_Triphosphate_(ATP)-01.jpg.

Fig. 15.4: Source: https://commons.wikimedia.org/wiki/File:Human_chromosomesXXY02.png.

Fig. 15.5: Source: https://commons.wikimedia.org/wiki/File:DNA-structure-and-bases.png.

Fig. 15.6: Copyright © 2010 by Miguelferig, (CC BY-SA 3.0) at https://commons.wikimedia.org/wiki/File:Phosphodiester_Bond_Diagram.svg.

Fig. 15.7: Source: https://commons.wikimedia.org/wiki/File%3ADNA_replication_en.svg.

Fig. 15.8: Source: https://commons.wikimedia.org/wiki/File%3ARNA-codon.png.

Fig. 15.9: Copyright © 2014 by Antonion Geremia, (CC BY-SA 4.0) at https://commons.wikimedia.org/wiki/File%3AAmminoacidi_e_proteine.png.

Fig. 15.10: Source: https://commons.wikimedia.org/wiki/File:D-glucose_D-glucopyranoside_TOLLENS.png.

Fig. 15.11a: Source: https://commons.wikimedia.org/wiki/File:Aldose.png.

Fig. 15.11b: Copyright © 2004 by Rob Hooft, (CC BY-SA 3.0) at https://commons.wikimedia.org/wiki/File:Skeletal_structure_of_ketose.png.

Fig. 15.12: Source: https://commons.wikimedia.org/wiki/File:Photosynthesis_equation.svg.

Fig. 15.13: Source: https://commons.wikimedia.org/wiki/File:D-Glucose_vs._D-Fructose_Structural_Formulae_V.1.svg.

Fig. 15.14: Source: https://commons.wikimedia.org/wiki/File:RiboseAndDeoxy.gif.

Fig. 15.15: Source: https://commons.wikimedia.org/wiki/File:Lactose_color.png.

Fig. 15.16: Source: https://upload.wikimedia.org/wikipedia/commons/d/d2/Sucrose_structure_formula_inkscape.svg.

Fig. 15.17a: Source: https://commons.wikimedia.org/wiki/File:Amylose_3Dprojection.corrected.png.

Fig. 15.17b: Source: https://commons.wikimedia.org/wiki/File:Amylopektin.svg.

Fig. 15.18: Copyright © 2009 by Boumphreyfr, (CC BY-SA 3.0) at https://commons.wikimedia.org/wiki/File:Glycogen_chain.png.

Fig. 15.19a: Copyright © 2013 by BluceBlaus, (CC BY 3.0) at https://commons.wikimedia.org/wiki/File:Blausen_0396_FattyAcid.png.

Fig. 15.19b: Source: https://commons.wikimedia.org/wiki/File:12-Hydroxyheptadecatrienoic_acid.svg.

Fig. 15.20: Source: https://commons.wikimedia.org/wiki/File:Phospholipid.png.

Fig. 15.21: Source: https://commons.wikimedia.org/wiki/File:Phospholipids_aqueous_solution_structures.svg.

Fig. 15.22: Source: https://commons.wikimedia.org/wiki/File:Sphingolipid.png.

Fig. 15.23: Copyright © 2005 by Samulili, (CC BY-SA 3.0) at https://commons.wikimedia.org/wiki/File:ATP_chemical_structure.png.

Fig. 15.24: Source: https://commons.wikimedia.org/wiki/File:Mitochondrial_electron_transport_chain_short_PL.svg.

Fig. 15.25: Source: http://www.georgiastem.com/wp-content/uploads/2017/11/toxicity-categories-used-for-human-poisons.jpg.

Fig. 15.26: Source: https://commons.wikimedia.org/wiki/File:Methylmercury.png.

Fig. 15.27: Copyright © 2009 by Bretwood Higman, (CC BY 3.0) at https://commons.wikimedia.org/wiki/File:MercuryFoodChain-01.png.

# Agricultural Chemistry

## Chemistry can't get more important than this...

## 16.1 History

The development of agriculture occurred around the world in ancient times, but the Middle East, especially the well-named Fertile Crescent, was the origin of many of the grains and domesticated animals we know so well today. The world as we know it today, and yesterday, would simply not exist without the hands-on development of agricultural of all kinds, animal and plant kingdoms together.

Chemistry has played and will continue to play a fundamental role in agriculture. The goal of this chapter is to summarize key atomic elements, alone and in combinations, and discuss their chemical importance, especially in the plant world. This is somewhat of an arbitrary choice as the area of animal husbandry is just as important and interesting; however, the plant side is more fundamental. After all, without plant husbandry there would be no animal husbandry.

The following sections are not intended to examine chemical mechanisms of what goes on in agriculture. Rather it presents an overview of what fundamental chemicals are crucial for making a very complex system function at all in the modern world. It is also worth noting the interplay and dependence of what is discussed in this chapter with what is presented in the others, especially the chapter on energy. For example, without fuels to run the tractors much of the world would be starving.

## 16.2 Soil Chemistry and Nutrients

There is a truism that says we all owe our existence to the presence of six inches of soil and the fact that it rains. Rain we've got down, but what about soil chemistry?

### Fundamentals

Soils are complex mixtures of inorganic and organic matter, and are quite variable in composition. The presence of adsorbed and absorbed water is crucial to plant growth as is the presence of gases, especially oxygen, carbon dioxide and nitrogen.

The inorganic or mineral (for example the geode of Figure 16.1) parts of soils are characterized by physical as well as chemical properties. Let's break these down a bit, beginning with the simpler one, physical:

FIGURE 16.1. Mineral—Geode—Brazil.

## Physical Properties of Minerals

The basic physical forms of soils are very familiar to us. Ranging in size from nanoparticles ($10^{-9}$ meters across) to microns ($10^{-6}$ m) up to small pebbles to rocks. We are more familiar with the names clays, sand and gravel. The differences between these common names is basically particle size. Clays are composed of very small or fine particles that tend to cling to one another tenaciously via very high surface areas and corresponding high between-particle bonding forces making them able to retain water well. Sand particles are larger may have similar bonding forces, but also much lower surface areas so total attraction forces will be correspondingly weaker.

In the absence of organic matter soils have rock-like properties in terms of hardness and tend to fracture in a brittle manner. Soils tend to be moderate to poor conductors of heat and electricity; however, when wet and full of ions the latter can improve markedly.

Addition of organic matter from plants and animals enriches the basic rocklike nature of soils. In combination with adsorbed water the rich substance needed to efficiently grow crops results.

## Chemical Composition

The structural components of soils, basically little rocks, are composed primarily of Si, O, Al and Fe. An interesting point is that silicon and aluminum are not important to the function of biological systems. Oxygen and iron are very important; however. There are many other elements present in soils at substantially lower levels which are important. Table 16.1 contains a list of the primary elements that occur in plants, and their biological function.

TABLE 16.1 Primary Plant Nutrients and Their Function

| Nutrient | Functions |
|---|---|
| Nitrogen (N) | Synthesis of proteins for growth |
| Phosphorus (P) | Cellular division, formation of energetic structures |
| Potassium (K) | Transport of sugars, formation of starch, stomata control, co-factor of enzymatic reactions, reduces susceptibility to plant diseases |
| Calcium (Ca) | Building block in cell walls, reduces susceptibility to plant diseases |
| Magnesium (Mg) | Part of chlorophyll molecule |
| Sulfur (S) | Synthesis of essential amino acids |
| Boron (B) | Cell wall formation, germination and elongation of pollen tube, metabolism and transport of sugars |
| Copper (Cu) | Influences the metabolism of nitrogen and carbohydrates |
| Iron (Fe) | Chlorophyll synthesis |
| Manganese (Mn) | Aides in photosynthesis |
| Molybdenum (Mo) | Component of nitrate-reductase and nitrogenase enzymes |
| Zinc (Zn) | Auxin synthesis |

Table 16.1 also lists the inorganic elements available. The organic fraction contains the big six elements that compose organic compounds; oxygen, carbon, hydrogen, nitrogen, sulfur and phosphorous. Without these six elements life would simply not exist on earth as we know it. Now let's look at the elements that are responsible for the success of agriculture.

## Basic Plant Structures

Chapter 15 discusses photosynthesis and the solar power driven formation of carbohydrates from $CO_2$ and water, or more fundamentally from the elements C, O and H. Two products of the biochemical processes are ***cellulose*** and ***lignins***: (Chemical Structures 16.1 and 16.2, respectively):

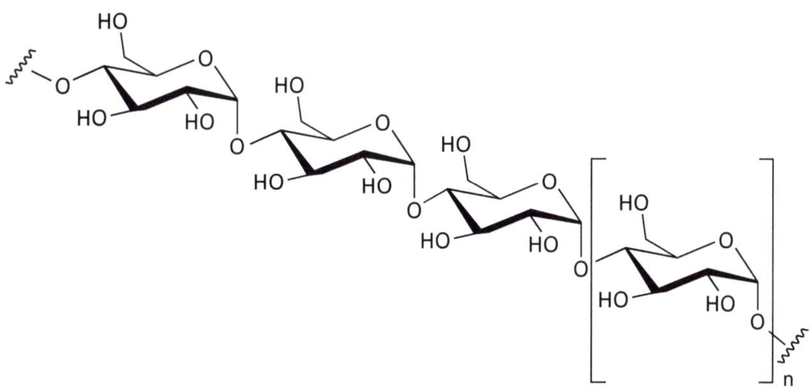


CHEMICAL STRUCTURE 16.1. Cellulose Chemical Structure.

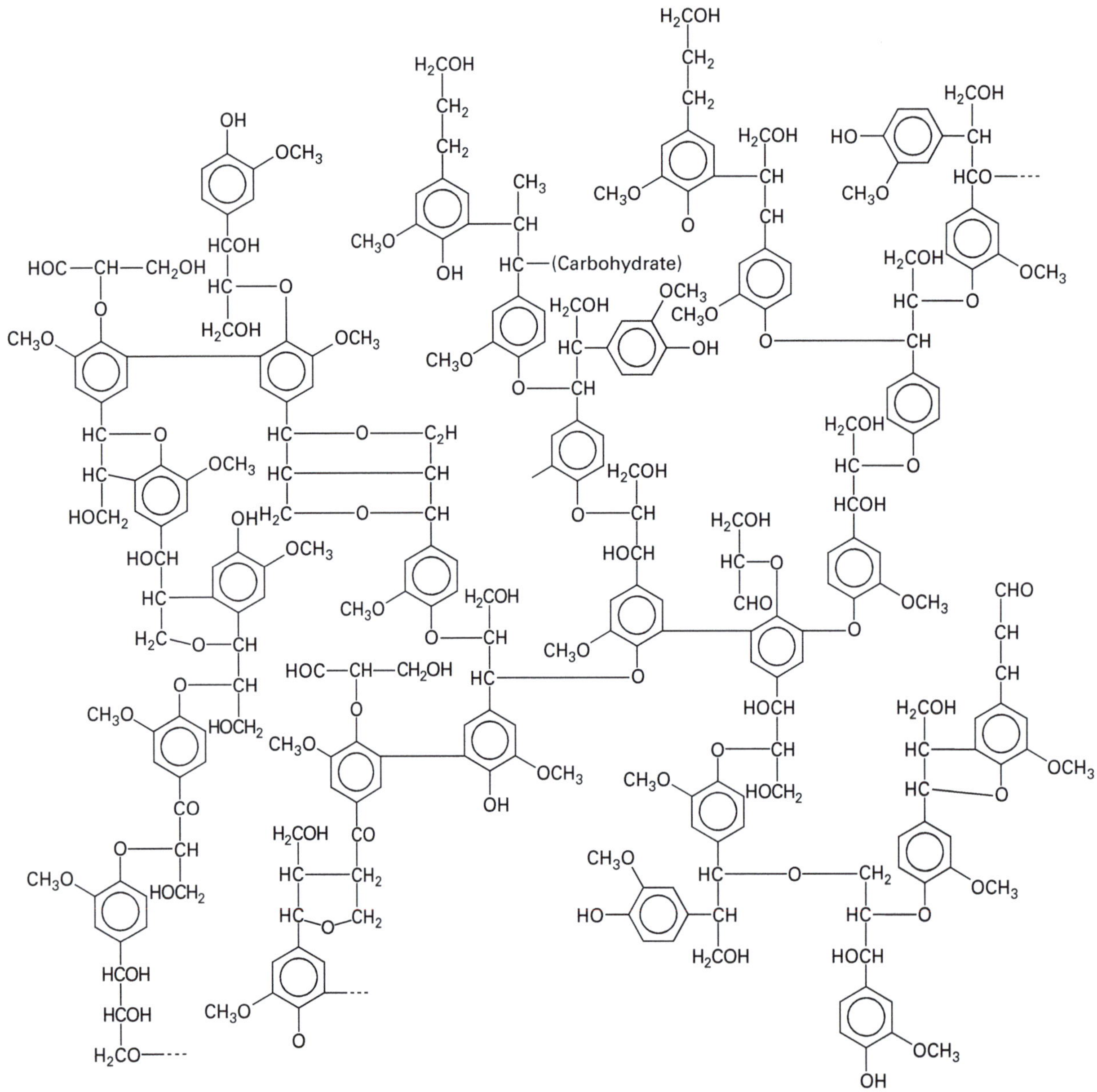

**CHEMICAL STRUCTURE 16.2.** Lignin—Generalized Chemical Structure.

Both of these substances serve structural purposes in a plant, with cellulose being a key component of cell walls, and lignins providing even stronger intercellular support as well as being key for efficient water transport. Notice how complex the lignin chemistry appears to be. In fact there is no exact structure for lignins as the molecular structures vary a great deal in nature. However, the basic unit is phenol:

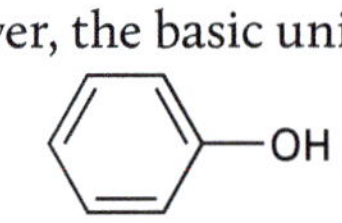

**CHEMICAL STRUCTURE 16.3.** Phenol.

Along with C, O and H; the elements N, S and P are especially important in the formation and function of proteins, DNA, RNA, amino acids, enzymes and many other biological systems as described in Chapter 15, Biochemistry.

## Primary Plant Nutrients

The three elements that are needed in large amounts for suitable plant growth and function to are nitrogen (N), phosphorus (P), and potassium (K). On fertilizer bags they are represented by N-P-K and a triad of numbers, for example 10-10-10. The latter means 10% N, 10% P and 10% K, by weight. Usually the nitrogen component is supplied using ammonium nitrate ($NH_4NO_3$) or urea, the phosphorus as $P_2O_5$, and potassium as $K_2O$. In the plant these elements are responsible and important for:

- ***Nitrogen:*** Leaf and seed production, N is a key component of chlorophyll.
- ***Phosphorus:*** Key to cell energetics. Encourages root development and flowering.
- ***Potassium:*** Disease prevention, fruit quality, protein synthesis aid.

Usually nitrogen is the limiting nutrient of plant growth.

***Nitrogen Fertilizer Chemistry:*** Since nitrogen comprises about 78% of the earth's atmosphere it is in essence free as a starting material; however, $N_2$ is a very stable molecule as the atoms are joined by a triple bond. There are a small number of organisms found in soil that can "fix" nitrogen gas from the air, but for accessing the large quantities of nitrogen needed for fertilizers a chemical process is required. The ***Haber Bosch*** process is deceptively simple considering its immense importance to humanity:

$$3H_2(g) + N_2(g) \underset{Pressure}{\overset{Heat}{\rightleftharpoons}} 2NH_3(g) \qquad (16.1)$$

It is not an exaggeration to state that without this reaction to supply ammonia, $NH_3$, the precursor to ammonium nitrate fertilizer, we simply could not come close to growing enough food to feed the earth's population.

Notice that this reaction requires hydrogen gas, which does not occur naturally in any substantial amounts. So first the $H_2$ must be synthesized. The reader is referred back to Chapter 12 where this topic is discussed.

Ammonium nitrate is then made from ammonia by reaction with nitric acid:

$$HNO_3 + NH_3 \longrightarrow NH_4NO_3 \qquad (16.2)$$

The nitric acid needed here is also made from ammonia by reaction with oxygen and water in the Ostwald process:

$$5O_2 + 4NH_3 \longrightarrow 4NO + 6H_2O \qquad (16.3)$$

$$O_2 + 2NO \longrightarrow 2NO_2 \qquad (16.4)$$

$$H_2O + 3NO_2 \longrightarrow NO + 2HNO_3 \qquad (16.5)$$

The NO produced as byproduct in the last reaction is recycled while the $HNO_3$ is in aqueous solution where it can be concentrated as desired using simple distillation.

## Phosphorus Fertilizer Chemistry

The long term supply of phosphorus is problematic. Most elements (C, N, O, H) that are needed in large amounts for biological processes are available in gaseous forms from the atmosphere. Those that are not are either available terrestrially in very large quantities (K and S) or are needed by plants in very small quantities (micronutrient section below). Phosphorus is different in that it is available only in the solid state, and worldwide supplies are limited both in volume and locations. This limitation in availability is somewhat misleading in the sense that phosphorus occurs in substantial amounts in all living organisms. However, this form of phosphorus, although widespread, is very dilute and thus impractical to access in large quantities. Before the advent of inorganic fertilizers phosphorus was replenished by recycling dead organisms, manure and crop waste (see section 13.5**). This technique is not practical for the large volume of food that must now be produced.

Phosphorus occurs naturally in substantial deposits primarily in the form of phosphate rock, especially apatite: $\mathbf{Ca_{10}(PO_4)_6(OH, F, Cl)_2}$, which is really shorthand for a mixture of closely related compounds: $\mathbf{Ca_{10}(PO_4)_6(OH)_2}$, $\mathbf{Ca_{10}(PO_4)_6(F)_2}$ and $\mathbf{Ca_{10}(PO_4)_6(Cl)_2}$.

Treatment of apatite with strong mineral acids results in the more soluble phosphate fertilizers. Here is this chemistry for another form of apatite:

$$\mathbf{7H2SO_4 + 3Ca_3(PO_4)_2 \cdot CaF_2 + 3H_2O \longrightarrow} \quad \mathbf{(16.6)}$$
$$\mathbf{3Ca(H_2PO_4)_2 . H_2O + 2H_2O + 7CaSO_4 + 2HF}$$

This Ca-$PO_4$ compound is known as superphosphate and contains about 10% P by weight. The 20% product (concentrated superphosphate) can be made by removal of the gypsum, $CaSO_4$.

## Potassium (Potash) Fertilizer Chemistry

The term potash is commonly used in connection with fertilizers, and refers to a mixture of a number of potassium containing salts, especially sylvite (**KCl**), sylvinite (**KCl.+ NaCl**), and langbeinite ($\mathbf{K_2SO_4 . 2MgSO_4}$). These minerals occur in large quantities worldwide and need minimal purification processing before an immediate use in fertilizers.

## Secondary Plant Nutrients

These are secondary nutrients only in the sense that they are required by plants in lesser amounts than those previously discussed. They are all still absolutely necessary for normal plant growth. The three elements here are:

### Sulfur

Sulfur is a key part of many biochemical compounds and processes. Key groups of compounds including proteins, enzymes, amino acids and many more. It is also a crucial part of photosynthetic processes.

Sulfur is available in vast quantities on earth, so there is no potential issue with supply. Currently the sulfur needed industrially is supplied as a byproduct of oil and natural gas extractions, and far more is produced annually than is needed. Greater than 80% of the sulfur used goes into the production of sulfuric acid. $H_2SO_4$ is the chemical produced industrially in the greatest quantity every year, and about 50% of this goes into the production of fertilizers (see phosphorus above).

### Magnesium

Without magnesium there would be no photosynthesis since it is a key component of the compound chlorophyll:

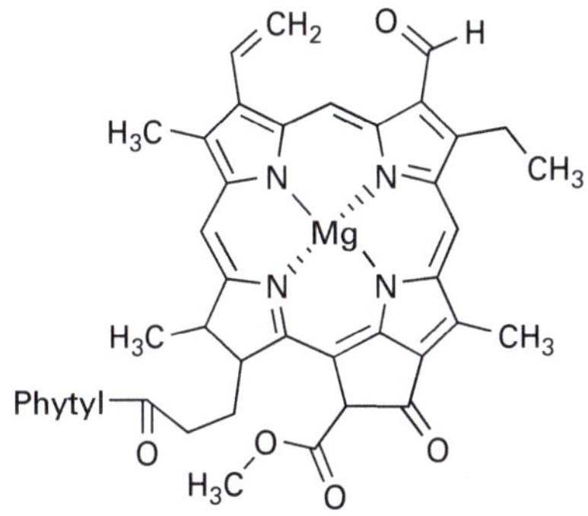


**CHEMICAL STRUCTURE 16.4.** Magnesium.

Magnesium is also important in enzymatic processes as well as electrolyte balances in organisms.

### Calcium

Calcium is widely available in the environment. It is involved a in many parts of plant biochemical systems, including nutrient uptake, enzymatic processes, photosynthesis, and cell wall structures, among others. $Ca^{+2}$ deficiencies show up in root development issues, as well ongoing solar conversion chemistries.

## Plant Micronutrients

There are (at least) 14 elements that have been shown to be a part of plant metabolisms or simply beneficial to the overall well being of plants. Since they are required by plants in trace amounts it is relatively rare that one of them will be unavailable in the environment.The plant micronutrients include:

***Iron*** which is essential for photosynthesis, as well as enzymatic processes.

***Molybdenums*** an enzymatic cofactor and involved in amino acid ***synthesis***.

***Borons*** involved in cell division, ion transport, flowering, and amino acid metabolisms.

***Coppers*** involved in photosynthesis, enzymatic processes, production of grains and lignin synthesis.

***Manganeses*** involved in photosynthesis and chloroplast construction.

***Sodium*** is ubiquitous and involved in a large number of biochemical processes.

***Zinc*** is involved in many enzymatic processes, as well as DNA transcription into RNA.

***Nickel*** in higher plants Ni is required by the enzyme urease to remove potentially toxic levels of urea.

***Chlorine (as chloride)*** is involved in photosynthesis, the maintenance of ionic balance, and cell osmotic mechanisms.

***Cobalt*** known to be beneficial to some plants, but is also essential for nitrogen fixation by bacteria.

#### Other Elements

Other Elements are also important. ***Selenium*** is not essential to plants, but it is to animals hence a deficiency in plants can be manifested upstream in animal health. ***Silicon*** functions to impart structural strength to plants. ***Aluminum*** is generally toxic to plants. ***Vanadium*** may be required by plants at extremely low levels, possibly as a useful substitute for cobalt.

In finishing this section it is important to note that for any of the plant nutrients listed above more is not necessarily better. There is an optimal range of use either side of which is likely toxic to the plant. Especially in the cases of the micronutrients erring on the side of using too little is likely better since these elements are needed in such minute quantities to begin with.

## 16.3 Pesticides

***Pesticides*** are substances that kill or inhibit insects or animals from damaging or destroying plants. Subgroups of this main category are insecticides and rodenticides. The use of pesticides is always a controversial topic, but it is probably not possible to feed a human population even half the size of what it is today without using pesticides.

The focus here is upon insecticide which can be split into a number of categories, two being naturally occurring and synthetic (man-made). There is an understandable inclination to believe that naturally occurring insecticides are inherently better or safer; however, this is not necessarily true. For example, why would anyone assume a synthetic rattlesnake venom was any more lethal than that of the rattlesnake itself?

Here we will look at the most important, broad classes of chemistries used as pesticides after a brief look at historical approaches.

### Historical Approaches to Insecticides

Before the advent of environmental concerns insect control was somewhat of a brute force chemical approach. Compounds containing arsenic, lead, and mercury were popular, as well as derivatives of thallium, boron, copper, and various heavy metals. Fluorides, such as sodium and potassium fluoride, were also used. Although they offer an effective approach, there are a number of problems with all of these compound families. First of all they tend to be broadly toxic to many organisms, including humans. Second they are persistent in the environment. For example, arsenic is arsenic. It never breaks down into simpler or safer materials. Third, they often require use in large amounts to be effective.

### Chemical Classes of Insecticides

There are examples of naturally occurring and synthetic insecticides within some of these families of compounds. At the end of the descriptions there is a table (Table 16.2) that reflects the persistence of various insecticides in the environment. This is followed by a very brief summary of what chemically happens to chemicals after exposure to the elements, mainly to water and oxygen.

### Organochlorine Compounds

Chemical Structure 16.5 shows the chemical structures of some of the most important members of this family of compounds, and there are many members.

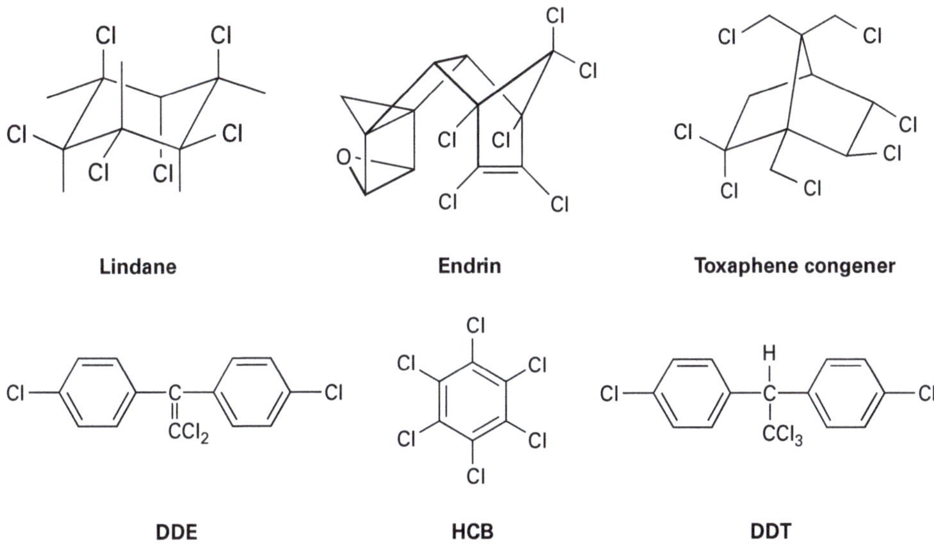


**CHEMICAL STRUCTURE 16.5.** Some Important Chlorinated Organic Pesticides.

The poster child here is DDT (dichlorodiphenyltrichloroethane). DDT is well known for its effect in thinning of the eggshells of birds of prey. Apparently it inhibits the transport of calcium ions that are required to make the shells physically strong. Since the banning of DDT's agricultural use in 1972 in the United States, birds of prey population numbers have rebounded dramatically. This is an excellent example of an unknown side effect from the usage of chemicals, naturally occurring or synthetic. There are many others.

This class of compounds is mainly of historical interest since few of them remain in the pesticide arsenal of chemicals.

**Mode of Action** How the chlorinated organic insecticides act as insecticides is not well understood. The balance of $Na^+$ and $K^+$ ions appears to be disrupted in the axons of neurons in some cases, but there is more to it than just that. Regardless of the details the neurochemistry of respiration is affected in a terminal manner for the insect.

## Organophosphates and Phosphonates

A charter member of this class of insecticides is diazinon, which especially toxic towards ants. It is just one of a family of closely related compounds, a few of which are shown in Chemical Structure 16.6.

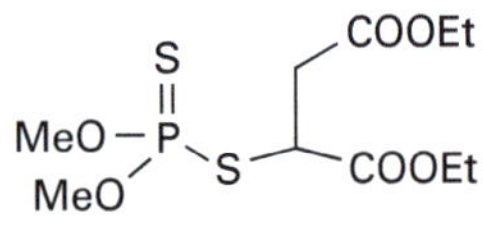

**Malathion**

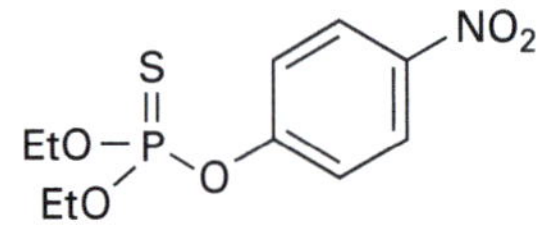

**Parathion**

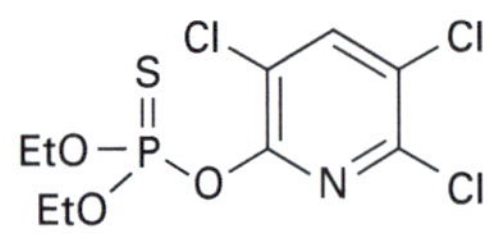

**Chlorpyrifos**

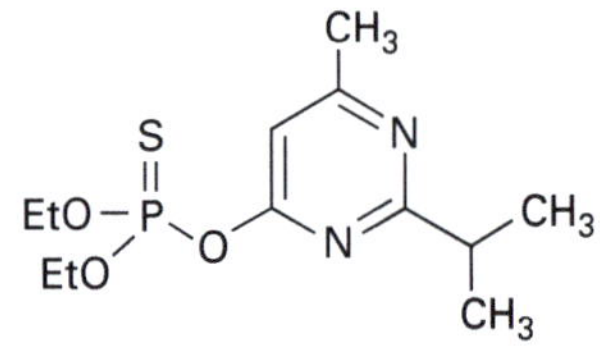

**Diazinon**

**CHEMICAL STRUCTURE 16.6.** Some Common Organophosphonate Pesticides.

The reader may note that doubly bonded C=S bond in some of these structures, basically replacing C=O. As alluded to previously in this book as members of Group 16 oxygen and sulfur have similar properties, but not identical. In general the S is exchanged for O upon exposure to water and oxygen in the environment.

Compared with the chlorinated organic class of compounds just described, the organophosphorus compounds are attractive because they are not persistent in the environment as they degrade quickly (see Table 16.2 below). A downside is that they are the most acutely toxic class to vertebrates. In fact, they are closely related structurally to known and deadly nerve gases such as *Sarin*. They were initially discovered to be good insecticides by the Germans in WWII, not as potential nerve agents, but rather as a replacement for the naturally occurring insecticide nicotine, which was in short supply.

### Mode of Action

These insecticides function by inhibiting crucial enzymes in the central nervous system, especially cholinesterase. Eventually this inhibition leads to paralysis and death of the organism.

## Carbamates

The compound carbaryl, made famous under the trade name of SEVIN, is the key representative of this broad class of insecticides. Carbaryl and many of its stablemates (Chemical Structure 16.7) are attractive because they have an unusually broad spectrum of active in terms of type of insect affected. In addition, they collectively have very low toxicities in mammals.

### Mode of Action

Similar to the phosphorus esters the carbamates function by inhibiting crucial enzymatic processes in the nervous system, although the exact enzymes affected are often different than in the phosphorus cases discussed for the prior categories.

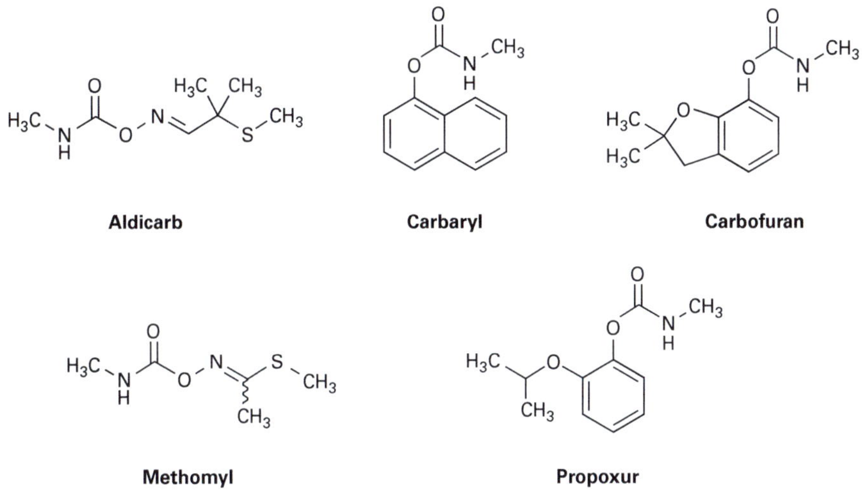


**CHEMICAL STRUCTURE 16.7.** Common Carbamate Insecticides.

## Pyrethroids

This interesting class of naturally occurring and synthetically derived chemicals have complex and interesting organic structures. The parent pyrethrin's structure is shown in Chemical Structure 16.8.

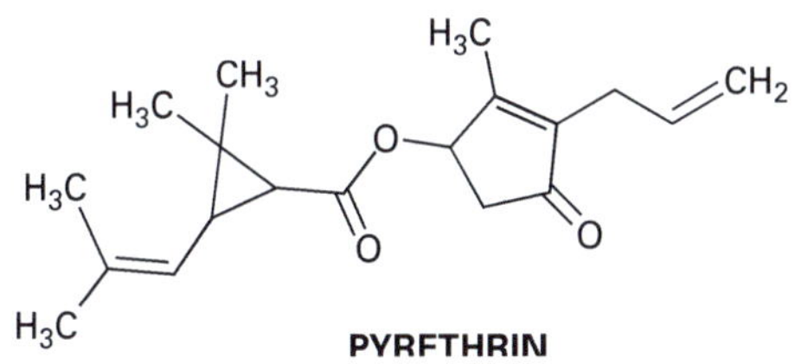


**CHEMICAL STRUCTURE 16.8.** Pyrethrin.

Since the late 1940's a number of different generations of members of this family have been synthesized. The latest examples are especially promising because they are effective at very low levels of application. They are non-persistent in the environment.

### Mode of Action

Pyrethroids act in a similar to DDT, that is, they interrupt the $Na^{+}$ ion channeling process as axonic neurochemical poisons.

## Nicotinoids

This relatively new class of insecticides is parented by the not so new compound nicotine of cigarette smoking infamy (Chemical Structure 16.9):

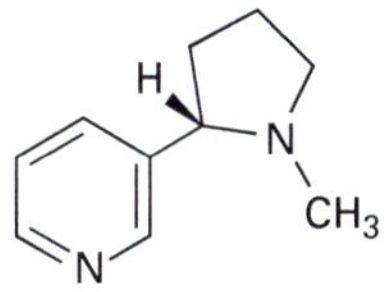

**CHEMICAL STRUCTURE 16.9.** Nicotine.

### Mode of Action

These compounds also act by disrupting the central nervous system basically by blocking important neurotransmitter sites.

There are many other insecticidal chemicals that have been and are been explored. It is outside the scope of this book to expand upon this subject, as interesting as it is.

## 16.4 Herbicide Chemistries

Chemicals that kill or inhibit the growth of plants. Probably the most famous and controversial member of this group is glyphosate, widely known under the trade name ROUNDUP (Chemical Structure 16.10).

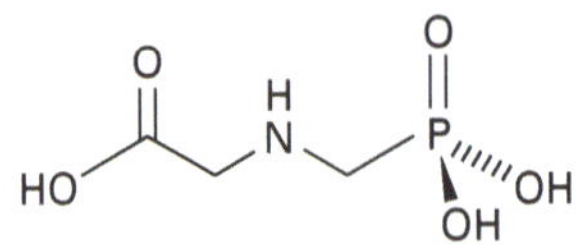


CHEMICAL STRUCTURE 16.10. Glyphosate.

### Half Lives of Selected Insecticides and Herbicides

One of the big issues with DDT that was not recognized until very late is its persistence in the environment. DDT has not been used in the United States for 45 years, and yet it still persists in the environment as well as fat tissues of living organisms, including humans. When organic pesticides and herbicides do degrade, quickly or slowly, they form smaller and smaller molecular pieces which may be more or less harmful than the starting material. Ideally, the final products are carbon dioxide, water and minerals. Persistence is a long term effect and is measured by ***half-life*** ($t_{1/2}$). Basically, half-life is defined as how long it takes for half of a quantity of a chemical to react and decompose in the environment (Figure 16.2).

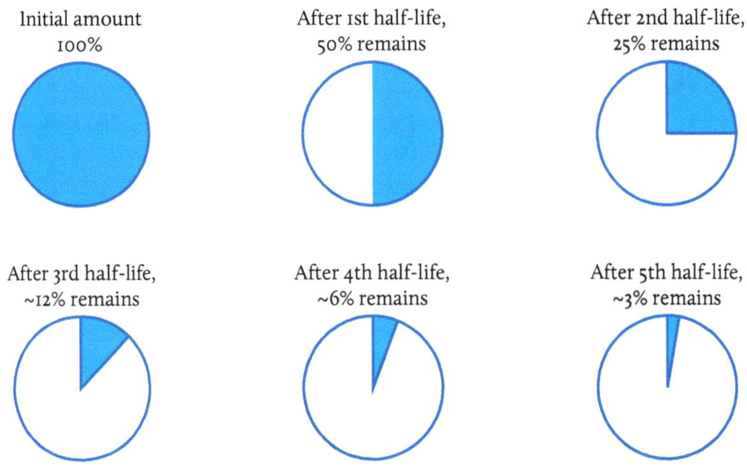


FIGURE 16.2. Half Life Degradation Example.

Table 16.2 contains the basic mechanistic fates of pesticides in the environment.

TABLE 16.2 **Fates of Pesticides and Herbicides in the Environment**

| *Environmental Factor* | *Effect* |
|---|---|
| *Microorganisms* | *Bacteria and fungi degrade organic compounds into smaller products* |
| *Sunlight* | *Photodegradation, primarily photoxidation coupled to hydrolysis* |
| *Water and Oxygen* | *Hydrolysis coupled with oxidation to break down molecules* |
| *Plant/Animal Metabolism* | *Plants can uptake, and animas ingest chemicals wherein the molecules are stored or metabolized.* |
| *Adsorption* | *Chemicals can be adsorbed* ***onto*** *the surfaces of minerals and organic materials* |
| *Absorption* | *Chemicals can be absorbed* ***into*** *the interior of minerals and organic materials* |

Table 16.3 gives a listing of the half-lives and water solubilities of a few common insecticides.

TABLE 16.3 **Soil Half Lives of Common Pesticides and Herbicides**

| ***Compound*** | ***Soil Half Life*** | ***Water Solubility (ppm)*** |
|---|---|---|
| *DDT* | *3-10 years* | *0.006* |
| *Chlordane* | *2-4 years* | *0.1* |
| *Parathion* | *5 days* | *57* |
| *Chlorpyrifos (Lorsban)* | *30-90 days* | *1* |
| *Aldicarb* | *70 days* | *6000* |
| *Carbaryl (Sevin)* | *10 Days* | *40* |
| *Permethrin* | *30 days* | *0.04* |
| *Atrazine* | *60 days* | *33* |
| *Alachlor* | *15 days* | *242* |

http://slideplayer.com/slide/5714325/18/images/18/Soil+Half-life+of+Insecticides+and+Solubility.jpg

It is unreasonable to expect any biologically active chemical to be active just upon a specific organism, therefore side effects are just part of the game. In addition there is no reason to think, a priori, that the decomposition products from a chemical are less harmful than the starting material. The situation is complex.

## Mechanisms of Insecticide and Herbicide Decompositions in the Environment

Over the past 50 years or so the fates of and effects upon the environment of chemicals has come to the fore, and with good reason. Using the term environment to encompass the whole earth, what happens to insecticides when they are released? This is a complex and difficult question to answer. However, working chemists, biochemists, and environmentalists can make predictions as to the likelihoods of mechanisms of degradation.

The reader is referred to Chapter 14, *Organic Chemistry—Reactions*, which describes the basic reaction mechanisms for organic molecules.

Two of the most common chemicals on earth are water and oxygen. They are both very reactive species. In addition, heat and sunlight are ubiquitous. Combine the four of them and it is sort of amazing any organic chemical can persist at all! Add in microorganisms which are always looking for food sources and you would seem to have an unstoppable quintet. And you do, geologically speaking, but our time frame as humans is very short.

In the environment organic compounds are especially reactive by two mechanisms, hydrolysis and oxidation. These mechanisms are usually happening in tandem and/or in combination. An example of a *possible* mechanism of each (there are many, many possibilities) will suffice as examples here.

### Hydrolysis of Carbaryl

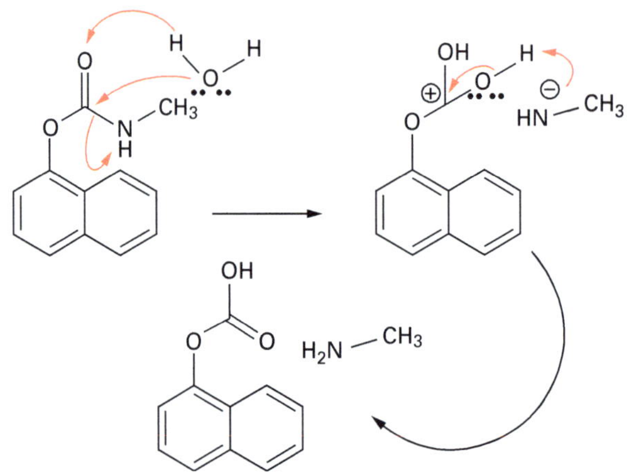


**CHEMICAL STRUCTURE 16.11.** Hydrolysis of Carbaryl.

Here the red arrows represent electron movements. The water oxygen nulceophilically attacks the carbonyl carbon of the amide in Carbaryl, followed by electron rearrangements and an H transfer to form the corresponding carboxylic acid and a primary amine. This is a case where the products may be more toxic than the starting material. Subsequent reactions to the newly formed compound can be imagined.

**PRACTICE** Looking at the structure of DDT, and referring back to Chapter 14 and nucleophilc reactions, what might be a first step in the hydrolysis of DDT?

### Glyphosate Oxidation

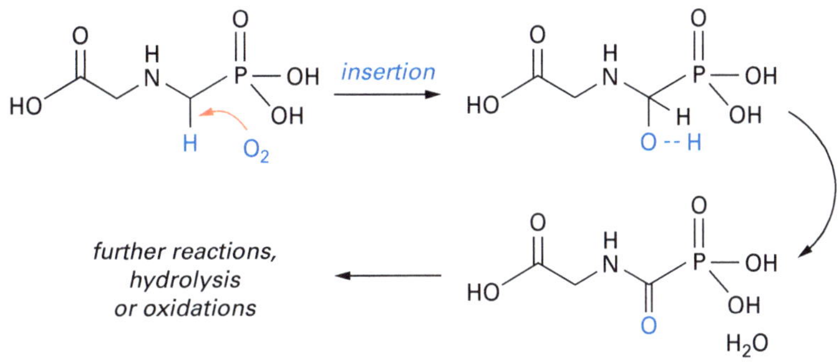


**CHEMICAL STRUCTURE 16.12.** Glyphosate Oxidation.

In this case an oxygen molecule inserts into a C – H bond to form an unstable hydroperoxide intermediate. This is followed by a transfer of a second hydrogen to the terminal O-H group to form a water molecule and a new carbonyl bond. This new molecule can then undergo more oxidation reactions, or more likely, hydrolyses.

**PRACTICE** Look up the structure of Atrazine. What might be a first step in its oxidation?

These are just examples as there are many more possible mechanisms of decomposition for glyphosate. The same is true for any organic compound. Once again as mentioned above, it is a mistake to assume a compounds decomposition products are more benign than the parent compound. The situation has a great many variables, most of which will result in affecting the overall properties of the compound under study.

## 16.5 Plant Hybridizations, Genetic Modifications, and Organic Foods

There has been a great deal of interest in organic foods of many different kinds. It is worth pointing out that virtually all foods are organic by definition.; However, the term is loosely connected to the production of foodstuffs with no or minimal amounts of manmade chemicals. Many times an associated suggestion is to avoid chemical treatments of any kind anywhere along the sequence of events from seed to consumption. This approach is commendable, but should be understood on the basis of risk versus reward. As mentioned above, it will simply not be possible to feed a world population of many billions without using chemicals.

### GMOs (Genetically Modified Organisms)

There are few topics involving food and agriculture as controversial as that of GMOs. Man has been performing genetic modifications of plants and animals in the interest of superior food production for thousands of years. Standard hybridization procedures by definition involves genetic manipulation. Where the topic gets more interesting is with the incorporation of genetic materials that cross species boundaries. Here the discussion becomes much more clouded, especially in terms of ethics and morals. There are no easy answers to this situation, but what will almost surely occur is the science. It always does. We just have to figure out what to do to control the various technologies are used.

# Chapter 16 Summary

*Agricultural Chemistry:* the study of the usage of chemicals in plant and animal husbandry, as well as the toxicological properties and fate of the same in the environment.

## Soil Chemistry and Nutrients

*Soils:* Complex mixtures of inorganic and organic materials. Adsorbed and absorbed water is crucial to plant growth as is the presence of gases, especially oxygen, carbon dioxide and nitrogen.

*Basic Plant Structures:* The photosynthetic cycle is complex, and results in many different carbohydrate products starting basically with carbon dioxide and water. Two of the primary products are **cellulose** and **lignins** high molecular weight polymers based upon structures 16.13a and b, respectively.

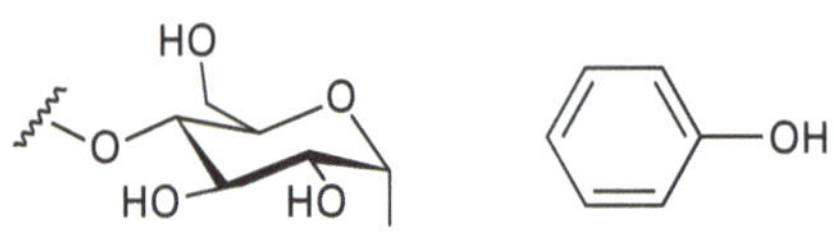


a b

CHEMICAL STRUCTURE 16.13A AND B. Basic Plant Structures.

*Primary Plant Nutrients:* N, P and K are needed in large quantities.

- *Nitrogen:* Leaf and seed production, N is a key component of chlorophyll.
- *Phosphorus:* Key to cell energetics. Encourages root development and flowering.
- *Potassium:* Disease prevention, fruit quality, protein synthesis aid.

*Secondary Plant Nutrients:* Needed in lesser amounts.

- *Sulfur:* A key part of many biochemical compounds and processes
- *Magnesium:* Key to photosynthesis, enzymatic processes and electrolyte balances.
- *Calcium:* Important to nutrient uptake, enzymatic processes, photosynthesis, and cell wall structures.

*Plant Micronutrients:* Needed in trace amounts.

- Iron, Molybdenum, Boron, Copper, Manganese, Sodium, Zinc, Nickel, Chlorine, Cobalt

*Pesticides:* Substances that kill or inhibit insects or animals from damaging or destroying plants.

Organochlorine Class

Organophosphate and phosphonate class

Crabamates

*Herbicides:* Substances that kill or inhibit the growth of plants.

*Half Lives of Chemicals:* How long they will last in the environment.

## Mechanisms of Insecticide and Herbicide Decompositions in the Environment

***Plant Hybridizations, Genetic Modifications, and Organic Foods***

# CHAPTER 16 EXERCISES

1. What comprises soil?
2. What is a mineral?
3. Two primary biological components of plants are cellulose and lignins. What simple chemical components underlie their basic structure? What are their functions in a plant?
4. The three primary plant nutrients are N, P and K.

   a. What is the function of N? Where does the N come from?

   b. What is the function of P? Where does the P come from?

   c. What is the function of K? Where does the K come from?
5. What are the secondary plant nutrients? What are their basic biological functions?
6. Silicon and aluminum are the $2^{nd}$ and $3^{rd}$ most common elements on earth, and yet they are not important in plant or animal biochemistry. In fact, Al can be toxic to plants. Consider their respective positions in the periodic table and related properties. Why might this be the case?
7. What is a pesticide?
8. What is an herbicide?
9. What is the most common mechanism of action of various groups of pesticides?
10. Phosphonates are an important group of pesticides and herbicides. What is the key structural component of this group of compounds?
11. Carbamates are an important group of pesticides and herbicides. What is the key structural component of this group of compounds?

12. Consider the structure of the long-known pesticide Malathion. Using the information contained in Chapter 14 predict mechanisms for the reactions of this compound in the environment. Also what are the products?

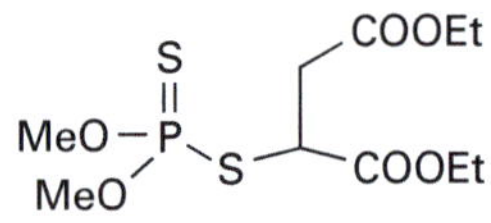


CHEMICAL STRUCTURE 16.14. Malathion.

13. Consider the structure of Aldicarb. Using the information contained in Chapter 14 predict mechanisms for the reactions of this compound in the environment. Also what are the products?

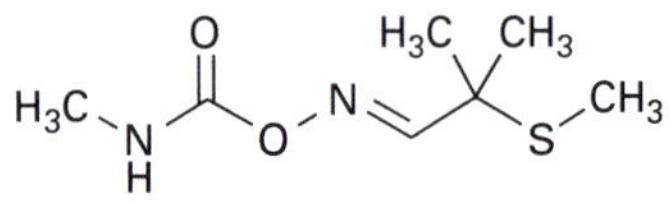


CHEMICAL STRUCTURE 16.15. Aldicarb.

14. Consider the structure of the Pyrethrin. Using the information contained in Chapter 14 predict mechanisms for the reactions of this compound in the environment. Also what are the products?

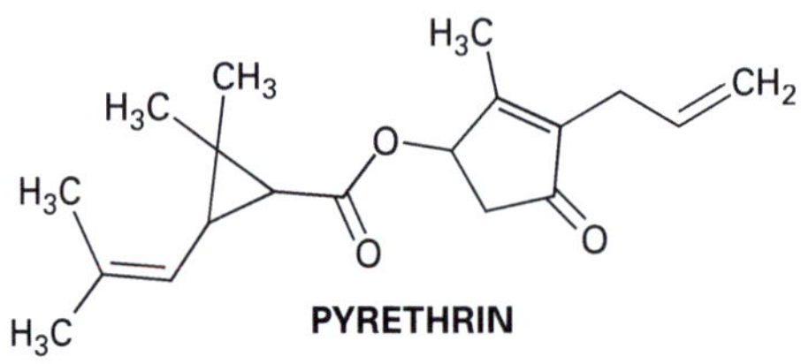


CHEMICAL STRUCTURE 16.16. Pyrethrin.

15. Of the three primary plant nutrient elements
16. Ammonia, $NH_3$, is one of the most important compounds prepared.
    a. How is it synthesized from $N_2$ gas?
    b. How is the key fertilizer component $NH_4NO_3$ prepared from $NH_3$? (Hint: there is more than one synthetic step.)
17. The second primary nutrient for plants is phosphorous. Where does this come from?
18. The third primary plant nutrient is potassium. Where does this come from?
19. Which of the primary and secondary plant nutrients are non-renewable resources?
20. Can any of the primary or secondary plant nutrients be safely replaced by another element?

21. Consider the long term environmental stability of DDT in relation to its molecular structure:

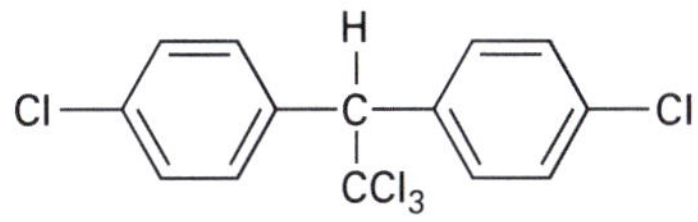


**CHEMICAL STRUCTURE 16.17.** Consider the Long-Term Environmental Stability of DDT in Relation to Its Molecular Structure.

22. Why might this structure be less sensitive to degradation that those shown above in problems 12-14? !Hint: think about the three dimensional characteristics of this molecule and the accessibility of sites that might be susceptible to reaction.}
23. The first degradation product from DDT is DDE:

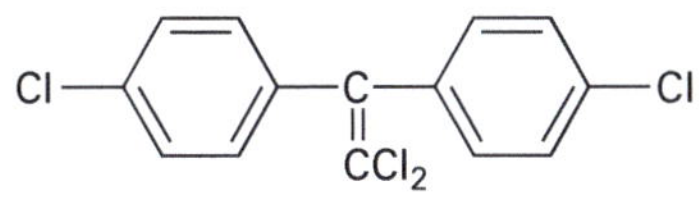


**CHEMICAL STRUCTURE 16.18.** The First Degradation Product from DDT is DDE.

24. Suggest a mechanism in getting from DDT to DDE.
25. What might be the next step in DDE's degradation in the environment?
26. What are the differences between classical plant hybridization and genetic modification techniques?
27. What are the CRISPR gene editing technologies based upon?

# Credits

Fig. 16.1: Source: https://commons.wikimedia.org/wiki/File:Geode_mineral.png.

Chemical Structure 16.1: Source: http://beerandwinejournal.com/wp-content/uploads/2014/10/Amylose_3Dprojection.corrected-1024x492.png.

Chemical Structure 16.2: Copyright © 2007 by Karol Głąb, (CC BY-SA 3.0) at https://commons.wikimedia.org/wiki/File:Lignin_structure.svg.

Chemical Structure 16.7: Source: http://pubs.rsc.org/services/images/RSCpubs.ePlatform.Service.FreeContent.ImageService.svc/ImageService/Articleimage/2016/NJ/c5nj03524d/c5nj03524d-f1_hi-res.gif.

Chemical Structure 16.8: Source: https://www.the-piedpiper.co.uk/graphics1/pyrethrin.jpg.

Chemical Structure 16.9: Source: http://www.apsnet.org/edcenter/intropp/topics/Article%20Images/defensesfig21.jpg.

Chemical Structure 16.10: Source: http://www.scielo.br/img/revistas/jbchs/v26n3//0103-5053-jbchs-26-03-0411-gf01.jpg.

Fig. 16.2: Source: http://npic.orst.edu/images/fsimages/halflife-pie.jpg.

Table 16.3: Source: http://slideplayer.com/slide/5714325/18/images/18/Soil+Half-life+of+Insecticides+and+Solubility.jpg.

# Fundamental Constants[1]

| | |
|---|---|
| **Avogadro's Number (NA)** | *$6.02 \times 10^{33}$ atm/mol* |
| **Gas Constant (R)** | *0.082 L.atm/mol.K =<br>1.99 cal/mol.K =<br>8.31 J/mol.K* |
| **Boltzman Constant (k)** | *$1.38 \times 10^{-23}$ J/K* |
| **Pi** | *3.14* |
| **Electron's charge €** | *$1.60 \times 10^{-19}$ C* |
| **Faraday Constant (F)** | *$9.65 \times 104$ C/mol* |
| **Planck's Constant (h)** | *$6.63 \times 10^{-34}$ J-s* |
| **Speed of light** | *$3.00 \times 10^{8}$ m/s* |

*[1]Fundamental constants are available from the National Institute of Standards (NIST): http// physics.nist/gov*

APPENDIX 2

# Conversion Factors

| | |
|---|---|
| ***Length:*** *Metric unit: meter (m)* | ***Mass:*** *Metric unit: gram (g), kilogram (kg)* |
| 1 km = 0.621 mi | 1 kg = 1000 g |
| 1 mi = 5289 ft = 1.61 km | 1 lb = 454 gm = 16 oz |
| 1 in. = 2.54 cm | 1 amu = $1.66 \times 10^{-24}$ g |
| 1 cm = 0.394 in | |
| 1Å = $1 \times 10^{-10}$ m | |
| ***Energy:*** *Metric unit: Joule (J)* | ***Temperature:*** *Metric units: Celsius ©; Kelvin (K)* |
| 1 J = 0.239 cal = 1 kg · $m^2/s^2$ | −273 °C = 0 K = −459 °F |
| 1 cal = 4.184 J | K = °C + 273 |
| 1 J = 1 Coulomb © × 1 Volt (V) | °C = 5/9 (°F − 32) |
| 1eV = $1.6 \times 10^{-19}$ J | °F = °C + 32 |
| ***Volume:*** *Metric unit: Liter (L) = 1000 mL* | ***Pressure:*** *Metric unit: Pascal (Ps)* |
| 1L = 1.06 qt | 1 Ps = 1 N/ $m^2$ = 1kg/m · $s^2$ |
| 1 gal = 4 qt = 3.79 L | 1atm = 101 Ps = 760 torr = 760 mm Hg |
| $1cm^3$ = 1 mL | 1atm = 1.01 bar = 14.7 psi |

# Aqueous Reduction Potentials (25°C)

| Acidic Solution | | | E° (V) |
| --- | --- | --- | --- |
| $F_2(g) + 2\ e^-$ | → | $2\ F^-(aq)$ | +2.87 |
| $Co^{3+}(aq) + e^-$ | → | $Co^{2+}(aq)$ | +1.82 |
| $Pb^{4+}(aq) + 2\ e^-$ | → | $Pb^{2+}(aq)$ | +1.8 |
| $H_2O_2(aq) + 2\ H^+(aq) + 2\ e^-$ | → | $2\ H_2O$ | +1.77 |
| $NiO_2(s) + 4\ H^+(aq) + 2\ e^-$ | → | $Ni^{2+}\ (aq) + 2\ H_2O$ | +1.7 |
| $PbO_2(s) + SO_4^{2-}(aq) + 4\ H^+(aq) + 2\ e^-$ | → | $PbSO_4(s) + 2\ H_2O$ | +1.69 |
| $Au^+(aq) + e^-$ | → | $Au(s)$ | +1.68 |
| $2\ HClO(aq) + 2\ H^+(aq) + 2\ e^-$ | → | $Cl_2(g) + 2\ H_2O$ | +1.63 |
| $Ce^{4+}(aq) + e^-$ | → | $Ce^{3+}(aq)$ | +1.61 |
| $NaBiO_3(s) + 6\ H^+(aq) + 2\ e^-$ | → | $Bi^{3+}(aq) + Na^+(aq) + 3\ H_2O$ | +~1.6 |
| $MnO_4^-(aq) + 8\ H^+(aq) + 5\ e^-$ | → | $Mn^{2+}(aq) + 4\ H_2O$ | +1.51 |
| $Au^{3+}(aq) + 3\ e^-$ | → | $Au(s)$ | +1.5 |
| $ClO^{3-}(aq) + 6\ H^+(aq) + 5\ e^-$ | → | $1/2\ Cl_2(g) + 3\ H_2O$ | +1.47 |
| $BrO^{3-} + 6\ H^+(aq) + 6\ e^-$ | → | $Br^-(aq) + 3\ H_2O$ | +1.44 |
| $Cl_2(g) + 2\ e^-$ | → | $2\ Cl^-(aq)$ | +1.36 |
| $Cr_2O_7^{2-} + 14\ H^+(aq) + 6\ e^-$ | → | $2\ Cr^{3+}(aq) + 7\ H_2O$ | +1.33 |
| $N_2H_5^+(aq) + 3\ H^+(aq) + 2\ e^-$ | → | $2\ NH^{4+}(aq)$ | +1.24 |
| $MnO_2(s) + 4\ H^+(aq) + 2\ e^-$ | → | $Mn^{2+}(aq) + 2\ H_2O$ | +1.23 |
| $O_2(g) + 4\ H^+(aq) + 4\ e^-$ | → | $2H_2O$ | +1.2 |
| $Pt^{2+}(aq) + 2\ e^-$ | → | $Pt(s)$ | +1.2 |
| $IO^{3-}(aq) + 6\ H^+(aq) + 5\ e^-$ | → | $1/2\ I_2(aq) + 3\ H_2O$ | +1.20 |
| $ClO_4^-(aq) + 2\ H^+(aq) + 2\ e^-$ | → | $ClO_3^-(aq) + H_2O$ | +1.19 |
| $Br_2(l) + 2\ e^-$ | → | $2\ Br^-(aq)$ | +1.07 |
| $AuCl_4^- + 3\ e^-$ | → | $Au(s) + 4\ Cl^-(aq)$ | +1 |

| Acidic Solution | | | E° (V) |
|---|---|---|---|
| $Pd^{2+}(aq) + 2\ e^-$ | ⟶ | $Pd(s)$ | +0.99 |
| $NO_3^-(aq) + 4\ H^+(aq) + 3\ e^-$ | ⟶ | $NO(g) + 2\ H_2O$ | +0.96 |
| $NO_3^-(aq) + 3\ H^+(aq) + 2\ e^-$ | ⟶ | $HNO_2(aq) + H_2O$ | +0.94 |
| $2\ Hg^{2+}(aq) + 2\ e^-$ | ⟶ | $Hg_2^{2+}(aq)$ | +0.92 |
| $Hg^{2+}(aq) + 2\ e^-$ | ⟶ | $Hg(l)$ | +0.86 |
| $Ag^+(aq) + e^-$ | ⟶ | $Ag(s)$ | +0.80 |
| $Hg_2^{2+}(aq) + 2\ e^-$ | ⟶ | $2\ Hg(l)$ | +0.79 |
| $Fe^{3+}(aq) + e^-$ | ⟶ | $Fe^{2+}(aq)$ | +0.77 |
| $SbCl_6^-(aq) + 2\ e^-$ | ⟶ | $SbCl_4^-(aq) + 2\ Cl^-(aq)$ | +0.75 |
| $[PtCl_4]^{2-}(aq) + 2\ e^-$ | ⟶ | $Pt(s) + 4\ Cl^-(aq)$ | +0.73 |
| $O_2(g) + 2\ H^+(aq) + 2\ e^-$ | ⟶ | $H_2O_2(aq)$ | +0.68 |
| $[PtCl_6]^{2-}(aq) + 2\ e^-$ | ⟶ | $[PtCl_4]^{2-}(aq) + 2\ Cl^-(aq)$ | +0.68 |
| $H_3AsO_4(aq) + 2\ H^+(aq) + 2\ e^-$ | ⟶ | $H_3AsO_3(aq) + H_2O$ | +0.58 |
| $I_2(s) + 2\ e^-$ | ⟶ | $2\ I^-(aq)$ | +0.54 |
| $TeO_2(s) + 4\ H^+(aq) + 4\ e^-$ | ⟶ | $Te(s) + 2\ H_2O$ | +0.53 |
| $Cu^+(aq) + e^-$ | ⟶ | $Cu(s)$ | +0.52 |
| $[RhCl_6]^{3-}(aq) + 3\ e^-$ | ⟶ | $Rh(s) + 6\ Cl^-(aq)$ | +0.44 |
| $Cu^{2+}(aq) + 2\ e^-$ | ⟶ | $Cu(s)$ | +0.34 |
| $HgCl_2(s) + 2\ e^-$ | ⟶ | $2\ Hg(l) + 2\ Cl^-(aq)$ | +0.27 |
| $AgCl(s) + e^-$ | ⟶ | $Ag(s) + Cl^-(aq)$ | +0.22 |
| $SO_4^{2-}(aq) + 4\ H^+(aq) + 2\ e^-$ | ⟶ | $SO_2(g) + 2\ H_2O$ | +0.2 |
| $SO_4^{2-}(aq) + 4\ H^+(aq) + 2\ e^-$ | ⟶ | $H_2SO_3(g) + H_2O$ | +0.17 |
| $Cu^{2+}(aq) + e^-$ | ⟶ | $Cu^+(aq)$ | +0.15 |
| $Sn^{4+}(aq) + 2\ e^-$ | ⟶ | $Sn^{2+}(aq)$ | +0.15 |
| $S(s) + 2\ H^+(aq) + 2\ e^-$ | ⟶ | $H_2S(aq)$ | +0.14 |
| $AgBr(s) + e^-$ | ⟶ | $Ag(s) + Br^-(aq)$ | +0.07 |
| **$2\ H^+(aq) + 2\ e^-\ H_2(g)$ (reference electrode)** | | | **0** |
| $N_2O(g) + 6\ H^+(aq) + H2O + 4\ e^-$ | ⟶ | $2\ NH_3OH^+(aq)$ | −0.05 |
| $Pb^{2+}(aq) + 2\ e^-$ | ⟶ | $Pb(s)$ | −0.13 |
| $Sn^{2+}(aq) + 2\ e^-$ | ⟶ | $Sn(s)$ | −0.14 |
| $AgI(s) + e^-$ | ⟶ | $Ag(s) + I^-(aq)$ | −0.15 |
| $[SnF_6]^{2-}(aq) + 4\ e^-$ | ⟶ | $Sn(s) + 6\ F^-(aq)$ | −0.25 |
| $Ni^{2+}(aq) + 2\ e^-$ | ⟶ | $Ni(s)$ | −0.25 |
| $Co^{2+}(aq) + 2\ e^-$ | ⟶ | $Co(s)$ | −0.28 |
| $Tl^+(aq) + e^-$ | ⟶ | $Tl(s)$ | −0.34 |
| $PbSO_4(s) + 2\ e^-$ | ⟶ | $Pb(s) + SO_4^{2-}(aq)$ | −0.36 |

| Acidic Solution | | | E° (V) |
|---|---|---|---|
| $Se(s) + 2\ H^+(aq) + 2\ e^-$ | $\longrightarrow$ | $H_2Se(aq)$ | –0.4 |
| $Cd^{2+}(aq) + 2\ e^-$ | $\longrightarrow$ | $Cd(s)$ | –0.40 |
| $Cr^{3+}(aq) + e^-$ | $\longrightarrow$ | $Cr^{2+}(aq)$ | –0.41 |
| $Fe^{2+}(aq) + 2\ e^-$ | $\longrightarrow$ | $Fe(s)$ | –0.44 |
| $2\ CO_2(g) + 2\ H^+(aq) + 2\ e^-$ | $\longrightarrow$ | $(COOH)_2(aq)$ | –0.49 |
| $Ga^{3+}(aq) + 3\ e^-$ | $\longrightarrow$ | $Ga(s)$ | –0.53 |
| $HgS(s) + 2\ H^+(aq) + 2\ e^-$ | $\longrightarrow$ | $Hg(l) + H_2S(g)$ | –0.72 |
| $Cr^{3+}(aq) + 3\ e^-$ | $\longrightarrow$ | $Cr(s)$ | –0.74 |
| $Zn^{2+}(aq) + 2\ e^-$ | $\longrightarrow$ | $Zn(s)$ | –0.76 |
| $2H_2O(l) + 2\ e^-$ | $\longrightarrow$ | $H_2(g) + 2OH^-(aq)$ | –0.83 |
| $Cr^{2+}(aq) + 2\ e^-$ | $\longrightarrow$ | $Cr(s)$ | –0.91 |
| $Mn^{2+}(aq) + 2\ e^-$ | $\longrightarrow$ | $Mn(s)$ | –1.18 |
| $V^{2+}(aq) + 2\ e^-$ | $\longrightarrow$ | $V(s)$ | –1.18 |
| $Zr^{4+}(aq) + 4\ e^-$ | $\longrightarrow$ | $Zr(s)$ | –1.53 |
| $Al^{3+}(aq) + 3\ e^-$ | $\longrightarrow$ | $Al(s)$ | –1.66 |
| $H_2(g) + 2\ e^-$ | $\longrightarrow$ | $2\ H^-(aq)$ | –2.25 |
| $Mg^{2+}(aq) + 2\ e^-$ | $\longrightarrow$ | $Mg(s)$ | –2.37 |
| $Na^+(aq) + e^-$ | $\longrightarrow$ | $Na(s)$ | –2.71 |
| $Ca^{2+}(aq) + 2\ e^-$ | $\longrightarrow$ | $Ca(s)$ | –2.87 |
| $Sr^{2+}(aq) + 2\ e^-$ | $\longrightarrow$ | $Sr(s)$ | –2.89 |
| $Ba^{2+}(aq) + 2\ e^-$ | $\longrightarrow$ | $Ba(s)$ | –2.9 |
| $Rb^+(aq) + e^-$ | $\longrightarrow$ | $Rb(s)$ | –2.93 |
| $K^+(aq) + e^-$ | $\longrightarrow$ | $K(s)$ | –2.93 |
| $Li^+(aq) + e^-$ | $\longrightarrow$ | $Li(s)$ | –3.03 |

| Basic Solution | | | E° (V) |
|---|---|---|---|
| $ClO^-(aq) + H_2O + 2\ e^-$ | $\longrightarrow$ | $Cl^-(aq) + 2\ OH^-(aq)$ | 0.89 |
| $OOH^-(aq) + H_2O + 2\ e^-$ | $\longrightarrow$ | $3\ OH^-(aq)$ | 0.88 |
| $2\ NH_2OH(aq) + 2\ e^-$ | $\longrightarrow$ | $N_2H_4(aq) + 2\ OH^-(aq)$ | 0.74 |
| $ClO_3^-\ (aq) + 3\ H_2O + 6\ e^-$ | $\longrightarrow$ | $Cl^-(aq) + 6\ OH^-(aq)$ | 0.62 |
| $MnO_4^-(aq) + 2\ H_2O + 3\ e^-$ | $\longrightarrow$ | $MnO_2(s) + 4\ OH^-(aq)$ | 0.59 |
| $MnO_4^-(aq) + e^-$ | $\longrightarrow$ | $MnO_4^{2-}(aq)$ | 0.56 |
| $NiO_2(s) + 2\ H_2O + 2\ e^-$ | $\longrightarrow$ | $Ni(OH)_2(s) + 2\ OH^-(aq)$ | 0.49 |
| $Ag_2CrO_4(s) + 2\ e^-$ | $\longrightarrow$ | $2\ Ag(s) + CrO_4^{2-}(aq)$ | 0.45 |
| $O_2(g) + 2\ H_2O + 4\ e^-$ | $\longrightarrow$ | $4\ OH^-(aq)$ | 0.4 |

| Basic Solution | | | E° (V) |
|---|---|---|---|
| $ClO_4^-(aq) + H_2O + 2e^-$ | $\longrightarrow$ | $ClO_3^-(aq) + 2\ OH^-(aq)$ | 0.36 |
| $Ag_2O(s) + H_2O + 2e^-$ | $\longrightarrow$ | $2\ Ag(s) + 2\ OH^-(aq)$ | 0.34 |
| $2\ NO_2^-(aq) + 3\ H_2O + 4\ e^-$ | $\longrightarrow$ | $N_2O(g) + 6\ OH^-(aq)$ | 0.15 |
| $N_2H_4(aq) + 2\ H_2O + 2\ e^-$ | $\longrightarrow$ | $2\ NH_3(aq) + 2\ OH^-(aq)$ | 0.1 |
| $[Co(NH_3)_6]^{3+}(aq) + e^-$ | $\longrightarrow$ | $[Co(NH_3)_6]^{2+}(aq)$ | 0.1 |
| $HgO(s) + H_2O + 2e^-$ | $\longrightarrow$ | $Hg(l) + 2\ OH^-(aq)$ | 0.10 |
| $O_2(g) + H_2O + 2\ e^-$ | $\longrightarrow$ | $O_2^-(aq) + OH^-(aq)$ | 0.08 |
| $NO_3^-(aq) + H_2O + 2\ e^-$ | $\longrightarrow$ | $NO_2^-(aq) + 2\ OH^-(aq)$ | 0.01 |
| $MnO_2(s) + 2\ H_2O + 2\ e^-$ | $\longrightarrow$ | $Mn(OH)_2(s) + 2\ OH^-(aq)$ | –0.05 |
| $CrO_4^{2-}(aq) + 4\ H_2O + 3\ e^-$ | $\longrightarrow$ | $Cr(OH)_3(s) + 5\ OH^-(aq)$ | –0.12 |
| $Cu(OH)_2(s) + 2\ e^-$ | $\longrightarrow$ | $Cu(s) + 2\ OH^-(aq)$ | –0.36 |
| $Fe(OH)_3(s) + e^-$ | $\longrightarrow$ | $Fe(OH)_2(s) + OH^-(aq)$ | –0.56 |
| $2\ H_2O + 2\ e^-$ | $\longrightarrow$ | $H_2(g) + 2\ OH^-(aq)$ | –0.83 |
| $2\ NO_3^-(aq) + 2\ H_2O + 2\ e^-$ | $\longrightarrow$ | $N_2O_4(g) + 4\ OH^-(aq)$ | –0.85 |
| $Fe(OH)_2(s) + 2\ e^-$ | $\longrightarrow$ | $Fe(s) + 2\ OH^-(aq)$ | –0.88 |
| $SO_4^{2-}(aq) + H_2O + 2\ e^-$ | $\longrightarrow$ | $SO_3^{2-}(aq) + 2\ OH^-(aq)$ | –0.93 |
| $N_2(g) + 4\ H_2O + 4\ e^-$ | $\longrightarrow$ | $N_2H_4(aq) + 4\ OH^-(aq)$ | –1.15 |
| $[Zn(OH)_4]^{2-}(aq) + 2\ e^-$ | $\longrightarrow$ | $Zn(s) + 4\ OH^-(aq)$ | –1.22 |
| $Zn(OH)_2(s) + 2\ e^-$ | $\longrightarrow$ | $Zn(s) + 2\ OH^-(aq)$ | –1.25 |
| $[Zn(CN)_4]^{2-}(aq) + 2\ e^-$ | $\longrightarrow$ | $Zn(s) + 4\ CN^-(aq)$ | –1.26 |
| $Cr(OH)_3(s) + 3\ e^-$ | $\longrightarrow$ | $Cr(s) + 3\ OH^-(aq)$ | –1.3 |
| $SiO_3^{2-}(aq) + 3\ H_2O + 4\ e^-$ | $\longrightarrow$ | $Si(s) + 6\ OH^+(aq)$ | –1.7 |

APPENDIX 4

# Standard Reduction Potentials (in Volts), 25°C

| Reaction | | | E° |
|---|---|---|---|
| $F_2 + 2e^-$ | → | $2F^-$ | +2.87 |
| $Co^{3+} + e^-$ | → | $Co^{2+}$ | +1.80 |
| $PbO_2 + 4H^+ + SO_4^{2-} + 2e^-$ | → | $PbSO_4(s) + 2H_2O$ | +1.69 |
| $MnO_4^- + 8H^+ + 5e^-$ | → | $Mn^{2+} + 4H_2O$ | +1.49 |
| $PbO_2 + 4H^+ + 2e^-$ | → | $Pb^{2+} + 2H_2O$ | +1.46 |
| $Cl_2 + 2e^-$ | → | $2Cl^-$ | +1.36 |
| $Cr_2O_7^{2-} + 14H^+ + 6e^-$ | → | $2Cr^{3+} + 7H_2O$ | +1.33 |
| $O_2 + 4H^+ + 4e^-$ | → | $2H_2O$ | +1.23 |
| $Br_2 + 2e^-$ | → | $2Br^-$ | +1.07 |
| $NO_3^- + 4H^+ + 3e^-$ | → | $NO + 2H_2O$ | +0.96 |
| $Hg^{2+} + 2e^-$ | → | $Hg$ | +0.85 |
| $Ag^+ + e^-$ | → | $Ag$ | +0.80 |
| $Fe^{3+} + e^-$ | → | $Fe^{2+}$ | +0.77 |
| $I_2 + 2e^-$ | → | $2I^-$ | +0.54 |
| $Cu^+ + e^-$ | → | $Cu$ | +0.52 |
| $Fe(CN)_6^{3-} + e^-$ | → | $Fe(CN)_6^{4-}$ | +0.36 |
| $Cu^{2+} + 2e^-$ | → | $Cu$ | +0.34 |
| $Cu^{2+} + e^-$ | → | $Cu^+$ | +0.15 |
| $Sn^{4+} + 2e^-$ | → | $Sn^{2+}$ | +0.15 |
| $2H^+ + 2e^-$ | → | $H_2$ | 0.00 |
| $Fe^{3+} + 3e^-$ | → | $Fe$ | –0.04 |
| $Pb^{2+} + 2e^-$ | → | $Pb$ | –0.13 |
| $Sn^{2+} + 2e^-$ | → | $Sn$ | –0.14 |
| $Ni^{2+} + 2e^-$ | → | $Ni$ | –0.25 |

| Reaction | | | E° |
|---|---|---|---|
| $Co^{2+} + 2e^-$ | → | Co | –0.29 |
| $PbSO_4 + 2e^-$ | → | $Pb + SO_4^{2-}$ | –0.359 |
| $PbI_2 + 2e^-$ | → | $Pb + 2I^-$ | –0.365 |
| $Cr^{3+} + e^-$ | → | $Cr^{2+}$ | –0.40 |
| $Cd^{2+} + 2e^-$ | → | Cd | –0.40 |
| $Fe^{2+} + 2e^-$ | → | Fe | –0.41 |
| $Cr^{3+} + 3e^-$ | → | Cr | –0.74 |
| $Zn^{2+} + 2e^-$ | → | Zn | –0.76 |
| $2H_2O + 2e^-$ | → | $H_2(g) + 2OH^-$ | –0.83 |
| $V^{2+} + 2e^-$ | → | V | –1.18 |
| $Mn^{2+} + 2e^-$ | → | Mn | –1.18 |
| $Al^{3+} + 3e^-$ | → | Al | –1.66 |
| $Mg^{2+} + 2e^-$ | → | Mg | –2.37 |

APPENDIX 5

# Dissociation Constants of Some Acids at 25°C[a]

| Acid[b] | HA | A⁻ | $k_a$ | $pk_a$ |
|---|---|---|---|---|
| Perchloric | $HClO_4$ | $ClO_4^-$ | $\sim 10^{+8}$ | ~−6 |
| Permanganic | $HMnO_4$ | $MnO_4^-$ | $\sim 10^{+8}$ | ~−8 |
| Chloric | $HClO_3$ | $ClO_3^-$ | $\sim 10^{+3}$ | ~−3 |
| Nitric | $HNO_3$ | $NO_3^-$ | | |
| Hydrobromic | HBr | $Br^-$ | | |
| Hydrochloric | HCl | $Cl^-$ | | |
| Sulfuric (1) | $H_2SO_4$ | $HSO_4^-$ | | |
| Hydrated proton or protonated solvent | $H^+$ | $H_2O$(solvent) | 1.00 | 0.00 |
| Trichloroacetic | $CCl_3COOH$ | $CCl_3COO^-$ | $2 \times 10^{-1}$ | 0.70 |
| Oxalic (1) | HOOC—COOH | $HOOC—COO^-$ | $5.9 \times 10^{-2}$ | 1.23 |
| Dichloroacetic | $CHCl_2COOH$ | $CHCl_2COO^-$ | $3.32 \times 10^{-2}$ | 1.48 |
| Sulfurous (1) | $H_2SO_3$ | $HSO_3^-$ | $1.54 \times 10^{-2}$ | 1.81 |
| Sulfuric (2) | $HSO_4^-$ | $SO_4^{2-}$ | $1.20 \times 10^{-2}$ | 1.92 |
| Phosphoric (1) | $H_3PO_4$ | $H_2PO_4^-$ | $7.52 \times 10^{-3}$ | 2.12 |
| Bromoacetic | $CH_2BrCOOH$ | $CH_2BrCOO^-$ | $2.05 \times 10^{-3}$ | 2.69 |
| Malonic (1) | $HOOC—CH_2—COOH$ | $HOOC—CH_2—COO^-$ | $1.49 \times 10^{-3}$ | 2.83 |
| Chloroacetic | $CH_2ClCOOH$ | $CH_2ClCOO^-$ | $1.40 \times 10^{-3}$ | 2.85 |
| Nitrous | $HNO_2$ | $NO_2^-$ | $4.6 \times 10^{-4}$ | 3.34 |
| Hydrofluoric | HF | $F^-$ | $3.53 \times 10^{-4}$ | 3.45 |
| Formic | HCOOH | $HCOO^-$ | $1.77 \times 10^{-4}$ | 3.75 |
| Benzoic | $C_6H_5COOH$ | $C_6H_5COO^-$ | $6.46 \times 10^{-5}$ | 4.19 |
| Oxalic (2) | $HOOC—COO^-$ | $^-OOC—COO^-$ | $6.4 \times 10^{-5}$ | 4.19 |

| Acid[b] | HA | $A^-$ | $k_a$ | $pk_a$ |
|---|---|---|---|---|
| Acetic | $CH_3COOH$ | $CH_3COO^-$ | $1.76 \times 10^{-5}$ | 4.75 |
| Propionic | $CH_3CH_2COOH$ | $CH_3CH_2COO^-$ | $1.34 \times 10^{-5}$ | 4.87 |
| Malonic (2) | $HOOC—CH_2—COO^-$ | $^-OCC—CH_2—COO^-$ | $2.03 \times 10^{-6}$ | 5.69 |
| Carbonic (1) | $CO_2 + H_2O$ | $HCO_3^-$ | $4.3 \times 10^{-7}$ | 6.37 |
| Sulfurous (2) | $HSO_3^-$ | $SO^2{}_3^-$ | $1.02 \times 10^{-7}$ | 6.91 |
| Hydrogen sulfide (1) | $H_2S$ | $HS^-$ | $9.1 \times 10^{-8}$ | 7.04 |
| Phosphoric (2) | $H_2PO_4^-$ | $HPO^2{}_4^-$ | $6.23 \times 10^{-8}$ | 7.21 |
| Ammonium ion | $NH_4^+$ | $NH_3$ | $5.6 \times 10^{-10}$ | 9.25 |
| Hydrocyanic | HCN | $CN^-$ | $4.93 \times 10^{-10}$ | 9.31 |
| Silver ion | $Ag^+ + H_2O$ | AgOH | $9.1 \times 10^{-11}$ | 10.04 |
| Carbonic (2) | $HCO_3^-$ | $CO^2{}_3^-$ | $5.61 \times 10^{-11}$ | 10.25 |
| Hydrogen peroxide | $H_2O_2$ | $HO^-{}_2$ | $2.4 \times 10^{-12}$ | 11.62 |
| Hydrogen sulfide (2) | $HS^-$ | $S^{2-}$ | $1.1 \times 10^{-12}$ | 11.96 |
| Phosphoric (3) | $HPO^2{}_4^-$ | $PO^3{}_4^-$ | $2.2 \times 10^{-13}$ | 12.67 |
| Water[c] | $H_2O$ | $OH^-$ | $1.8 \times 10^{-16}$ | 15.76 |

[a] HA is the acid form, with acid strength decreasing down the table. $A^-$ is the conjugate base, with base strength increasing down the table. The equilibrium is $HA \rightleftharpoons H^+ + A^-$ and the equilibrium-constant expression is

$$K_a = \frac{[H^+][A^-]}{[HA]} \qquad pK_a = -\log_{10} K_a$$

[b] The notation (1) indicates a first dissociation or proton-transfer reaction; (2) indicates a second dissociation; and (3) indicates a third dissociation.

[c] This $K_a$ value for water explicitly uses $[H_2O] = 55.6$ moles liter$^{-1}$ in the denominator, for the sake of consistency with the other entries in the table. The standard $K_w$ is obtained by noting that $55.6 \times 1.8 \times 10^{-16} = 1.0 \times 10^{-14} = K_w$.

# Credit

# INDEX

## A



## B



## C

## D



## E



## F



## G



## H



## I

## J



## K



## L



## M



## N



## O



## P

9 781516 541041

Printed by Libri Plureos GmbH in Hamburg,
Germany